AF602095

VERTEBRATE ZOÖLOGY

VERTEBRATE ZOÖLOGY

HORATIO HACKETT NEWMAN, PH.D.

Professor of Zoology and Embryology in the
University of Chicago

Chennai Trichy Tirunelveli New Delhi

ISBN 978-81-8094-142-9

MJP Publishers
No. 44, Nallathambi Street,
Triplicane, Chennai 600 005

MJP 124

First Published : 1920

MJP's First Print : 2012

THIS VOLUME IS DEDICATED
TO MY FATHER
ALBERT HENRY NEWMAN

PREFACE

This volume is intended for use as a text-book in college courses in vertebrate zoölogy such as are required of premedical students and others who have had a course in general zoölogy. The aim of the book is to present those aspects of the subject which are not adequately brought out by laboratory work in comparative anatomy. It is taken for granted that the student who uses this as a text in connection with the lecture and recitational part of a course, shall also pursue a laboratory course in comparative anatomy, using a laboratory manual. It is also believed advisable to use as laboratory references the various text-books on comparative anatomy.

The book is avowedly dynamic in tone, emphasizing the physiological, developmental, phylogenetic, and ecological aspects of vertebrates. Structural features must of course be dealt with extensively, but purely anatomical details are as a rule subordinated to physiologic and evolutionary considerations.

The vertebrates are, moreover, viewed not merely as a group of animals belonging to the present, but, *historically*, as a very ancient assemblage of related forms, that arose from simple beginnings many millions of years ago and have passed through many vicissitudes involved in the mighty world changes of ancient times. Hence more than the usual attention is given to earlier chapters in the ancestral history of the vertebrate classes, chapters that are often of more dramatic interest than those of the present and that give to the student a new conception of the significance of modern end-products of evolution which, in themselves, are often relatively unattractive and devoid of interest.

The writer has for some years been much impressed with the far-reaching applicability to problems of animal morphology, of the *axial gradient* conception of his colleague, Professor Child, and one of the features of the present book is the attempt to interpret vertebrate structures in terms of this conception. In some cases it is probable that the theory has been stretched beyond the limits its author would consider justified; hence the present writer takes entire responsibility for the applications of Child's theory brought out in this volume.

The axial gradient conceptions have appeared to the writer to be strictly in accord with the principles of racial senescence as presented by H. F. Osborn and R. S. Lull in their recent volumes on evolution. The present volume has much to say about *adaptive radiation* and the various degenerative and senescent conditions so common among vertebrates. The writer has freely made use of the material found in Osborn's and in Lull's books. Many figures have been borrowed from both authors, for which the writer is deeply indebted.

Much of the data used has been taken from the several volumes on vertebrates of the Cambridge Natural History, and from the various comparative anatomies such as those of Wiedersheim, of Kingsley, and of Wilder. Kellicott's " Chordate Embryology " has been found excellent for much of the embryological data. A list of about a hundred references to which the writer has had access is presented in an appendix.

The illustrations have been borrowed to a considerable extent from Macmillan Company publications, but the great majority of the figures have been redrawn by Mr. Kenji Toda to whom I herewith express my hearty thanks. Many of the figures are made up of several related illustrations arranged in such fashion that they readily may be put into chart form. In our own laboratory we are already using a considerable proportion of these compound figures as charts. The sources of these redrawn figures are various and the author wishes to acknowledge with thanks the courtesy of those who have permitted their originals to be thus modified and used. The greatest care has been taken properly to acknowledge the source of each borrowed illustration. If in any case the figure has been incorrectly attributed to an author, information regarding the error will be gratefully received.

The author is not unaware of the shortcomings of the present book. Some critics will doubtless feel that much valuable, and to their minds necessary, data has been omitted. Others will probably believe that much that has been included, especially in connection with the notes on the natural history of certain types, might better have been omitted. But the selection of what to present and what to omit has been carefully canvassed and what appears to be a workable compromise has been decided upon. The teacher may readily omit what he feels to be superfluous, and will be glad to supply the data that he feels should be included.

Doubtless, errors of various kinds have crept into the text in spite of our scrupulous efforts to avoid them. Although the writer has been indebted to several competent readers of his manuscript, he holds himself entirely responsible for the book with all of its defects and whatever virtues it may have. He will be grateful for criticism and correction on the part of his readers.

H. H. Newman.

Chicago, Ill.
May 1, 1919.

TABLE OF CONTENTS

CHAPTER I. PRINCIPLES OF VERTEBRATE MORPHOLOGY

PAGES

The fundamental architectural plan of vertebrates....................1–8

A physiological interpretation of the fundamental structural plan of vertebrates....................8–12

Generalized, specialized, senescent, and retarded (pædogenetic) types of vertebrates....................12–22

Vertebrate phylogeny....................22–33

CHAPTER II. THE PHYLUM CHORDATA

Cephalochordata....................33–48

Urochordata....................48–60

Hemichordata....................60–68

Craniata (Vertebrates)....................68–70

CHAPTER III. THE ORIGIN AND EVOLUTION OF THE VERTEBRATES

The Amphioxus Theory....................72–75

The Annelid Theory....................77–79

The Arthropod Theory....................79–82

Minor Theories....................82–86

CHAPTER IV. CYCLOSTOMATA

Myxinoidea....................89–91

Petromyzontia....................91–96

CHAPTER V. PISCES (FISHES)

Introduction to Pisces....................97–112

Elasmobranchii....................112–127

Teleostomi....................127–153

Dipneusti....................153–159

Generalized and specialized types of fishes and the axial gradient theory of structural relations....................159–163

Eggs, reproduction, and breeding habits of fishes....................163–168

Appendix to fishes (Ostrachodermi, etc.)....................168–172

Chapter VI. Amphibia

PAGES
Introductory.........173–179
Extinct Amphibia.........180–183
Present-day Amphibia.........183–203
The development of the frog.........203–209

Chapter VII. Reptilia

Introductory.........210–213
Extinct reptiles.........213–224
Modern reptiles.........225–259

Chapter VIII. Aves (Birds)

Introductory.........260–276
Extinct birds.........276–281
Birds of to-day.........281–312
Ratitæ.........281–288
Carinatæ.........288–312
Development of birds.........314–323

Chapter IX. Mammalia

Introductory.........324–336
Extinct mammals.........336–345
Mammals of the present.........345–359
Prototheria.........347–352
Eutheria (Didelphia).........352–359

Chapter X. Mammalia (continued) Placental Mammals

Unguiculata.........360–363
Dermoptera.........363
Chiroptera.........363–364
Carnivora.........364–371
Rodentia.........371–373
Edentata.........373–377
Tubulidentata.........377–378
Primates.........378–388
Artiodactyla.........389–392
Perissidactyla.........392–396
Sirenia.........396–398
Hyracoidea.........398–400

PAGES
Odontoceti...400–403
Mystacoceti...403–404
The development of mammals.................................404–411

Partial List of Literature.................................413–415

Index...417–432

VERTEBRATE ZOÖLOGY

CHAPTER I

PRINCIPLES OF VERTEBRATE MORPHOLOGY

Definition.—The vertebrates may be defined as animals having:—pronounced antero-posterior, dorso-ventral, and bilateral axes; internal metameric segmentation, especially of the mesoblast; a central nervous system dorsal in position and tubular in structure, with a well-defined central canal or neurocoel; a well-defined head, characterized by highly specialized sense organs and by a concentration of nervous tissue into a complex brain; the alimentary tract opening by anterior mouth and posterior anus, and provided with paired pharyngeal clefts; a notochord derived from the primitive endoderm and situated between the central nervous system and the alimentary tract; an open cœlom, at first segmented, but later the segmental cœlomic cavities unite to form the large pericardial, peritoneal, and, in the mammals, thoracic cavities; a closed circulatory system quite distinct from the cœlom; a post-anal prolongation of the body into a metamerically segmented tail, without cœlomic cavity; usually paired appendages, pectoral and pelvic. These characters and a few others serve to mark off the vertebrates quite sharply from all other groups. Several of the most fundamental of these characteristics must now be discussed. The diagrams on the following page (Fig. 1) illustrate most of these characters.

THE FUNDAMENTAL ARCHITECTURAL PLAN OF VERTEBRATES

The Three Morphological Axes

The Axis of Polarity (Primary Axis).—A typical vertebrate has an elongated form, with head and tail ends clearly defined. An imaginary line drawn from the extreme anterior to the extreme posterior end indicates the primary structural and functional axis of the body, which is designated the *antero-posterior* or *apico-basal* axis. *The or-*

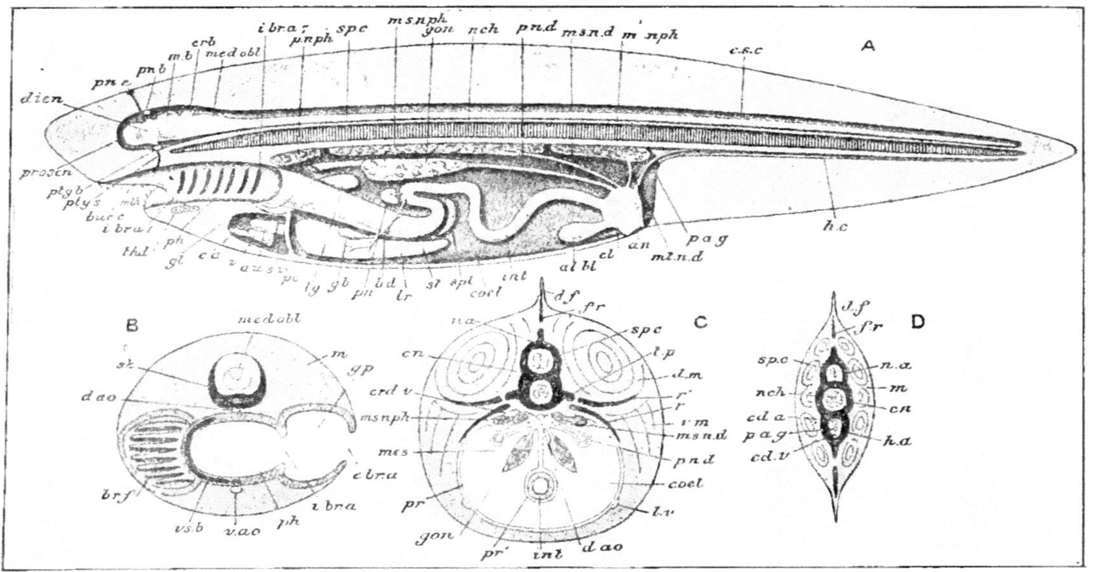

FIG. 1.—A, sagittal section of an ideal craniate, to show among other things the axial organization; B, transverse section of the head; C, of the trunk; D, of the tail; *al. bl*, allantoic bladder; *an*, anus; *au*, auricle; *b. d*, bile-duct; *br. f*, branchial filaments; *buc. c*, buccal cavity; *c. a*, conus arterio; *cd. a*, caudal artery; *cd. v*, caudal vein; *coel*, cœlom; *crd. v*, cardinal vein; *cn*, centrum; *crb*, cerebellum; *c. s. c*, cerebro-spinal cavity; *d. ao*, dorsal aorta; *dien*, diencephalon; *d. f*, dorsal fin; *d. m*, dorsal muscles; *e. br. a*, external branchial aperture; *f. r*, fin-ray; *g. b*, gall bladder; *gl*, glottis; *gon*, gonad; *g. p*, gill-pouch; *h. a*, hæmal arch; *h. c*, hæmal canal; *i. br. a*, internal branchial aperture; *int*, intestine; *lg*, lung; *lr*, liver; *l. v*, lateral vein; *m*, muscles; *m. b*, mid-brain; *med. obl*, medulla oblongata; *mes*, mesentery; *ms. n. d*, mesonephric duct; *ms. nph*, mesonephros; *mth*, mouth; *mt. n. d*, metanephric duct; *mt. nph*, metanephros; *n. a*, neural arch; *nch*, notochord; *p. a. g*, post-anal gut; *pc*, pericardium; *ph*, pharynx; *pn*, pancreas; *pn. b*, pineal body; *pn. d*, pronephric duct; *pn. e*, pineal eye; *p. nph*, pronephros; *pr*, peritoneum, parietal layer, and *pr'*, visceral layer; *prosen*, prosencephalon; *pty. s*, pituitary sac; *r*, sub-peritoneal rib; *r'*, intermuscular rib; *sk*, skull; *sp. c*, spinal cord; *spl*, spleen; *st*, stomach; *s. v*, sinus venosus; *thd*, thyroid; *t. p*, transverse process; *v*, ventricle; *v. ao*, ventral aorta; *v. m*, ventral muscles; *vs. b*, visceral bar. (From Parker and Haswell.)

gans of highest dynamic activity and most pronounced sensitivity are at the apical or anterior end and the organs of lowest dynamic activity and least sensitivity are at the basal or posterior end of this axis. Between these extremes, or opposite poles, of the axis the remaining organs or functions are arranged, at least primitively, in a graded series of diminishing dynamic activity and sensitivity. These geometrical relations serve as an index of an inherent spatial orderliness in the arrangement of the functions with reference to one another, and demonstrate that the organism is based on a single plan—is a coherent entity.

From a purely physiological point of view this gradient represents a linear series of functions, ranging from dominant or controlling functions to subordinate or controlled functions, a series which, broadly speaking, runs somewhat as follows:—olfactory and visual, the most anterior and dominant functions, entirely sensory in character; motor functions associated with movement of the eyes, and motor centers for most voluntary functions; sensory and motor activities associated with feeding, including the sense of taste, and the motor activities of jaws and tongue; sensory and motor activities associated with hearing and equilibrium; the active functions of respiration and circulation, which are closely correlated; the most anterior locomotor functions, associated with the pectoral appendages; the most active phases of the alimentary or digestive functions, associated with the stomach and the larger glands; the excretory and lower alimentary functions, associated with the kidneys, the lower intestine and rectum; the reproductive functions, associated with ovaries and testes, their accessory ducts and copulatory organs; the functions of the tail or post-anal body, which may be considered as a developmental afterthought and as more or less beyond the limits of the original primary axis. The tail has a gradient of its own and does not belong to the primary gradient. This is only a rough outline of the real physiological gradient, but is clear enough for our purposes. The true gradient no longer exists in modern vertebrates because there has been a great deal of secondary concentration at various levels, especially at the anterior end, where the original metameric arrangement of the functional series has been profoundly disturbed and distorted. The primary gradient is further obscured by the fact that various systems of organs, such as the heart and blood vessels, the brain, the alimentary tract, etc., have developed secondary axes of functional activity and structural differentiation that are largely independent of the primary axis and have little reference to the polarity of the body as a whole. In the embryo, however, the axis is less distorted than in the adult.

The Dorso-Ventral Axis (Secondary Axis).—If we take a cross section at any level of the primary axis, we at once perceive that a line drawn from the mid-dorsal to the mid-ventral line represents another main architectural axis. In the diagram (Fig. 2) it will be noted that the central nervous system occupies the high point or apical end of the axis and that, at this level, one loop of the intestine occupies the ventral or basal point of the axis. The gradient of functions in between the two poles has been so decidedly disturbed by lateral foldings and secondary displacements, that any attempt to construct a list of graded functions would be futile. We know in general, however, that the dorsal side is the dominant part of the axis and that, during embryonic development, the various functions differentiate almost exactly in the order of their relative dynamic activity.

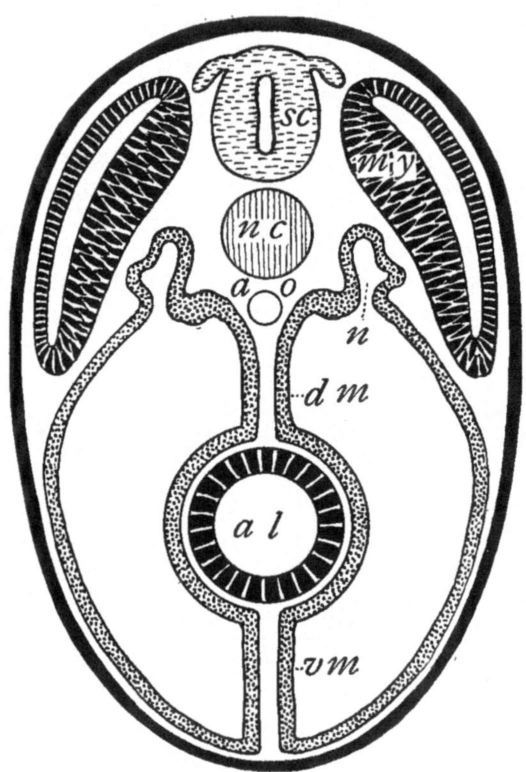

Fig. 2.—Diagrammatic transverse section of a vertebrate to illustrate the dorso-ventral (secondary) and bilateral (tertiary) axes of organization. *al*, alimentary tract; *ao*, aorta; *d. m*, dorsal mesentery; *my*, myotome; *nc*, notochord; *n*, nephrotome; *o*, omentum; *sc*, spinal chord; *v. m*, ventral mesentery. (Modified after Kingsley.)

The Bilateral Axis (Tertiary Axis).—The dorso-ventral axis divides any level of the antero-posterior axis into two mirror-image halves (Fig. 2), so that each side is the reversed counterpart of the other. It would appear then that the bilateral axis is a necessary consequence of the dorso-ventral axis. The axis has a single median or apical point and two lateral or basal points. Any vertical level in the dorso-ventral axis may constitute the apical point of a double bilateral axis. A good example of this type of axial organization is seen in connection with the differentiation of the segmented mesoblast of the chick (Fig. 3). The median dorsal part of the somite forms the myotome (segmental voluntary muscles), the most highly dynamic of all mesodermal structures; next comes the dermatome which forms the deeper skin and many complex sensory and glandular structures, etc.; next comes the nephrotome or primordium of the excretory system; next,

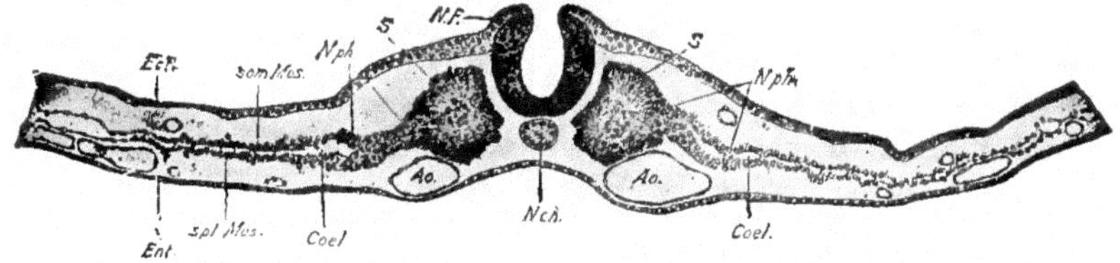

Fig. 3.—Transverse section across the primary axis of a 10 somite chick embryo, to illustrate the bilateral (tertiary) axis of vertebrate organization. *Ao*, aorta; *Coel*, cœlom; *ect*, ectoderm; *ent*, entoderm; *nch*, notochord; *N. F.*, neural furrow; *N. ph*, nephrotome; *s*, somite; *som. Mes*, somato-pleure; *spl. Mes*, splanchopleure. (From Lillie's "Development of the Chick" [Henry Holt and Co].)

but only at posterior levels, the gonatome or primordium of the reproductive system; next, the embryonic cœlom, from which are derived mainly such passive structures as the peritoneal lining of the body cavities and the mesenteries; and finally, the extra-embryonic body cavity, which has to do primarily with the formation of embryonic membranes that serve as protective and respiratory organs during the life of the embryo, but play no rôle in the organization of the adult.

It should be understood that the appendages are, as the name indicates, truly added structures and are not to be considered as part of the original bilateral gradient. Each appendage has its own three gradients, with the free end representing the apical end and the fixed end, the basal. It need hardly be said that the main specializations of the appendages take place at the apical or free end, while the basal parts remain comparatively conservative and undifferentiated.

Metamerism

The segmental organization, metameric structure, is secondary in importance only to the primary axiate organization and is undoubtedly one of the developmental consequences of the latter. The vertebrates constitute but one of three great metameric groups, the other two being the annelids and the arthropods. **Annelids,** the lowest of metameric types, show metamerism in its most generalized condition. In them segmentation involves equally both internal and external structures, and there is little or no obliteration of primary metamerism by fusion of groups of contiguous metameres into larger regions. In the **arthropods** the external metamerism to a large extent retains its primitive condition, especially in regions back of the head or cephalothorax, but the primitive internal metamerism is obscured through the almost complete obliteration of the cœlom by the system of venous sinuses.

Metamerism in the vertebrates, especially in the higher groups, is confined largely to certain internal structures and is scarcely at all expressed in external characters. Even the internal metamerism is to a large extent obscured by the compacting of the anterior metameres into a head, and by the fusion of the original segmental body cavities into a few large compound cœlomic chambers. The segmentation is best expressed in the structures derived from the embryonic mesoblast: the myotomes, the nephric tubules, the vertebral units, and their skeletal accessories. The central nervous system also retains its primitive metameric segmentation in post-cephalic regions, but in the cephalic region the metameric elements are very difficult to make out. There are believed to be primitively five (possibly six) metameres, three or four in front of the otic vesicle and one behind it.

Back of the head, however, the vertebrate is composed of a linear series of body segments or metameres, each of which has, theoretically, its own complete set of body organs. There is, however, never an ideal or complete development of all the characteristic segmental organs in any one metamere. In the anterior metameres the organs of lowest dynamic activity, such as those of excretion and reproduction, do not appear; and in the posterior metameres the organs of highest dynamic activity, such as the organs of special sense and the higher coördinative faculties, do not appear. And there is a graded series between these extremes.

As a rule, a series of two to ten or more metameres unites into a functional region and a concentration of these metameres occurs that makes it difficult to discover the original segmental conditions involved. Even the most primitive vertebrates exhibit wide departures from the primitive metameric plan seen in some of the lower annelids, and these modifications of the generalized segmental arrangement become progressively more pronounced from the lowest to the highest vertebrate classes.

Cephalization

One of the most significant advances over ancestral conditions that the vertebrates have to their credit has to do with the evolution of intelligence, and this is intimately bound up with the pronounced specialization of the anterior metameres into a head, which is principally a brain and a group of sense organs. This foreshortening, condensation, and specialization of the anterior metameres has been called *cephalization.* The course of evolution from the lowest to the highest vertebrates has been one of more and more pronounced cephalization involving a progressively larger number of metameres, an increased dominance of the head over the rest of the body, and closer integration of all the functions. The climax of the process of cephalization is Man, who is essentially a dominating intelligence and a set of accessory organs. Various attempts have been made to decipher the highly modified metamerism of the vertebrate head. One of these attempts gave rise to the classic "Vertebral Theory" of the skull by Goethe and by Oken. It was thought that the skull was made up of a modified series of vertebral units and that an analysis of these elements would give the number of metameres. This theory was proven by Huxley to be untenable. The cranial nerves have been taken as evidences of metamerism, but the old arrangement of ten cranial nerves is no longer taken to indicate ten metameres, since some of these nerves are the motor and others the sensory components of single metameric divisions of the neuron. Perhaps the most reliable index of the number of metameres in the head is seen in the mesoblastic somites of some of the fishes and cyclostomes. There are but three somites in front of the auditory capsule, which forms a convenient landmark in that it seems to mark off the original head from the postcephalic region. Possibly then the primitive craniate had a head of three metameres, a number characteristic of the larva of Balanoglossus, and possibly that of the echinoderms. The most characteristic stage in certain annelid larvæ has also three metameres. The higher vertebrates have in the process of their evolution gradually appropriated to the head, one after the other, the adjacent body metameres.

The Backward Retreat of the Lower Functions.—Accompanying the concentration of the chief nervous elements in the head there is a progressive series of changes in the opposite direction, consisting of a steady retreat toward the posterior end of the respiratory, digestive, excretory, and reproductive functions. These functions are, in the more primitive vertebrates, present in the metameres just back of the head, but in the higher forms they are steadily pushed back into the posterior metameres by becoming atrophied in the anterior metameres and by the development of new sets of more or less equivalent organs further back. An excellent example of this type of process is seen in the evolution of the kidneys. In the most primitive vertebrates the embryonic kidney is located well toward the anterior part of the cœlom, and in the hag-fishes this anterior kidney, or pronephros, functions in the adult. In vertebrates of the fish and amphibian grades the kidney of the adult is a mesonephros, located farther back in the body cavity; while in the land vertebrates the functional kidney is a metanephros, situated still farther toward the posterior end.

The farther these organs of lower dynamic activity retreat from the dominant anterior end the more readily they appear to differentiate and the more specialized and condensed they become, just as though they were steadily growing out from under an influence that tends to suppress their full developmental possibilities. The evolution of the urogenital organs of the vertebrates affords an interesting example of this. In the lowest vertebrates the gonads and their ducts are simple, are located far forward in the body cavity, and are to a large extent separate from the urinary elements. In the highest vertebrates, however, these organs have retreated far to the posterior end of the primary axis and have been intimately involved with the urinary system.

A PHYSIOLOGICAL INTERPRETATION OF THE FUNDAMENTAL STRUCTURAL PLAN OF VERTEBRATES

Professor C. M. Child has developed what appears to the writer to be the most scientific and far-reaching dynamic interpretation of animal structure that has yet appeared. His "axial gradient theory" is a statement in energistic terms of the axiate organization of animals and of the relations of dominance and subordination that are so obvious in such animals as the vertebrates. He has shown experimentally "that the apical or anterior region is primarily the region of greatest dynamic or metabolic activity in the individual. The apical end becomes the most highly specialized and differentiated region of the body, and in those forms which possess a central nervous system, the cephalic ganglion or brain and the chief sense organs usually arise in this region; . . . in other words it becomes the head, and in motile forms usually precedes in locomotion." "The basal or posterior region, on the other hand, is primarily the least active region and in motile forms its activity is more or less under the control of the apical or anterior region."

The Axial Gradient in Development

There is a remarkable parallelism between the orderly spatial arrangement of structures and functions down the polar axis and the orderly sequence in time of these structures during the development of the individual from the egg. The first structures to differentiate are those of the head; in fact the early vertebrate embryo is nearly all head and the body grows out from it as though it were a mere axial outgrowth of the head. The last part of the body to complete its differentiation is the tail or basal part of the axis of polarity. There is thus an important relation between the degree of dominance and the developmental age of the various levels along the axis. The more anterior part always differentiates before the more posterior part and dominates it at least for a time. Secondary alterations in the relationships of dominance and subordination are of frequent occurrence and result from functional adjustments and through the action of external factors. The same parallelism between spatial arrangement and sequence in time exists for the dorso-ventral axis. In vertebrates, the first structures to differentiate are those in the medullary plate region; in fact, in the higher vertebrates, the early embryo consists almost entirely of the anterior median dorsal region, which is destined to form the central nervous system of the head. The last structures fully to differentiate are the reproductive structures which, functionally at least, belong to the basal part of the gradient. Structures also develop mesio-laterally, differentiating first in the middle and proceeding laterally, so that, in the case of such structures as the paired appendages, the last parts to develop are the terminal elements.

The Nervous System as a Mechanism for Maintaining the Relations of Dominance and Subordination in the Organism.—The method of maintaining the dominance of the apical parts over basal parts is the method of conduction over nerve paths. There seems to be little doubt now that communications between an apical part and a basal part are electroid in character, so that changes in the chemical state of the apical part are transmitted speedily to the basal part and excite it to functional activity or inhibit its action, according to the character of the apical change. Thus a stimulated condition of a particular segment of the central nervous system involves a change of electric potential between it and the basal organ with which it is connected. The reaction of the basal part is in the nature of an adjustment to meet the change in the apical part so as to bring about a return to equilibrium between the two terminal points of the system. Thus, throughout the whole body there is a very intricate system of controlling and controlled parts, depending on the relative intensity or relative rate of metabolism of the various parts. It has been shown by the most exact types of apparatus, *e. g.*, the Tashiro biometer, which measures exceedingly minute differences in carbon dioxide production in small objects such as nerves, that the apical part of a nerve has a higher rate of metabolism than the basal part.

Relative Susceptibility to Inhibiting Agents of the Apical and Basal Parts of the Axes. Child has demonstrated differences in rate of metabolism between apical and basal parts of an organism by the so-called "susceptibility method." He finds that apical parts are more susceptible than basal parts to agents that retard metabolic activity (such as anæsthetics, potassium cyanide, etc.). If lethal concentrations are used, the apical part dies first and there is a progression of death changes down the axes until the last parts to retain life are the basal parts. These experiments demonstrate that the axial gradient is essentially a gradient in the rate or intensity of metabolic activity. From this we may conclude that the relationship of functional dominance and subordination is one depending on the relative rates of metabolism, the part with the more rapid or intense metabolism stimulating through conduction regions of less rapid metabolism, and exciting them to perform their peculiar functions.

The Inhibiting Influence of Dominant Regions Upon the Development and Degree of Differentiation of Subordinate Regions. Among the invertebrates that have the same three fundamental axial gradients possessed by vertebrates, experiments have shown that during development a dominant region exercises an inhibiting influence over a subordinate region. This may be demonstrated by experiments in regeneration. If a planarian worm is cut transversely across the axis of polarity at almost any level the posterior piece will grow a new head. Evidently then, while the posterior part was organically connected with the anterior piece, head formation was inhibited. As soon as the organic connection with the original head or dominant part has been severed, the inhibition is removed and a head develops. In another flatworm, *Microstomum,* the dominance of the head becomes less with age so that at a point about halfway down the axis new brain and eyes form. This means that a new head is established, but there is not an immediate isolation of the new individual from the old. As long as the new individual remains a part of the old, it remains a subordinate individual, not only functionally in following the lead of the original head, but structurally in that the eyes do not fully differentiate and the brain remains small. Each of the two individuals divides again into two and in each case the posterior individual is subordinate to the anterior.

In general therefore it would appear that *the presence of a dominant part inhibits the full differentiation of subordinate parts in proportion to their relative metabolic rates or intensities and the proximity of the two regions.*

Let us now apply these principles to vertebrate development. In the first place it will be recalled that vertebrates are metameric animals in which each metamere, at least back of the primitive head, is serially homologous with all the rest. Each metamere has therefore potentially the developmental capacity of any other. Whatever systems or structures that develop in any one metamere should, *ex hypothese,* be latent in all of the others. Any failure, therefore, on the part of a given metamere to realize the full extent of its differentiational possibilities must be attributed to some sort of inhibition. Let us examine the main regions of the body of a higher vertebrate with this idea in mind.

The head metameres are characterized by a great specialization of brain tissue and an almost complete suppression of the lower functions.

In the neck region there is very little differentiation of subordinate elements belonging to the ventral ranges of the dorso-ventral axis; muscles, glands, and skeletal elements are present, but no excretory or reproductive organs differentiate in that region. In the thoracic region, where the dominance of the anterior end of the dorsal structures is less intense, the organs of respiration, circulation, and locomotion differentiate fully; but there are no digestive, excretory, or reproductive specializations. Still further back in the upper abdominal region the digestive functions reach their maximum importance and the uro-genital functions begin to appear. And, finally, in the lower abdominal region, the excretory and genital functions develop fully; but the genital tissues do not fully differentiate till comparatively late in the life cycle.

The tail of the vertebrate appears to be a developmental afterthought. It is the last part to develop and in many specialized vertebrates scarcely develops at all, as though the developmental momentum slowed down to such an extent that there was not enough force left to push out the tail. When the tail does form, however, it is usually, though not always, the most primitive part of the body. A tailless condition is very common in highly specialized races with prolonged developmental period.

GENERALIZED, SPECIALIZED, SENESCENT, AND RETARDED (PÆDOGENETIC) TYPES OF VERTEBRATES

A **generalized vertebrate** of any class is one in which the axial relations are well in balance; the primary axis distinctly dominant over the secondary and tertiary axes. Such animals as a dog-shark, a salamander, a lizard, a shrew, are typical generalized vertebrates. They have in common certain characteristics of which the following are the most important:—the body is rather elongated and cylindrical in shape; with head, trunk, and tail in normal balance; the fore and hind limbs of approximately equal value; and no pronounced specializations either externally or internally. Types such as these are usually looked upon as prototypic of the groups to which they belong and therefore as affording a close approximation to the ancestral stock from which the group in question has been derived.

The history of vertebrate evolution has been closely associated with a series of radical geographic and climatic changes, that have had the effect of periodically eliminating large numbers of specialized groups which have become adapted to a peculiar and limited environment, and of permitting only a few plastic, generalized types, capable of adjusting themselves to the changed world conditions, to persist. These generalized forms have then become the ancestral stock from which the specialized types of the new period have arisen.

There has usually been a period of struggle on the part of the persisting generalized species to gain a foothold; then they have multiplied rapidly, and have entered upon a period of adaptive radiation, which has resulted in the development of terrestrial, arboreal, aquatic, fossorial, and volant types. Whether an animal is a fish, amphibian, reptile, bird, or mammal, it meets a given set of life conditions in much the same manner. The fish-like form, for example, has been adopted not only by fishes, but by amphibians (several of the persistently aquatic urodeles), by reptiles (notably by the extinct ichthyosaurs), by birds (penguins and the extinct Odontolcæ), and by mammals (whales and dugongs). These all have certain features in common (Fig. 4) that are the fundamental adaptations for active life in the water:—the spindle-shaped body, fin-like appendages, smooth, water-shedding exterior, tail-fin or hind limbs modified to act as a propeller, and usually dorsal fins (in fish, amphibia, ichthyosaurs, and some whales). Such structures that serve a similar function in adaptation to a given environmental complex are, for the most part, not homologous, but merely analogous. They are technically called "homoplastic," which implies that they have been molded out of diverse materials into like forms.

Every successful vertebrate class has had a period of youth, when the members were all comparatively generalized; a period of maturity, characterized by adaptive deployment into all of the available life zones; and a period of old age or senescence, characterized by the development of bizarre, overspecialized types, incapable of weathering a world crisis. The reptiles, for example, arose in the Palæozoic and were at first generalized lizard-like forms. Before the close of the Palæozoic there had arisen many specialized and a few precociously senescent types, most of which became extinct during the troublous climatic disturbances that ushered in the Mesozoic. Only a few of the more generalized reptilian stocks survived the crisis and adjusted themselves to the new conditions of the Mesozoic, the Golden Age of the reptiles. This age saw the great second adaptive radiation of the reptiles and the production of many overspecialized or senescent types, including the dinosaurs, ichthyosaurs, plesiosaurs, and pterosaurs, all of which became extinct before the close of the Mesozoic. Only a few of the more generalized stocks lived over into the Cenozoic to found the rather conservative modern reptilian classes.

The generalized forms are like the conservative germ-plasm in heredity, that passes on from generation to generation, progressing slowly and steadily along definitely directed lines, and largely uninfluenced by the somatic specializations that are the result of environmental or functional adaptations. Just as the germ-plasm is a continuous, unbroken series of cell generations that gives off tangentially the successive somatic generations with all of the accompanying specializations, so is the generalized stock of animal forms a continuous series of races that sprays off at intervals the specialized types. These specialized groups go their ways and become extinct, while the slowly evolving generalized types form the stock from which future specialized races may arise.

Generalized forms are never equally generalized in all particulars. As a rule, while retaining a fundamentally generalized structure, they become superficially specialized in one or more particulars. While the ideally generalized type is only approximated in any class of vertebrates, some one or more forms stand out from their fellows as more nearly prototypic than the rest. Such forms are of great phylogenetic interest and will subsequently claim a large share of our attention.

Not infrequently forms that appear to have some of the ear-marks of prototypes prove on examination to be pretenders, in that they are either secondarily simplified by parasitic or sedentary life, or are retarded forms that have failed to complete the life cycle normal for the group to which they belong. Usually the degenerate forms may readily be recognized by the study of their life history; for their larval stages show more advanced conditions of various structures than do the definitive forms. The retarded forms, on the other hand, simply cease to develop somatically in a larval or juvenile stage, while the germinal cycle completes itself. As a consequence these juvenile forms become sexually mature and reproduce their kind, a phenomenon known as *pædogenesis* or *neoteny*.

Specialized types should be carefully discriminated from senescent types, though few specialized forms are free from the evidences of senescence. Specializations are to be looked upon as adaptive in character, and their origin as a sort of response to the demands of certain environmental conditions. Senescent characters, although sometimes apparently adaptive, are frequently valueless or distinctly disadvantageous to their possessors, and have often been responsible for racial extinction. The various types of adaptive complexes and the several criteria of racial senescence must receive separate consideration.

Specializations and Adaptations

It has been pointed out that a primitive group, after weathering a radical world change and thus surviving its more highly specialized relatives, tends to begin a new adaptive deployment and to occupy all of the available life zones. In order to enter specialized life zones or live upon very restricted types of food, adaptive specializations must occur that fit various offspring of the primitive stock to occupy these restricted life conditions.

In general it may be said that the active, predaceous life is primitive as compared with the more passive herbivorous life and that, in the case of vertebrates at least, the generalized types are for the most part active and predaceous, characterized by quick action, keen sensitivity, sharp grasping teeth, claws as opposed to hoofs, and normal balance of head, trunk, and tail. Bottom-feeding types in the waters and browsing, herbivorous types on land are characterized by slow action (though many herbivorous types have become secondarily cursorial), cutting and grinding teeth, hoofs instead of claws, and lack of balance in bodily proportions.

Many of the senile races are characterized by the possession of armatures of various sorts that are unquestionably of great defensive value to sluggish and otherwise defenseless creatures. In spite of the value of such structures to their possessors, they are rather to be thought of as the products of racial aging than as responses on the part of the organism to life needs. Only in a very limited sense, then, can such structures be classed as adaptations, for many bony and scaly excrescences of senile types have gone far beyond the bounds of usefulness and have been a mere burden and useless incumbrance to their owners.

In the body of this volume many examples of adaptive specializations and equally numerous illustrations of the structures resulting from racial senescence will be cited and commented upon. In this place we shall concern ourselves with a study of the laws of adaptation, following closely the analysis given by Osborn.

The Laws of Adaptation

"The form evolution of the back-boned animals, beginning with the pro-fishes of Cambrian and pre-Cambrian time, extends over a period estimated at not less than 30,000,000 years. The supremely adaptable vertebrate body type begins to dominate the living world, overcoming one mechanical difficulty after another, and passes through the habitat zones of water, land, and air. Adaptations in the motions necessary for the capture, storage, and release of plant and animal energy continues to control the form of the body and its appendages, but simultaneously the organsim through mechanical and chemical means protects itself either offensively or defensively and also adapts itself to reproduce and protect its kind." Osborn finds that there are two great laws of adaptation:

1. *The law of convergence or parallelism of form in locomotor, offensive, and defensive adaptations.* There is a widespread tendency for members of totally unrelated groups to develop similar body form and similar external characters in adaptation to similar habitat zones. Several adaptive types are readily distinguished and are herewith listed:

Aquatic or swimming animals. Among the thoroughly marine types of vertebrates, whether they be fish, reptile or mammal, there is in general the same fish-like body form, with the same sorts of locomotor structures, paired and median fins. Apart from their general aquatic adaptations the shark, ichthyosaur (reptile), and dolphin (mammal) are profoundly different (Fig. 4).

"The three mechanical problems of existence in the water habitat are: first, overcoming the buoyancy of water either by weighting down or increasing the gravity of the body or by the development of special gravitating organs, which enable animals to rise and descend in this medium; second, the mechanical problem of overcoming the resistance of water in rapid motion, which is accomplished by means of warped surfaces and well-designed entrant and re-entrant angles of the body similar to the 'stream-lines' of the fastest modern yachts; third, the problem of propulsion of the body, which is accomplished, first, by sinuous motion of the entire body, terminating in powerful propulsion by the tail fin; secondly, by supplementary action of the four lateral fins;

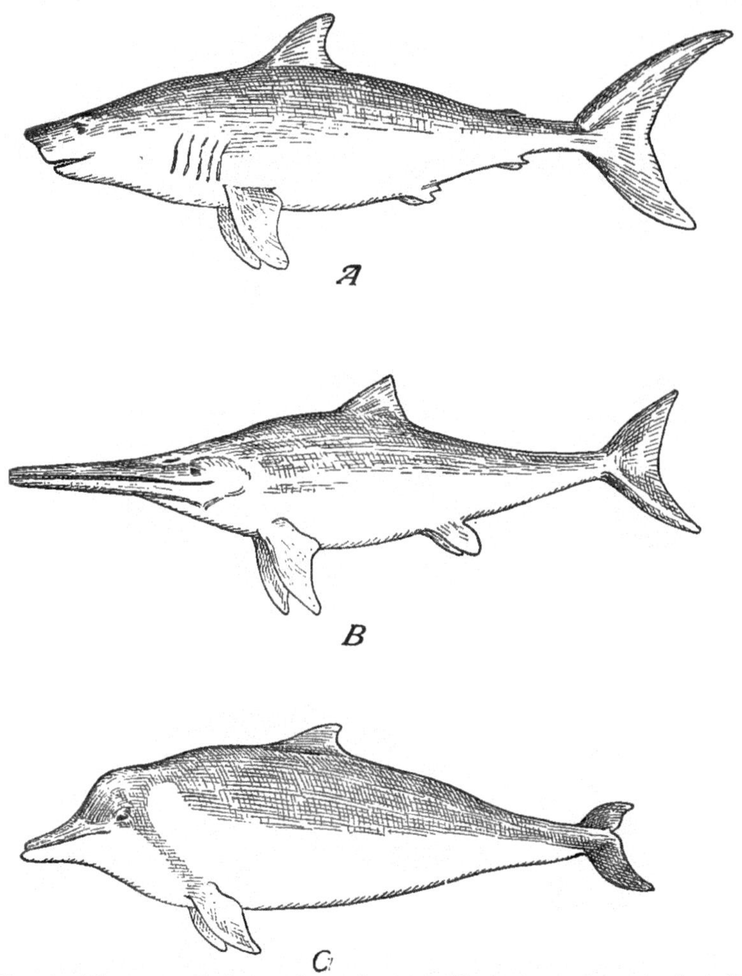

FIG. 4.—Three aquatic types of vertebrate, to illustrate convergent adaptation of three wholly unrelated forms for marine life. All three show the fusiform body, median and paired fins, though the skeletal structures are radically different. A, shark (Pisces); B, ichthyosaur (Reptilia); C, porpoise (Mammalia). (After Osborn's "Origin and Evolution of Life" [Charles Scribner's Sons].)

third, by the horizontal steering of the body by the median system of fins."

Flying or parachuting animals, belonging to all of the vertebrate classes except the cyclostomes, exhibit similar parallelisms which consist of: first, planes for the support of the weight in the air; and second, tapering body-form for decreasing the resistance of the body in rapid motion through the air; third, a rigid framework for the planes.

Arboreal or climbing (scansorial) animals of all vertebrate classes have one prime requisite—clinging appendages. Some cling by means of prehensile fingers or tail, others by the use of adhesive pads, others by air-suction pads, and still others by hook-like claws.

Running (cursorial) animals are very much alike in their mechanical adaptations. These are primarily foot modifications. The prime requisite is a long limb with plenty of spring to it. This requisite is met by standing high on the toes with the legs well under the body. Usually there is a progressive loss of some of the digits, a reduction that reaches its climax in the one-toed horses.

Digging, burrowing (fossorial) animals are usually long and narrow so as to require only a narrow-bore burrow. The fore limbs are strong and have a heavy shoulder girdle for the attachment of the heavy digging muscles. Eyes and ears are usually reduced and the tail is usually poorly developed.

Desert-dwellers.—The prime requisites for these animals are: first, coverings of some sort that prevent undue loss of moisture, such as heavy scales, spines, or armor; second, protection against the extremes of temperature. Many of them meet the latter requirement by burrowing in sand or in the soil at night. It is also very common for desert creatures to be venomous. This venom may be a chemical end-product of life under arid conditions.

Cave and deep sea animals.—These animals may be viewed largely as products of lowered vitality. It can scarcely be claimed that their characters are on the whole adaptive. One character very commonly found in abyssmal creatures is phosphorescence. Light-producing organs of all sorts are developed by these creatures that appear to be definitely adapted to life in the dark. Possibly phosphorescence may be one of the physiological accompaniments of life under abyssmal conditions. Deep-sea vertebrates (fishes) are also almost invariably forms exhibiting radical distortions of the generalized fish form. Two principal types are common: those with suppressed heads and those with exaggerated heads. This condition is discussed elsewhere.

Ant-eating adaptations.—Ants for a long time have been extremely numerous and have been an important factor in the environment of all land vertebrates since the establishment of the reptilian orders. Wheeler in his classic treatise on ants claims that many vertebrate integumentary structures, such as hair, feathers, and scales are primarily for protection against ants. Whether this position is justified or not, it is certainly true that ant-eaters among all of the classes of land vertebrates are heavily armored or densely covered with integumentary structures. Besides this most obvious adaptation, ant-eaters have long slender snout, with small terminal mouth, long sticky tongue, teeth reduced or absent, strong digging feet, and internal nostrils far forward and in such relation to the glottis that ants could not possibly crawl into the wind-pipe.

2. *The law of divergence of form; the law of adaptive radiation.* This law is the antithesis of the "Law of Convergence"; for, instead of a similarity of adaptive character acquired by unrelated groups, we have diversity of adaptive characters acquired by members of a single related stock, which tend to radiate into all of the available life zones and to develop the various adaptive complexes. For example, a primitive stock of cursorial reptiles splits up into fossorial, aquatic, arboreal, volant, and giant herbivorous and carnivorous types. A type once arboreal may become a volant type, and that seems to be the usual sequence. A volant type does not become fossorial, but may return to a cursorial habit, as for example, the running birds. An aquatic type may become terrestrial and then secondarily return to the life in the water, carrying back with it, however, an air-breathing mechanism. When the aquatic vertebrates developed adaptations for land life they lost their gills, their lateral-line organs, etc. When once a character is lost it cannot be regained, so the aquatic reptiles and mammals must be dependent on lungs, although they would be much better off with gills. This illustrates the irreversibility of evolutionary changes, and especially of adaptive specializations. Osborn, in his book on "The Origin and Evolution of Life," takes the position that the irreversibility of evolution is due to the progressive chemical evolution of the "heredity chromatin." Its changes, he believes, are orderly and progress step by step toward more and more narrow specialization. Once the chromatin has acquired factors for specialized characters, it cannot reverse and return to the generalized condi-

tion. All it can do is to lose certain old characters and develop new ones of a different sort. A too narrowly specialized creature is at the mercy of a changing geologic age. Rather sudden climatic and geographic changes have been the rule in geologic history, and whenever such changes have occurred there has been a rapid extinction of the most highly specialized types, races that are no longer plastic enough to adjust themselves by adaptations to the new conditions. They are in a "cul-de-sac of structure," says Osborn, "from which there is no possible emergence by adaptation to a different physical environment or habitat. It is these two principles of too close adjustment to a single environment and of the non-survival of characters once lost by the chromatin which underlie the law that the highly specialized and most perfectly adapted types become extinct, while primitive, conservative, and relatively unspecialized types invariably become the centres of new adaptive radiation."

Racial Senescence and Degeneration

Just as the individual grows old and suffers a retardation of all vital activities, so races age and show similar evidences of lowered vitality and diminished activity. In young individuals and in young races the rate of metabolism is high and the expressions of a high rate of metabolism (a more intense vitality) are seen in their comparatively active life, predaceous habits, structures on the whole generalized, moderate size, and lack of heavy excrescences. When the individual or the race is young, the products of its metabolism are used up largely in motor activities of various sorts and there is little deposition of inert materials such as armor, spines, heavy bones, fats, or massive flesh.

A senescent race, on the other hand, is characterized by sluggish behavior, by herbivorous habits or feeding habits involving little exertion, by structures on the whole specialized or degenerate, often by giant size or bulky build, and by accumulations of inert materials such as armor, spines, heavy bones or flesh. These and other characters are now very generally recognized as criteria of racial senescence.

Structural Criteria of Racial Senescence

Large Size.—It has been found that the highly specialized races of the past have usually grown to giant size as compared with their less specialized relatives. Good examples of this phenomenon are seen in the great dinosaurs, and in the monster mammals of the past and of the present, such as the proboscidians and the whales. In these forms growth has run riot, probably because of the lack of growth-inhibiting factors.

Spinescence.—In both vertebrates and invertebrates the development of spines, horns and other chitinous or bony excrescences are generally believed to be evidences of racial old age. As the general metabolism slows down there appears to be a tendency for hard or dead substances to be deposited in regions of lowest metabolic rate, just as débris collects in an eddy. Good examples of this phenomenon are found in *Stegosaurus*, in *Edaphosaurus*, in the horned toads, in many teleost fishes, in porcupines, and in the deer family. In the last-named group the extinct Irish elk stands out as a classic example of a type that went a step too far in the elaboration of excrescences, and became extinct.

Degeneracy.—One of the systems most commonly found in a degenerating condition is the dentition. Many of the most highly specialized groups have undergone a partial or total loss of teeth. This is true of the edentates (archaic types of placental mamals), of the turtles (among the oldest and most specialized among reptilian orders), of the sturgeon (a degenerate end product of a long line of chondrostean fishes), and of modern birds (possibly the most highly specialized vertebrates). Loss of teeth is equally common among the extinct groups of the past, such as the beaked dinosaurs, and the great sauropods. Degeneration of the tail is almost as common among highly specialized and senescent races as is the loss of teeth. Examples of taillessness are too numerous to list. Loss of limbs is also very common in senescent types.

The Eel-like Form.—Lull also recognizes as a criterion of racial old age great bodily elongation, usually accompanied by partial or total loss of limbs. Gregory lists forty-four distinct types of eel-like creatures among the vertebrates, including:—three groups of cyclostomes, one of sharks, a lung fish, three teleostome groups, three amphibian groups, five reptilian groups, and one group of mammals. In all cases the criteria are about the same, though the degree of elongation and reduction of limbs varies greatly.

Elaborate Coloration.—Most primitive or generalized groups of animals have dull colors and indistinct patterns; but one of the most striking features of highly specialized climax groups of vertebrates, such as the teleost fishes and the birds, is their remarkably high coloration. The more brilliant pigments may be looked upon as end products of certain chemical processes in metabolism, and therefore appropriate in groups that represent end products of long lines of specialization.

Possible Internal Causes of Some Phases of Senescence.—The effects of atrophy or hypertrophy of such endocrine glands as the thyroid, pituitary body (hypophysis), ovaries or testes, upon growth and differentiation, are well known. Disturbances in the normal functionality of these various glands underlie giantism, dwarfism, degeneracy, failure to complete the differentiation of the secondary sexual characters, etc. It is not unlikely that racial senescence, like individual senescence, may be the result of a progressive deterioration of the germinal determiners of these important balance-wheels of organic growth. It is also not unlikely that most of the changes in bodily proportion, which constitute a vast preponderance of specific and racial differences, are largely due to relative racial atrophy or hypertrophy of these growth-regulating glands. The giant races may be those in which thyroid abnormality has unloosed the restraints to growth, and size has reached the limit of mechanical possibility. Similarly racial thyroid or hypophysis atrophy may account for degenerate types. An excellent example of the effectiveness of thyroid in controlling differentiation is seen in the case of frog development. The bull-frog larva in northern waters takes several years to reach the stage of metamorphosis; but if fed upon thyroid there is a precocious metamorphosis of the first-year larva while it is still only about one-fourth the size normally reached in nature. Thus giantism in frog larvæ is the result, presumably, of deficient or belated thyroid secretion. The converse of this experiment has been made by extirpation of the hypophysis in tadpoles. Allen has found that these operated larvæ do not develop a thyroid and are unable to metamorphose. The bearing of these experiments on the phenomenon of neoteny or pædogenesis is obvious.

VERTEBRATE PHYLOGENY

Phylogeny may be defined as the science of ancestries or genealogies, and, as such, is one of the chief concerns of the present book. In attempting to discover the origin, ancestry, and relationships of a given group of animals, we employ mainly three sets of criteria, which for the sake of brevity we may term: homologies, ontogenies, and fossils. These three phenomena form the chief subject-matter of the sciences of comparative anatomy, embryology, and palæontology.

Homologies have already been dealt with incidentally in the foregoing discussion of adaptations; for the second law of adaptation, "the law of divergence of form," implies homology, since the variously modified adaptive structures are the products of a single generalized ancestral prototype, and are strictly comparable in fundamental structure and mode of origin. Thus the wing of a bird, the fore leg of a horse, and the flipper of a whale are homologous in spite of their profound departures, both structurally and functionally, from the generalized vertebrate fore limb. Similarly, the rudimentary hind limbs of a whale and the splint bones of a horse are recognized as homologues of the fully functional structures of allied groups.

We must be constantly on our guard against the all too common error of mistaking for homologies "convergences of form" such as are dealt with in Osborn's first law of adaptation. Thus, while the shark, ichthyosaur, and porpoise appear to possess homologous adaptations in their general form and locomotor organs, such structures as the caudal and dorsal fins in these three types are more truly *analogous* than homologous, for they do not represent the same embryonic rudiments, nor are they made of the same structural materials. Before we can be certain about homologies we must put them to the test by means of a study of their embryonic development.

Embryological Aspects of Phylogeny

The Recapitulation Theory

This theory has often been called Hæckel's Biogenetic Law, and may be paraphrased as follows: The life history of the individual gives a brief condensed resumé of the evolutionary history of its ancestors. Some one has said that "the individual climbs its own ancestral tree." In a somewhat restricted and modified sense this theory is still accepted by most embryologists, though there are some who have cast it aside as worthless. Embryology has been compared to "an ancient manuscript with many sheets lost, other displaced, and with spurious passages interpolated by a later hand." The fact that the history is tremendously abbreviated precludes the recapitulation of every ancestral stage. The necessity for the embryo to prepare quickly to meet a difficult environment and to obtain its own livelihood causes the pushing back into the earlier stages of some of the characters that may have had a later phylogenetic origin, and also calls forth the development of new characters that are merely adaptations for larval life. Characters that are considered truly ancestral are known as *palingenetic*, and those that are mere interpolations for purposes of larval adaptation are known as *cænogenetic*. A good example of a palingenetic life history is that of the common frog, but even here there are certain larval characters that are purely cænogenetic. In some of the non-aquatic frogs in which the eggs are not laid in the water the typical cycle is greatly foreshortened through the omission of the larval stages; the young hatches out as a little frog with only a mere stump of a tail left. This type of foreshortening the life history is known as *tachygenesis*.

It will appear from what has just been said that some ontogenies may be fairly reliable recapitulations of phylogenies, but that others are entirely unreliable as guides in phylogenetic matters. The question may well be asked as to whether we are in a position to decide in any given case whether the embryonic history is truly palingenetic or not. In answer to this we may say that the evidence of embryology is valuable only when it is supported by a study of comparative anatomy and palæontology.

One point about the recapitulation theory that is seldom clearly apprehended has been clearly stated by Lillie:

> "If phylogeny is to be understood to be the succession of adult forms in the line of evolution, it cannot be said in any real sense that ontogeny is a brief recapitulation of phylogeny, for the embryo of a higher form is never like the adult of a lower form, though the anatomy of embryonic organs of higher species resembles in many particulars the anatomy of homologous organs of the adult of the lower species. However, if we conceive that the whole life history is necessary for the definition of a species, we obtain a different basis for the recapitulation theory. The comparable units are then entire ontogenies, and these resemble one another in proportion to the nearness of relationship, just as the definitive structures do. Thus in nearly related species the ontogenies are very similar; in more distantly related species there is less resemblance, and in species from different classes the ontogenies are widely divergent in many respects."

It must not be forgotten, however, that the ontogenies of closely related species may be made to differ markedly from one another by cænogenetic adaptive changes, as in the case of some of the frogs that have adopted the habit of laying the eggs out of water. In these cases the ontogenies are much more widely divergent than one would expect in forms so closely related. So the statement of Lillie that "ontogenies resemble one another in proportion to the nearness of relationship," might be emended by adding the clause:—*unless the ontogenies have been secondarily disturbed by adaptive cænogenetic modifications.* In all of our attempts to make use of ontogenies as evidences of phylogenetic relationships we must be on our guard against the various pitfalls that have been discussed, and must not fail to recognize that in phylogetic research palæontology is a more reliable guide than is embryology.

The Geologic Aspects of Vertebrate Phylogeny

In order to appreciate the historical side of vertebrate evolution it is necessary to know something about the geologic succession of the various classes. It is estimated that at least 25,000,000 years have elapsed (some authors estimate ten times this) since the deposition of the strata in which we find the earliest vertebrate fossils. Very likely at least 5,000,000 years of evolutionary progress had been made since the first chordate ancestors of these early armored vertebrates had made their appearance. The accompanying chart (Fig. 5) shows in compact form the great eras of geologic time and the number of years that each is believed to have occupied. Another excellent chart by Osborn (Fig. 6) indicates the characteristic vertebrates of the various great geologic ages and the successive geologic appearance, adaptive radiations and diminutions of the five vertebrate classes. It will be noted that all of the classes arise in the Palæozoic, that each higher class arises, not late in the developmental history of the ancestral class, but very near its base, when the earlier race was young. Thus the Amphibia are shown coming off from fishes of the Devonian; the reptiles from the Lower Carboniferous Amphibia before the Amphibia themselves had made any progressive adaptive radiation; while both birds and mammals came off near the base of the reptilian trunk, the birds probably somewhat later than mammals. In addition, it will be noted that the first vertebrate remains occur far down in the Palæozoic, in Ordovician times, which were characterized primarily by invertebrate forms. The student will find it

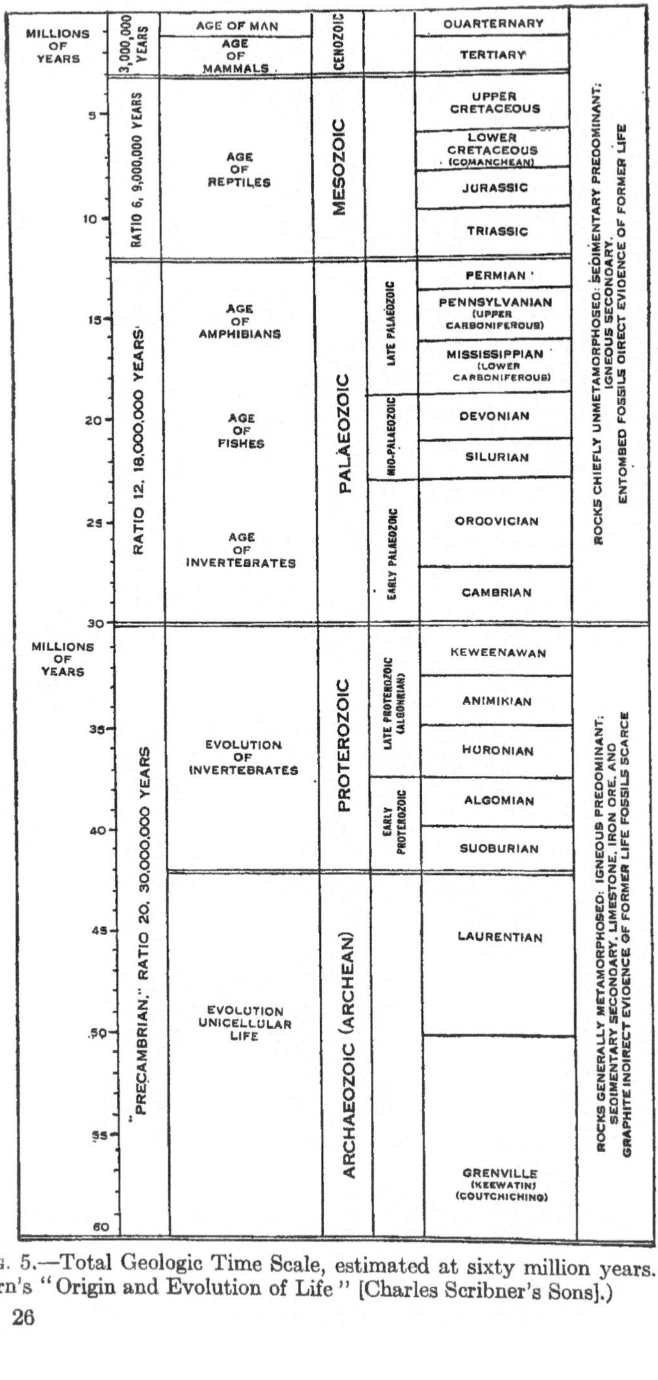

MILLIONS OF YEARS						
0–3	3,000,000 YEARS	AGE OF MAN	CENOZOIC		QUARTERNARY	ROCKS CHIEFLY UNMETAMORPHOSED: SEDIMENTARY PREDOMINANT; IGNEOUS SECONDARY. ENTOMBED FOSSILS DIRECT EVIDENCE OF FORMER LIFE
		AGE OF MAMMALS			TERTIARY	
5	RATIO 6, 9,000,000 YEARS	AGE OF REPTILES	MESOZOIC		UPPER CRETACEOUS	
					LOWER CRETACEOUS (COMANCHEAN)	
					JURASSIC	
10					TRIASSIC	
15	RATIO 12, 18,000,000 YEARS	AGE OF AMPHIBIANS	PALAEOZOIC	LATE PALAEOZOIC	PERMIAN	
					PENNSYLVANIAN (UPPER CARBONIFEROUS)	
					MISSISSIPPIAN (LOWER CARBONIFEROUS)	
20		AGE OF FISHES		MIO-PALAEOZOIC	DEVONIAN	
					SILURIAN	
25		AGE OF INVERTEBRATES		EARLY PALAEOZOIC	ORDOVICIAN	
30					CAMBRIAN	
MILLIONS OF YEARS	"PRECAMBRIAN," RATIO 20, 30,000,000 YEARS	EVOLUTION OF INVERTEBRATES	PROTEROZOIC	LATE PROTEROZOIC (ALGONKIAN)	KEWEENAWAN	ROCKS GENERALLY METAMORPHOSED: IGNEOUS PREDOMINANT; SEDIMENTARY SECONDARY. LIMESTONE, IRON ORE, AND GRAPHITE INDIRECT EVIDENCE OF FORMER LIFE FOSSILS SCARCE
35					ANIMIKIAN	
					HURONIAN	
40				EARLY PROTEROZOIC	ALGOMIAN	
					SUDBURIAN	
45		EVOLUTION UNICELLULAR LIFE	ARCHAEOZOIC (ARCHEAN)		LAURENTIAN	
50–60					GRENVILLE (KEEWATIN) (COUTCHICHING)	

FIG. 5.—Total Geologic Time Scale, estimated at sixty million years. (After Osborn's "Origin and Evolution of Life" [Charles Scribner's Sons].)

of great advantage to memorize the geologic epochs and ages in their order, for it will be necessary in dealing with the extinct representatives of the vertebrate classes to assign them without explanation to their appropriate geologic periods.

The study of geology teaches us that the earth's outer zones have undergone within the period of vertebrate history numerous profound changes which in general we may term climatic changes.

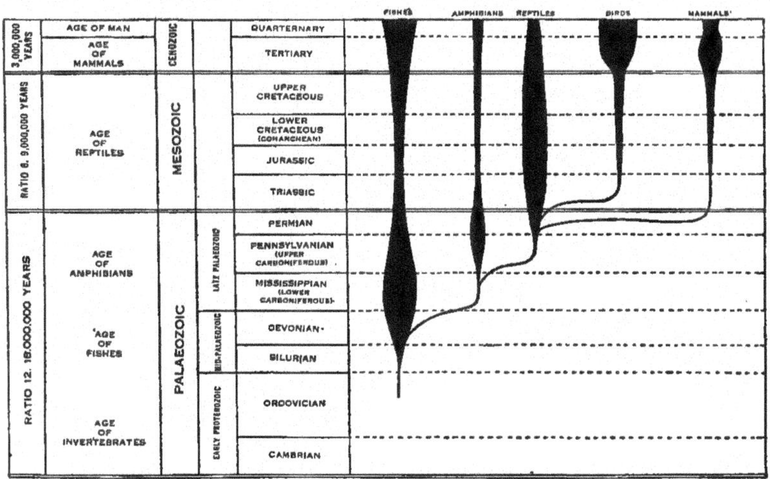

FIG. 6.—Chronological chart of vertebrate succession. Successive geologic appearance and epochs of maximum radiation (expansion) and diminution (contraction) of the five classes of vertebrates, namely, fishes, amphibians, reptiles, birds, and mammals. (After Osborn's "Origin and Evolution of Life" [Charles Scribner's Sons].)

There have been periods of continental subsidence, accompanied by ocean floor elevations, during which the great continental plains have been covered by comparatively shallow seas. The marine faunas of the seas have migrated into these shallows, and representatives of them have been buried in sediment. When the reverse change has occurred and the continental plains have been again elevated, the sedimentation of the shallow-sea period forms a great rocky stratum laden with marine fossils. Between periods of subsidence millions of years elapsed, and therefore a break in the continuity of the entombed fossils is to be expected. Discontinuity of fossil-bearing strata is the rule. If it were not for this periodicity of subsidence and elevation there would be no boundaries between consecutive geologic strata.

There have also been rhythms of alternating aridity and humidity which have been associated with marked evolutionary changes in land life. The principal periods of mountain uplift are shown in Fig. 7. One of the most significant of such periods was the great Permo-Triassic arid period which saw the earliest true land animals, the reptiles, make their first adaptive radiation and also gave rise to the first mammalian and perhaps avian stocks.

The elevations of mountain ranges have from time to time vastly altered the land habitats, especially those of the central continental plains, which were rendered arid by being cut off from the moisture of the sea. Finally, periodic eras of cold (glacial epochs) have alternated with tropical or semi-tropical eras, and these have naturally exercised a profound effect upon the character of the faunas and floras. In general the coal measures serve as a marker for the tropical periods, among which the Upper Carboniferous and Upper Cretaceous were the most striking. After the warm, moist climate of the Upper Carboniferous there followed rather suddenly the Permian glacial period accompanied by continental elevation and increased aridity. There was no such sudden change at the end of the Cretaceous; but during that period there were extensive continental elevation, considerable increase in aridity, and a period of cold, not severe enough however to be termed glacial.

A consideration of palæogeography and a knowledge of the climatic changes during the vast period of vertebrate evolution is absolutely essential for an intelligent understanding of the organic evolutionary processes.

It seems quite clear that the great geologic changes have been accompanied by equally marked changes in the floras and faunas of the earth. In general the changes in the organic world have been of such a kind as to bring about adjustments of the animals and plants to the changed conditions. What the exact mechanics employed by the organism in making these responses has been, has never been definitely determined. The mechanics of adaptation remains one of our unsolved problems. One theory, that of Lamarck, is that the organism responds directly to the changed environment and that the germ-plasm reflects the somatic response, so that the change becomes a permanent racial asset. Another theory is that of parallel

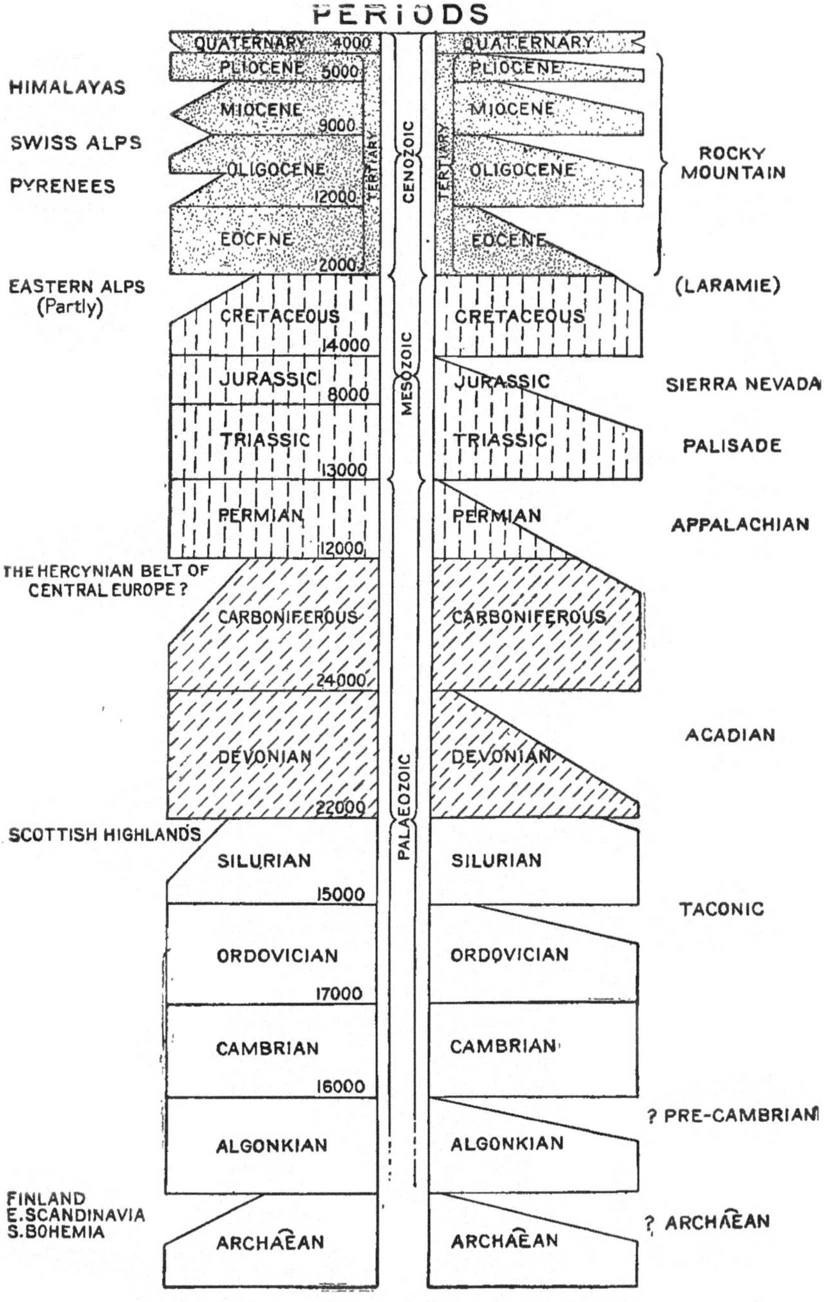

Fig. 7.—Main divisions of geologic time. The notches at the sides of the scale represent chiefly the periods of mountain uplift in the northern hemisphere of the Old World (left) and the New World (right). (From Osborn's "The Age of Mammals.")

induction, which means that both soma and germ are affected simultaneously and in the same way. Perhaps it would be best to think of the organism as a whole and view the response as a general organismal change that becomes fixed only after oft-repeated exposure generation after generation. In some cases it may be said that external factors simply accelerate or retard processes that were already under way in the germ-plasm, so that the response appears to be something new in kind when it is only the result of a sudden acceleration of a character evolution already under way. Whatever be the underlying mechanism involved in adaptive changes, there is no hope of explaining adaptations on the Darwinian basis, through the selection of the best out of a vast array of purely fortuitous variations; for if the historical study of vertebrate evolution reveals one thing more clearly than any other, it is that evolutionary changes are orderly, progressive, and determinate in character, and that in many respects these orderly processes of evolution are independent of each other and of environmental changes.

CHAPTER II

THE PHYLUM CHORDATA

The characteristics of the vertebrates have been sufficiently stated in the previous chapter. A vertebrate proper may be readily recognized by applying to it the definition on page 1, but there is a considerable assemblage of creatures, which, though falling short of being vertebrates in some respects, are obviously vertebrate-like in others, and are therefore classed with the Vertebrata or Craniata in the phylum *Chordata.*

These creatures range all the way from worm-like, burrowing forms such as *Balanoglossus* and the sessile, tubicolous *Cephalodiscus,* through the degenerate, sedentary, flask-like tunicates, to the pro-fish *Amphioxus.* When we include the vertebrates, the range of structural diversity and degree of specialization, with *Rhabdopleura* at one extreme and Man or the whales at the other, is so great that one doubts the advisability of making a single phylum so all-inclusive.

If, however, we adopt as the specifications of a chordate the possession of a *notochord, pharyngeal clefts,* and a *medullary plate,* we arbitrarily throw together all animals that possess these characters. It therefore becomes essential that we have exact criteria for recognizing these characters in whatever guise they may be presented.

Notochord.—The notochord is recognized by its position, by its histological structure, by its function, and by its embryonic derivation. Typically it is a stiff hyaline rod (Fig. 8) covered with a connective tissue sheath, lying ventral to the neural tube and dorsal to the alimentary tract. It is composed of vacuolated cells that histologically resemble pith. It is derived from a median dorsal strip of tissue cut off from the primitive endoderm or *archenteron.*

Pharyngeal Clefts (**Gill-slits**).—The term "pharyngeal clefts" is preferred to "gill-slits" because it is more accurately descriptive and has no dubious functional implications. The pharyngeal cleft is simply an opening through the body wall in the pharyngeal region that allows water to pass out of the pharynx to the exterior. The opening is made embryonically by an out-pouching of the pharynx and an in-pouching of the ectoderm, which meet and break through. There is therefore always an ectodermal and an endodermal part of a pharyngeal cleft. When gills are formed they are derived sometimes from the ectodermal and sometimes from the endodermal part of the cleft. In terrestrial vertebrates no gills, except minute transitory rudiments of gill filaments, ever develop, though the pharyngeal clefts may appear and either break through or remain closed. The number of pharyngeal clefts ranges from upwards of fifty in *Amphioxus* to one pair in *Cephalodiscus* and none in *Rhabdopleura*. Usually a framework of cartilage or bones constitutes the *branchial skeleton* and serves to support the clefts, their accompanying branchial tissues, and their blood supply.

Medullary Plate (Neural Tube).—The term "*medullary plate*" is perhaps somewhat more widely applicable than "*neural tube*," because in some of the more primitive chordates the plate never reaches the tubular condition. The medullary plate is a dorsal area of ectoderm that typically becomes first folded in longitudinally into a *neural groove* and then converted into a tube with a central canal or *neurocoel*. All of the vertebrates proper at some period have the central nervous system external (in the medullary plate condition), but some of the more primitive chordates have merely a diffuse plexus of nerve cells in the skin with a tendency for them to concentrate along the dorsal side. Such individuals can hardly be credited with a medullary plate, much less a neural tube. Some of their immediate relatives, however, have the beginnings at least of an infolding that makes it probable that the structure concerned is a medullary plate.

In dealing with the various groups that lay claim to membership, along with the vertebrates, in the chordate fraternity, it will be necessary to test their claims by a rigid examination of their credentials: *notochord*, *pharyngeal clefts*, and *medullary plate*.

The heterogeneous collection of types that has been assembled by comparative anatomists and called chordates is customarily subdivided into four sub-phyla, which, beginning with the group that shows most certain affinities to the vertebrates and followed by those forms that have a less valid claim to vertebrate relationship, are as follows:

Sub-Phylum I. Cephalochordata (Adelochorda).

This includes but a single family of fish-like creatures, of which there are about twelve species. The type form is *Amphioxus* (more correctly known as *Branchiostoma*).

Sub-Phylum II. Urochordata.

Order 1. Larvacea (Appendicularia), free-swimming forms with permanent tail.

Order 2. Ascidiacea (Tunicates or Sea-Squirts), fixed forms without tail in the adult.

Order 3. Thaliacea (Salpians), free swimming forms without tail in the adult.

Sub-Phylum III. Hemichordata.

Order 1. Enteropneusta, including worm-like forms such as *Balanoglossus*.

Order 2. Pterobranchiata, sessile, tube-dwelling forms—*Cephalodiscus and Rhabdopleura.*

Order 3. Phoronidia, tubicolous forms—*Phoronis.*

Sub-Phylum IV. Vertebrata (Craniata).

Class 1. Cyclostomata (round mouth eels).

Class 2. Pisces (true fishes with jaws).

Class 3. Amphibia (vertebrates with aquatic larvæ, but usually air-breathing in the adult condition).

Class 4. Reptilia (cold-blooded, air-breathing vertebrates).

Class 5. Aves (birds, feathered vertebrates).

Class 6. Mammalia (beasts or quadrupeds).

SUB-PHYLUM I. CEPHALOCHORDATA

The Cephalochordata are considered first because their claims to vertebrate relationship are stronger than those of the other provertebrate sub-phyla. They are rather small, marine, fish-like animals, usually called "lancelets" on account of their sharply pointed ends. *Amphioxus* was first described by Pallas in 1778. On account of its resemblance to a slug it was given the name of *Limax lanceolatus*, the implication being that it was a mollusk. In 1804 Costa, an Italian naturalist, redescribed it as a fish, allied to the lampreys and hagfishes, and, because he erroneously diagnosed the oral tentacles or

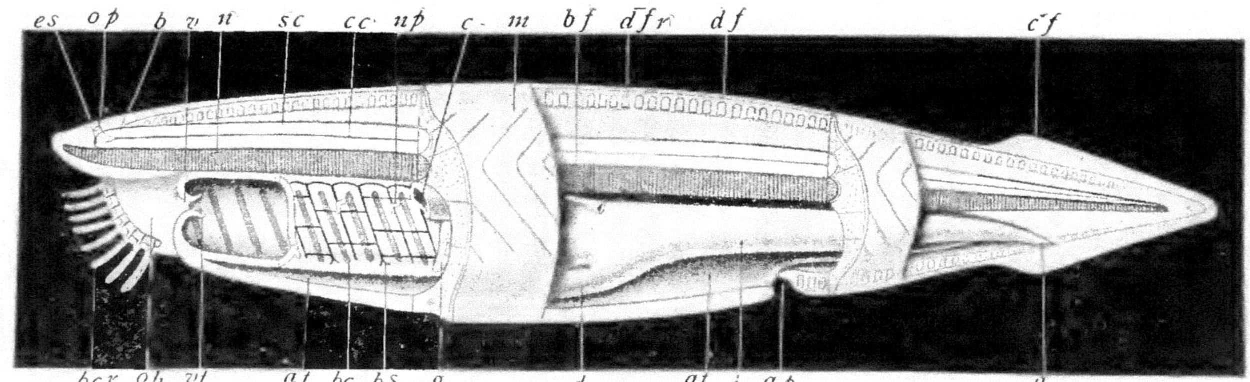

FIG. 8.—General anatomy of Amphioxus. *a*, anus; *ap*, atriopore; *at*, atrium; *b*, brain; *bcr*, buccal cirri; *bc*, branchial cleft; *bf*, brown funnel; *bs*, branchial skeletal rod; *cc*, central canal of neural tube; *c*, cœlom; *cf*, caudal fin; *d*, liver diverticulum; *df*, dorsal fin; *dfr*, dorsal fin ray; *es*, eye spot; *g*, gonad; *m*, myotome; *n*, notochord; *np*, nephridium; *oh*, oral hood; *op*, olfactory pit or funnel; *sc*, spinal cord or neural tube; *v*, velum; *vt*, velar tentacles. (Redrawn after Parker and Haswell.)

cirri as gills, he applied to it the name of *Branchiostoma* (gill-mouthed), a name still retained by experts in nomenclature. The name *Amphioxus*, however, though given a year or so later by Yarrel, has been in general use for so long that it will be difficult to displace.

Amphioxus has been studied extensively for over a century and few details of its structure or development have escaped analysis. It has come to be rather generally believed, following Willey, that this form, "though specialized in some particulars and degenerate in others, represents a grade of organization not far removed from that of the main line of early chordate ancestors." Whether Amphioxus is primitively simple or secondarily simplified by degeneration, the fact remains that in its structure and development it shows in a strikingly diagrammatic way the essential characters of the chordates. On this account Amphioxus has become a favorite laboratory type throughout the educational institutions of the civilized world and is studied annually by thousands of students.

The Cephalochordata consist of several closely similar species grouped into two genera, *Branchiostoma* (*Amphioxus*) and *Asymmetron*. The group is cosmopolitan in distribution, occurring almost everywhere in the temperate zone where sloping sandy sea-shores exist. This wide distribition and slight variability are taken by some writers to mean that the group is extremely archaic.

The writer is inclined to look upon Amphioxus as a form which has lost through sedentary life most of its head parts and is therefore partially acephalic, a view that is based on the following considerations. Typically, the chordate notochord runs only up to the head proper, and the fact that in Amphioxus the notochord extends to the anterior end of the body (Fig. 8) may mean that the head is degenerate—has retreated from the anterior end. This interpretation of Amphioxus is in accord with the fact that the tunicate larva has a much better head than has Amphioxus. Strictly speaking, Amphioxus is not headless; it has merely a reduced or degenerate head which does not extend in front of the trunk, but has come to lie back of the most anterior parts of the trunk. In the tunicates also the reduced brain lies posterior to the mouth and parts of the pharynx. The ancestral Amphioxus probably had a head with paired eyes, otic vesicles, and a considerably larger brain than is found in the modern representatives. Possibly also the pharynx is more specialized than primitive, in that it has such a very large number of pharyngeal clefts. In other respects Amphioxus may be accepted as an approximate prototype of the ancestral chordate.

Sessile Life and Method of Feeding of Amphioxus

Along the sandy shores of the Mediterranean and other temperate seas the "lancelet" leads a semi-sedentary life, burrowing rapidly

Fig. 9.—A group of lancelets (*Amphioxus lanceolatus*) in normal habitat, some in the sedentary position with only the anterior end protruding from the sand burrow, one in the foreground beginning to dig a new burrow, and others swimming about in fish-like fashion. (Redrawn from indistinct photograph after Willey.)

head first in the sand, leaving the head end protruding (Fig. 9) with its oral hood and buccal tentacles outspread to test and draw in the "sea-soup," as the food-laden water has been called. From time to time the burrow is left and another made. The method of food gathering is essentially that of a sedentary organism. There is no active searching for food, but the water which is swept in quantities through the pharynx and down through the pharyngeal clefts gives up its minute organic particles, such as protozoa and pelagic larvæ. The mechanism for concentrating this dilute food is somewhat complex. The pharynx is lined with ciliated epithelium and the cilia beat inward so as to cause a current of water to be drawn into the mouth and out through the gill-slits. The cilia beat also downward, so as to force the current to the floor of the pharynx, where there is a *hypopharyngeal groove* or *endostyle* (Fig. 10) filled with sticky mucus to which food particles adhere. The endostyle is provided with strong cilia which whip the mucus into a rope-like mass and drive it forward with its burden of food particles to the anterior end of the pharnyx. There the endostyle bifurcates around the mouth in the form of two semicircular grooves, the *peripharyngeal bands,* which unite again above the mouth and form the *dorsal* or *hyperpharngeal groove* (Fig. 10). The mucous rope travels backward in the dorsal groove till it reaches the stomach and intes-

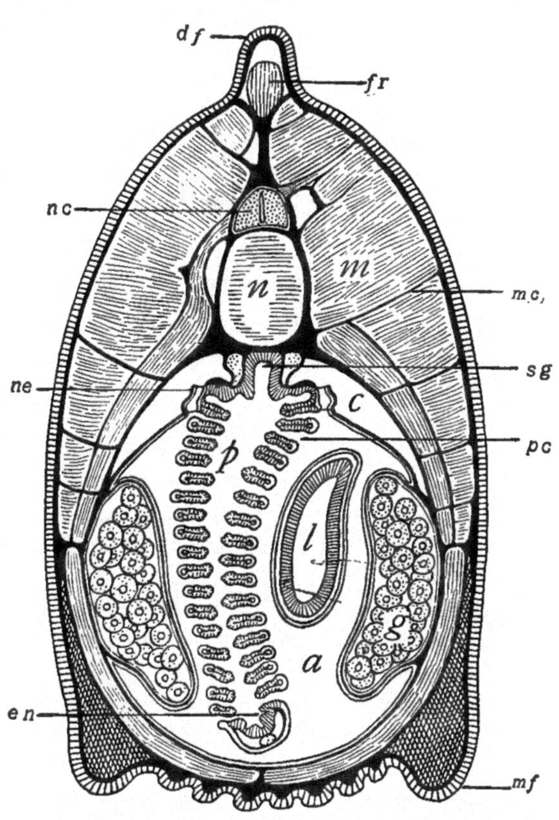

FIG. 10.—Transverse section through the pharyngeal region of Amphioxus. *a*, atrial cavity; *c*, cœlomic cavity; *df*, dorsal fin fold; *en*, endostyle; *fr*, fin ray; *l*, liver diverticulum; *m*, myotome; *mc*, myocomma; *mf*, metapleural fold; *n*, notochord; *nc*, nerve cord or spinal chord; *ne*, nephridium; *p*, pharynx; *pc*, pharyngeal cleft; *sg*, supra-pharyngeal or dorsal groove. (Redrawn and modified after Lankester and Boveri.)

testine. There its load of food-particles is digested and the mucus itself passes out of the anus.

It is of considerable interest to note in this connection that this complex food-concentrating mechanism is found in the urochordates and in the *Ammocœtes* larva of the lamprey eels, but nowhere else in the animal kingdom. Thus it furnishes a point of connection between Amphioxus and the lower chordates, on the one hand, and between Amphioxus and the vertebrates, on the other. It should also be noted that the primary function of the pharyngeal apparatus, including pharyngeal clefts, appears to be that of food concentration rather than that of respiration.

Characters of Amphioxus

1. **Characters Associated with the Reduced Head.**—The *brain* is extremely small, hardly as large in diameter as the rest of the neural tube, (Figs. 11, A and B). There are but two pairs of *cranial nerves* which have been called *olfactory* and *optic;* but in so reduced a brain homologies are uncertain. The sense organs consist of a *median olfactory funnel* opening into the neurocœl, a median rudimentary *eye-spot* on the anterior end of the brain, representing probably the last rudiments of the ancestral paired eyes. The *notochord* extends the entire length of the body, projecting in front of the brain. This may mean that the brain has retreated from a primitive anterior position. There is *no cranium.* Possibly the ancestral chordate had some sort of cranium.

2. **Characters that make up the Food-Concentrating Mechanism.**—Since the method of feeding has been described these characters need only be listed. The mouth is an *oral funnel* bounded by ciliated *buccal tentacles* with cartilaginous supports that serve to funnel the water into the pharynx. Separating the oral funnel from the pharynx is a *velum* composed of a membrane with sphincter muscles and a set of *velar tentacles* that serve as a grating and strain out the larger particles. The *pharnyx* has sometimes upwards of fifty pairs of clefts that are separated by partitions in which lie cartilaginous *skeletal rods* connected across with one another, forming a sort of *branchial basket.* The *endostyle, peripharyngeal,* and *hyperpharyngeal* grooves secrete mucus and propel the food to the stomach by means of the *mucous rope* food carrier. The *atrium* is a sort of mantle composed of folds of the body wall that incloses the whole branchial apparatus in

a voluminous water-filled chamber, the *atrial cavity*. The atrium is lined with ectoderm and has but one opening to the exterior, a posteriorly directed *atriopore*, which carries off the water that comes through the pharyngeal clefts. The atrium is a protection for the delicate pharynx while the animal is in its sandy burrow and helps to maintain an uninterrupted current of water.

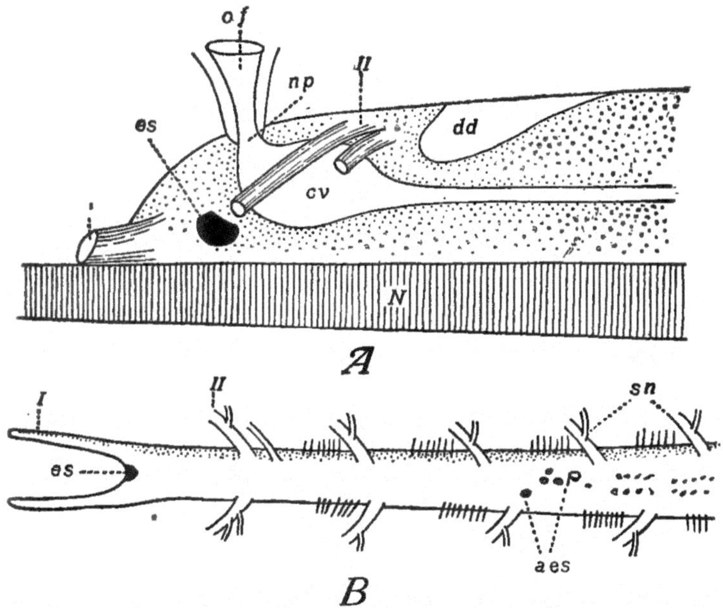

FIG. 11.—A. Lateral view of brain of Amphioxus. *cv*, central vesicle; *dd*, dorsal dilatation of the neural canal; *es*, eye spot; *of*, olfactory funnel; *np*, neuropore; *I*, first cranial nerve, olfactory; *II*, second (optic) cranial nerve, showing dorsal and ventral roots.

B. Dorsal view of brain and spinal cord of Amphioxus. *aes*, accessory dorsal eye spots, some median, some paired, *es*, eye spot; *I* and *II*, first and second cranial nerves; *sn*, spinal nerves. (Redrawn from Willey, after Hatschek and Schneider.)

3. **General Characters.**—The **alimentary** system consists of the *pharnyx*, a short, straight *stomach intestine* terminating in a ventral *anus*, which opens to the left of the ventral fin. The stomach gives off a ventral *diverticulum* or *liver*, which is directed forward.

The **circulatory system** (Fig. 14) consists of a *ventral pulsating vessel* with no specialized heart enlargement, which pumps the colorless

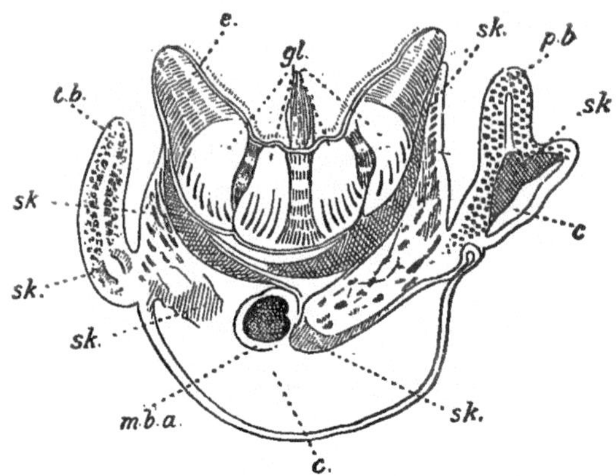

Fig. 12.—Transverse section of the ventral part of the pharynx of Amphioxus. *c*, cœlom; *e*, endostyle; *gl*, endostylar glands; *mba*, median branchial artery; *pb*, primary bar; *sk*, endostylar and branchial rods and skeletal plates; *tb*, tongue bar. (From Herdman, after Lankester.)

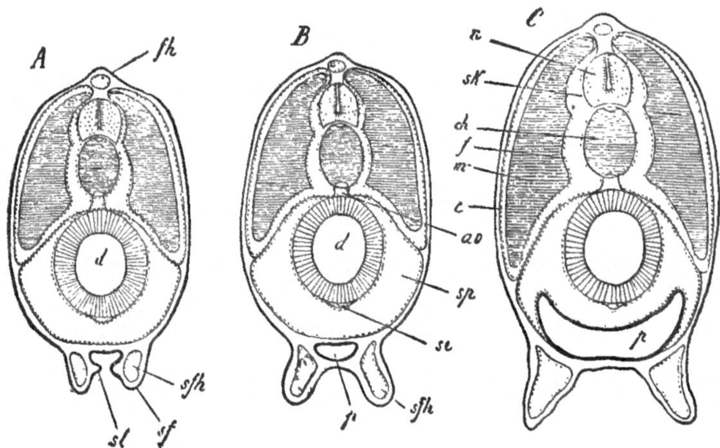

Fig. 13.—Diagrammatic transverse sections of Amphioxus to show three stages (A, B, C) in the development of the atrium. *ao*, aorta; *c*, dermis; *d*, intestine; *f*, fascia; *fh*, cavity for dorsal fin ray; *m*, myomere; *n*, neural tube; *p*, atrium; *sfh*, metapleural folds; *si*, subintestinal vein; *sk*, sheath of notochord and neural tube; *sl*, sub-atrial ridge; *sp*, cœlom. (From Parker and Haswell, after Lankester and Willey.)

blood forward and through the *branchial arches*, where it is aërated. The blood collects again in paired *dorsal aortæ*, which unite back of

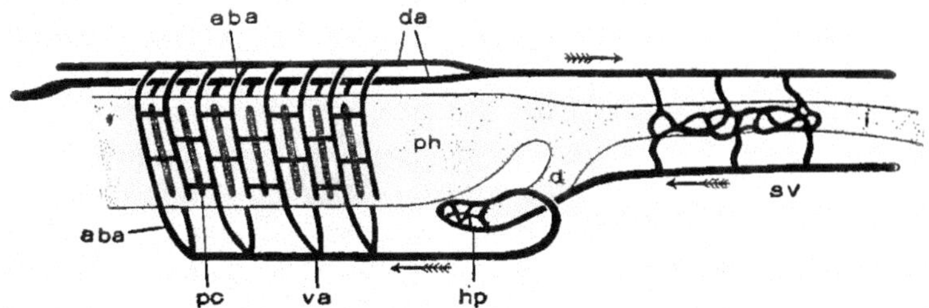

Fig. 14.—Diagram of the circulatory system of Amphioxus. *aba*, afferent branchial artery; *d*, liver diverticulum; *da*, dorsal aortæ; *e*, efferent branchial arteries; *hp*, hepatic portal system; *pc*, pharyngeal clefts; *ph*, pharynx; *sv*, sub-intestinal vein; *va*, ventral aorta. (Modified after Parker and Haswell.)

the pharnyx into a single dorsal aorta, that in turn carries the blood to the various systems. The *ventral vessel* takes a loop about the diverticulum and this loop is interpreted as a simple *hepatic portal system*.

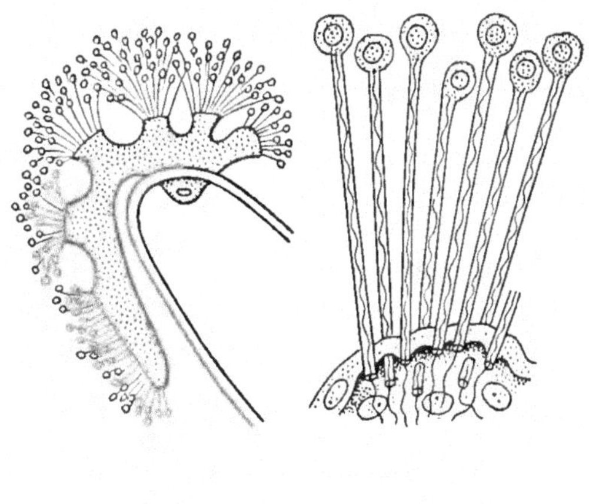

Fig. 15.—A. Nephridium of Amphioxus with incurrent funnels crowned with solenocytes. B. Enlarged view of a portion of one nephridial funnel, showing solenocytes. (After Boveri and Goodrich.)

The **excretory system** consists of paired *nephridia* (Fig. 15, A) with ciliated *nephrostomes* from which protrude knobbed cells called *solenocytes* (Fig. 15, B). The nephridia are true cœlomoducts, leading from the greatly reduced *cœlom* to the atrium. They are segmental organs and occur typically a pair to a metamere in the pharyngeal region. The resemblance between these structures and those of annelids is fairly close. The **muscular system** consists of segmental *myotomes* separated by connective tissue *myocommata.* The myotomes are chevron-shaped as seen from the side, and resemble those of the fishes.

The **spinal cord** has **spinal nerves** (Fig. 11) that alternate on the two sides and have dorsal and ventral roots. The **fin system** is very primitive, consisting of a low continuous *median fin-fold* running uninterruptedly about the tail and ending back of the atriopore. The fin-folds are supported by short connective tissue *fin-rays.* Paired ridges, called *metapleural folds,* run along the ventro-lateral portions of the body; they have been thought of as the primordia of paired appendages. The integument consists of a single layer of ectodermal cells and several layers of dermal cells.

The **gonads** are simply metameric pouches of the cœlom in the branchial region. The eggs and sperm escape by rupture of the body wall into the atrium, and fertilization is external. The **sexes** are separate. The **egg** is small and practically yokeless.

Embryology of Amphioxus

> "As an introduction to the study of embryology, and as an indispensable aid to a reasonable appreciation of the value of embryological facts, the life-history of Amphioxus provides an object, which for its capacity of application to almost every branch of zoölogical discussion is perhaps unrivaled. All the fundamental structures of the body are laid down with schematic clearness." (Willey.)

The ovum is microscopic and resembles those of many marine invertebrates. Spawning occurs at sun-down when simultaneously females and males discharge ova and spermatozoa into the sea-water, where fertilization occurs. Maturation phenomena resemble those of invertebrates, as do also the cleavage stages, the first two cleavages being from pole to pole (meridional) and the third equatorial, producing a tetrad of *micromeres* and a tetrad of *macromeres* (Fig. 16). The micromere cells divide more rapidly than the macromere cells and a *blastula* (Fig. 16, F) is formed with smaller ectodermal cells at the animal pole and larger endodermal cells at the vegetal pole. The endoderm cells soon cave into the dome of ectoderm cells and form an incipient *gastrula* as in Fig. 17, A, B, C. This invagination continues until a typical *embolic gastrula* is formed as in Figs. 17, D and E. A

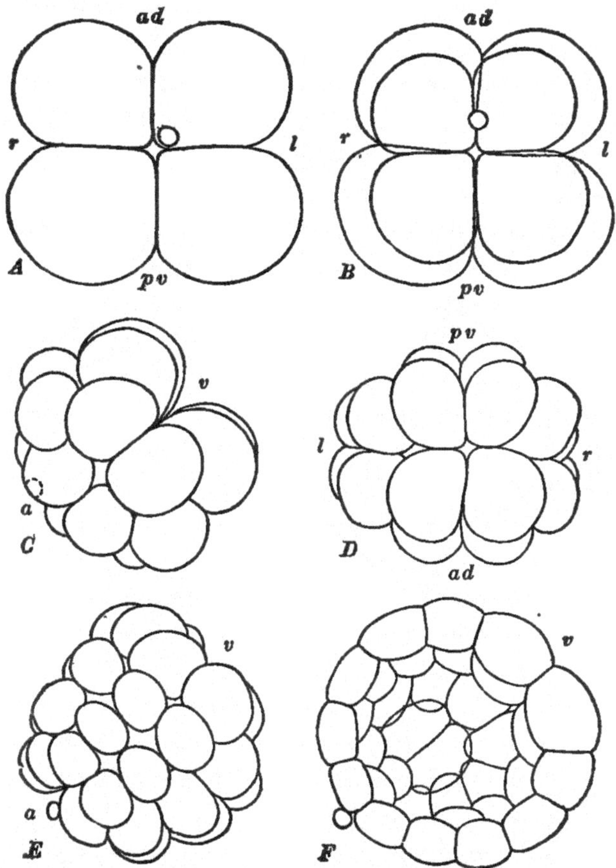

Fig. 16.—Cleavage of Amphioxus. A. Four-cell stage viewed from the animal pole. The two antero-dorsal cells are the smaller. B. Eight-cell stage viewed from the animal pole showing the four sizes of cells. C. Sixteen-cells viewed from the left side. D. Thirty-two cells viewed from the vegetal pole. E. Thirty-two passing into sixty-four cells, viewed from the antero-dorsal region. F. Optical section of right half of young blastula. About 128 cells. *a*, animal pole; *ad*, antero-dorsal; *l*, left; *pv*, posterior ventral; *r*, right; *v*, vegetal pole. (From Kellicott's "Outlines of Chordate Development" [Henry Holt and Company] after Cerfontaine.)

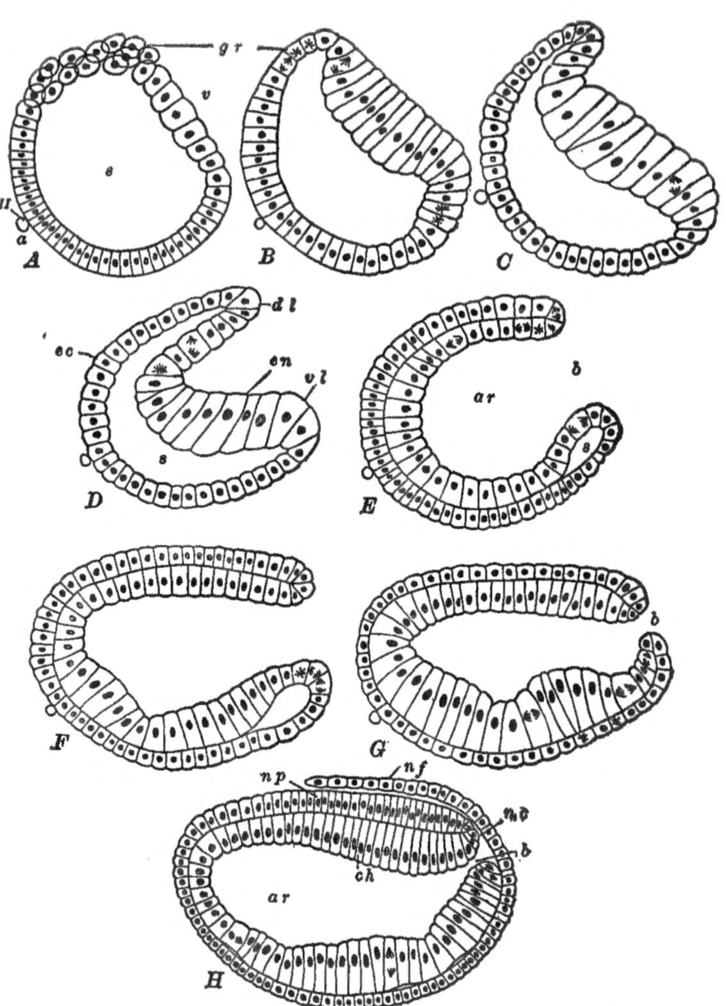

Fig. 17.—Gastrulation of Amphioxus. A. Blastula showing flattening of the vegetal pole and rapid proliferation of cells in the posterior region (germ ring). B. Flattening more pronounced; mitosis in cells of germ ring. C. Commencement of the infolding (invagination) of the cells of the vegetal pole. C. Continued infolding, and inflection, or involution, of ectoderm cells in the dorsal lip of the blastopore. The blastocœl becoming obliterated and the archenteron being established. E. Invagination complete. Continued involution of the dorsal lip of blastopore. Mitoses in germ ring. F. Constriction of blastopore and commencement of elongation of the gastrula. Remnants of blastocœl in ventral lip of blastopore. H. Neurenteric canal established by overgrowth of neural folds. Continued mitosis in germ ring. *a*, animal pole; *ar*, archenteron; *b*, blastoporal opening; *ch*, rudiment of notochord; *dl*, dorsal lip of blastopore; *ec*, ectoderm; *en*, endoderm; *gr*, germ ring; *nc*, neurenteric canal; *nf*, neural fold; *np*, neural plate; *s*, blastocœl or segmentation cavity; *v*, vegetal pole; *vl*, ventral lip of blastopore; *II*, second polar body. (From Kellicott's "Outlines of Chordate Development" [Henry Holt and Company] after Cerfontaine.)

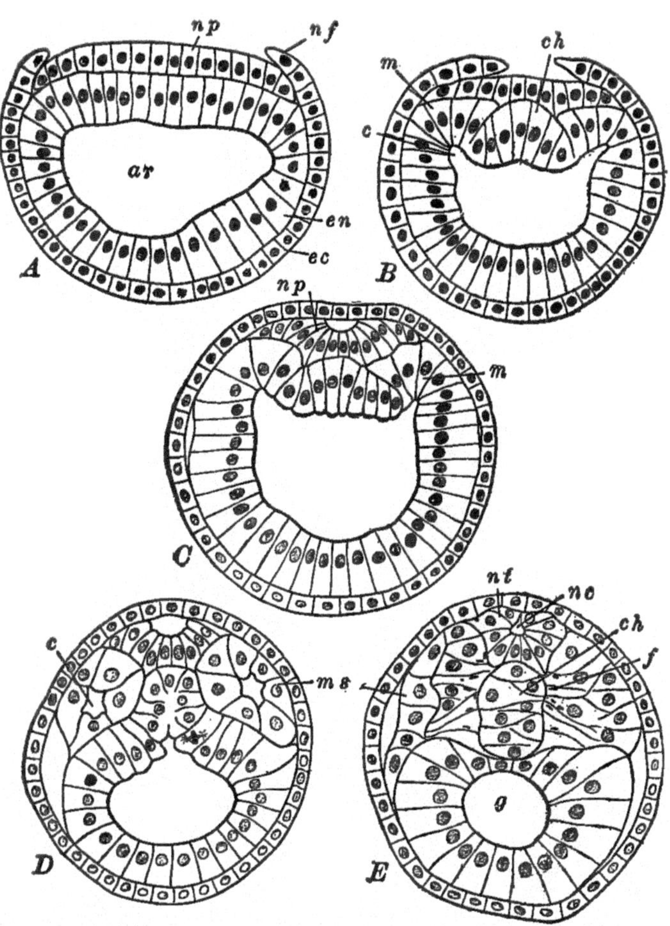

FIG. 18.—Transverse sections through young embryos of Amphioxus, showing formation of nerve cord, notochord, and mesoderm. A. Commencement of the growth of the superficial ectoderm (neural folds) above the neural plate (medullary plate). B. Continued growth of the ectoderm over the neural plate. Differentiation of the notochord, and first indications of mesoderm and enterocœlic cavities. C. Section through middle of larva with two somites. Neural plate folding into tube. D. Section through first pair of mesodermal somites now completely constricted off. E. Section through middle of larva with nine pairs of somites. Neural plate folded into a tube. Notochord completely separated. In the inner cells of the somites muscle fibrillæ are forming. *ar*, archenteron; *c*, enterocœl; *ch*, notochord; *ec*, ectoderm; *en*, endoderm; *f*, muscle fibrillæ; *g*, gut cavity; *m*, unsegmented mesoderm fold; *ms*, mesodermal somite; *nc*, neurocœl; *nf*, neural fold; *np*, neural plate; *nt*, neural tube. (From Kellicott's, "Outlines of Chordate Development" [Henry Holt and Company] after Cerfontaine.)

late gastrula shows a distinct bilaterality, a dorsal and ventral aspect, and anterior and posterior lips to the blastopore. The gastrula is a

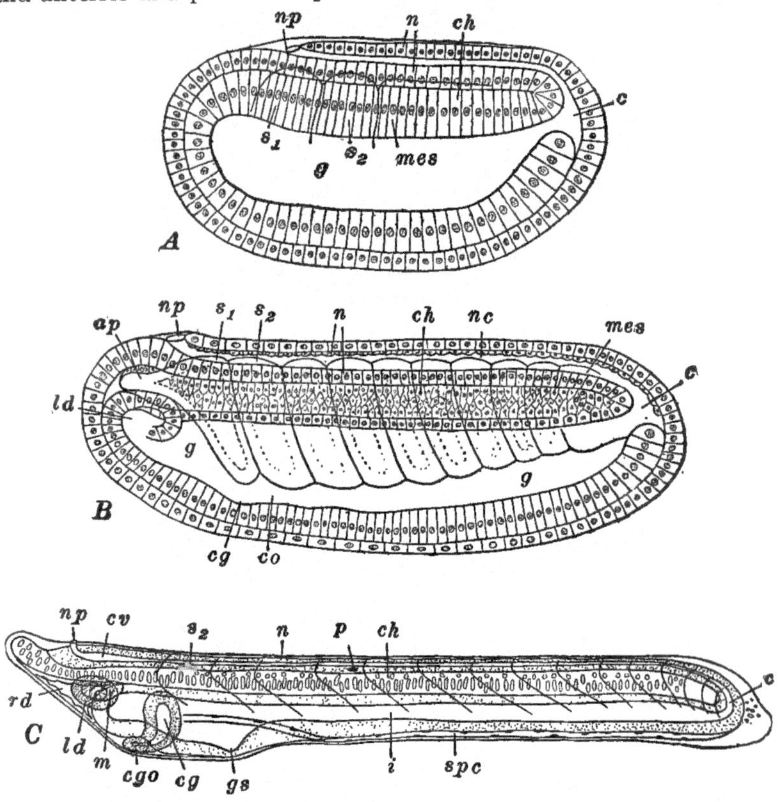

FIG. 19.—Optical section of young embryos of Amphioxus. The cilia are omitted. A. Two-somite stage, approximately at the time of hatching, showing relation of neuropore and neurenteric canal. B. Nine-somite stage, showing origin of anterior gut diverticula. C. Fifteen-somite stage. End of the embryonic period. *ap*, anterior process of first somite; *c*, neurenteric canal; *ch*, notochord; *cgo*, external opening of club-shaped gland; *co*, cœlomic cavity of somite; *cv*, cerebral vesicle; *g*, gut cavity; *gs*, rudiment of first gill slit; *i*, intestine; *l*, left anterior gut diverticulum; *m*, mouth; *mes*, unsegmented mesoderm; *n*, nerve cord; *p*, pigment spot (eye spot); *rd*, right anterior gut diverticulum; s_1, s_2, first and second mesodermal somites; *spc*, splanchnocoel (body cavity). (From Kellicott's "Outlines of Chordate Development" [Henry Holt and Company] after Hatschek.)

ciliated embryo that moves about within its vitelline membrane very much like that of an echinoderm or annelid. After about eight hours the gastrula emerges from the vitelline membrane as a *free-swimming larva.* It is now an elongated gastrula and has begun to develop the rudiments of definitive structures, such as notochord, cœlom, and medullary plate. The first signs of metamerism are seen in connection with a series of paired lateral pouches derived from the archenteron, (Fig. 19 A) beginning near the anterior end and proceeding posteriorly until fourteen pairs are formed. Additional segments arise as a direct outgrowth of the hind end of the body.

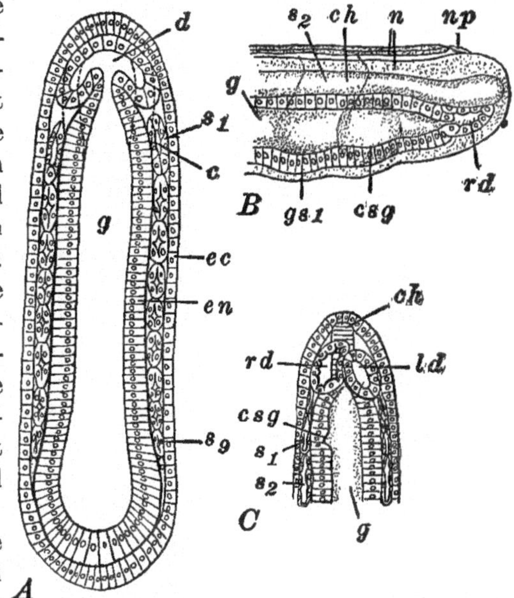

Fig. 20.—Sections through young Amphioxus embryos showing the origin of the anterior gut diverticula. A. Frontal section through embryo with nine pairs of somites. (See Fig. 19, B). The dotted line marks the course of the gut wall ventral to the level of the section. B. Optical sagittal section through anterior end of embryo with thirteen pairs of somites showing position of right anterior gut diverticulum. C. Same in ventral view. *c*, cœlomic cavity of somite; *ch*, notochord; *csg*, rudiment of club-shaped gland; *d*, rudiment of anterior gut diverticula; *ec*, ectoderm; *en*, endoderm; *g*, gut cavity; *gs1*, rudiment of first gill slit; *ld*, left anterior gut diverticulum; *n*, nerve cord; *np*, neuropore; *rd*, right anterior gut diverticulum; s_1, s_2, s_9, first, second, ninth mesodermal somites. (From Kellicott's "Outline of Chordate Development" [Henry Holt and Co.] after Hatschek.)

The medullary plate (Fig. 18) is cut off from the ectoderm of the body wall in a peculiar way by the coöperation of two factors. The ectodermal ridges at the side of the plate arch over the middle parts of the plate and the ectoderm of the ventral lip of the blastopore grows like a sheet over the blastopore and closes the latter so that instead of opening to the outside it communicates with the neural tube and has become a *neurenteric canal,* the homologue of which is found in all vertebrates and constitutes one of the most peculiar characters of the group.

A detailed study of the organogeny of Amphioxus would not fall within the scope of the present volume, but brief mention should be made of a few of the later stages, and of the more significant changes involved. A stage of much interest is shown in Fig. 20. This is an embryo showing nine pairs of primitive somites and in process of budding off the right and left head cavities. Up to this point the embryo is strictly bilaterally symmetrical. The first disturbance of bilaterality is seen in connection with the head-cavities, for the right one grows large and becomes the cavity of the snout or *pre-oral body cavity*, lying beneath the notochord, and the left becomes a rudimentary structure, the *pre-oral pit.* In Fig. 20, C is shown the beginning of the overgrowth of the right head cavity. This crowding over of the right side of the head toward the left disturbs the primitive location of the mouth, so that it opens to the left side and only secondarily adjusts itself to the nearly median postition characteristic of the adult. The gill-slits, myotomes, spinal nerves, nephridia, and other metameric structures are thrown out of the primitive paired arrangement and alternate on the two sides throughout life. In the genus *Asymmetron* the asymmetry of the whole body is more pronounced than in the more typical members of the Amphioxus family; hence the name. This twisting about of the body is a characteristic of sessile animals and it is significant to find in Amphioxus a condition comparable with that seen in the echinoderms, which are entirely headless creatures, and in Balanoglossus in which the head is greatly reduced.

More advanced larvæ of Amphioxus show the migration of the mouth from the left side to the middle, the method of uniting the primary gill-slits into the definitive double type separated by tongue bars, the formation of the oral *hood* and *buccal cirri*. There is no sudden change equivalent to metamorphosis, but at this time the larva gives up its free-swimming habit and begins to lead the semi-sedentary, sand-burrowing life of the adult. The period of adolescence is long and slow and culminates in the maturing of the gonads.

SUB-PHYLUM II. UROCHORDATA (TUNICATES)

Just as there is no doubt about the affinities of Amphioxus to the vertebrates, so there is no question but that the tunicates are related to Amphioxus.

In contrast with the cephalochordates, in which one family of two genera and only a few species exist, the urochordates constitute a large assemblage of highly diverse forms, ranging from purely pelagic to purely sessile forms, and from solitary forms of large size to colo-

Fig. 21.—Sketch of the chief kinds of Urochordata found in the sea showing their distribution and habits. The dotted lines on the left indicate the life zones of the sea: the surface or pelagic zone; the middle zone; and the sea-bottom zone. (From Herdman in the Cambridge Natural History, Vol. VII.)

nial forms of comparatively small size. They are plentiful all over the oceans from the shallow shore waters to the deepest abysses. By far the majority of them are sessile in habit in adult life, remaining permanently rooted to one spot. Some of the free-swimming forms (*Larvacea*) are interpreted as pædogenetic forms, which have retained the larval condition throughout life; others are evidently derived secondarily from sessile ancestors and now live a free pelagic life. The composite illustration (Fig. 21) shows the general appearance of many common types of urochordates. The sessile type seems to be representative of, and will serve to illustrate, the essential features of the whole sub-phylum.

Order I. Ascidiacea

A TYPICAL ASCIDIAN

An external view of an ascidian (Fig. 22) reveals little of interest. It does not even look like a living creature; much less a chordate. It is a dull brown object resembling a leather bag or bottle with two short necks, one terminal and one somewhat on the side, the former being the *oral funnel* and the latter the *atrial funnel* or *atriopore*. If one watches the water currents carefully he will observe that the water enters the *oral funnel* in a steady stream and exits from the atriopore. The wrinkled brown covering or *tunic* (which gives the name Tunicata to the group) is merely a lifeless protective layer composed of animal cellulose, a substance almost identical with wood. The body wall within the tunic is composed largely of glands and connective tissue and shows neither myotomes nor any other segmented structures. The tunic and body wall of one side removed, we get a view of the interior (Fig. 23). What we find is little more than an exaggerated *food-concentrating apparatus*, closely comparable with that of Amphioxus. The circular oral funnel opens through a *velum*, with *velar tentacles*, into a great sac-like *pharynx* that occupies far more than its share of the available space. This pharynx is an elaborate sieve with countless small slit-like openings, *stigmata*, which are subdivided pharyngeal clefts. On the ventral side there is an

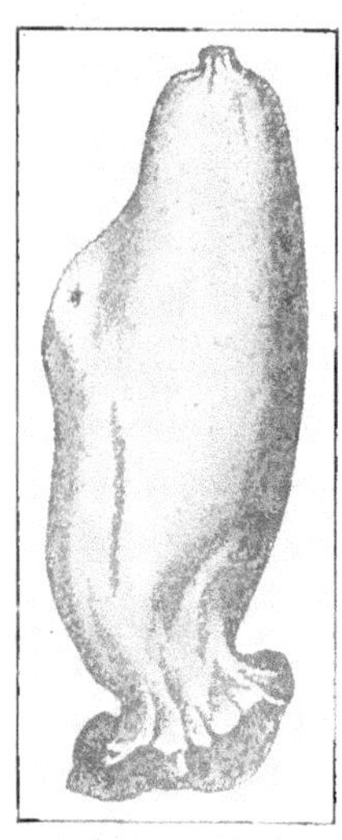

Fig. 22.—External appearance of a typical Ascidian (*Ascidia*) seen from the right side. (From Parker and Haswell, after Herdman.)

endostyle, around the mouth *peripharyngeal grooves*, and on the dorsal side a *dorsal lamina* (corresponding to the dorsal groove of Amphioxus). The method of food concentration and transportation is similar to that of Amphioxus, but the machine seems to be of a much improved type, appropriate for purely sedentary life. An *atrial cavity* surrounds the pharynx, which is inclosed by a *mantle* that surrounds the whole body. The ectoderm of this mantle it is that secretes the tunic. The *atriopore*, instead of being posterior in position and backwardly directed, is close to the mouth and forwardly directed. The stomach opens near the bottom of the pharynx and the intestine takes a complete turn and opens forward into the atrium. There is *no notochord, no neural tube*; indeed almost none of the structures characteristic of the dorsal side are present. A poor excuse for a *brain* occurs on the dorsal side of the mouth embedded in the body wall between the two funnels. It is nothing but a ganglion associated with a *sub-neural gland* and with a duct entering the pharynx. This last structure has been compared to the vertebrate *hypophysis*. The ascidian comes as near being an animated pharynx as could well be, only certain absolutely essential elements of the other parts of the body being retained.

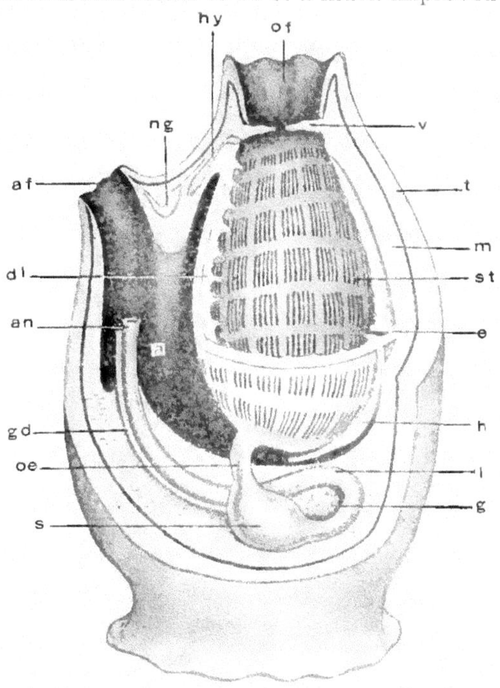

FIG. 23.—Internal anatomy of a typical Ascidian. *a*, atrial cavity; *af*, atrial funnel or atriopore; *an*, anus; *dl*, dorsal lamina; *e*, endostyle; *g*, gonad; *gd*, duct of gonad; *h*, heart; *hy*, hypophyseal duct; *i*, intestine; *m*, mantle, *ng*, neural gland; *oe*, oesophagus; *of*, oral funnel or mouth; *s*, stomach; *st*, stigmata or subdivided pharyngeal clefts; *v*, velum. (Considerably modified after Hertwig.)

The *heart* is no less than an oddity. It is a pulsating tube, lying ventral to the stomach, open at both ends into sinuses. It works by peristaltic contractions first in one direction for a few beats and then in the other. The *colorless blood* is thus driven through *sinuses* first toward the anterior and then toward the posterior organs. No other such heart is known among animals.

All ascidians are without exception *hermaphroditic,* as befits their sessile life, though there is a high degree of self-sterility. Usually the eggs of one individual meet the sperm of another in the sea-water.

An adult ascidian does not therefore appear to be an impressive candidate for the honor of being classed as a chordate, but it makes a much better showing when we consider the entire ontogeny.

Development and Metamorphosis of an Ascidian

The ascidian egg is small and almost yolkless. The cleavage stages, blastula, and early gastrula are essentially similar to those of Amphioxus, except that the embryonic life is longer. The gastrula is not a ciliated larva but a passive embryo; in fact the embryonic development is greatly prolonged and the larval life does not commence until the egg hatches and the advanced *tadpole* larva emerges.

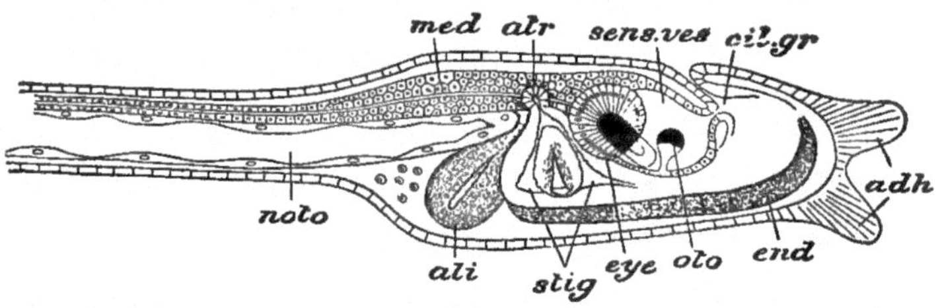

Fig. 24.—Anterior end of the "tadpole" larva of an ascidian (*Ascidia mammillata*) in optical section. *adh,* adhesive papillæ; *ali,* rudimentary alimentary tract; *atr,* atrial aperture; *cil. gr,* ciliated diverticulum, becoming ciliated funnel; *end,* endostyle; *eye,* eye; *med,* nerve-cord; *noto,* notochord; *oto,* otocyst; *sens. ves,* sense-vesicle (cavity of brain); *stig,* earliest stigmata or pharyngeal clefts. (From Parker and Haswell, after Kowalevsky.)

The Ascidian Tadpole.—When we examine the larva of the tunicate (Fig. 24) we understand why the Urochordata have been classed with the chordates. The notochord is quite typical but extends only as far as the pharynx, or, as the name Urochordata implies, is confined to the tail. The tail is also characterized by having dorsal and ventral fins and a neural tube. The *brain* or sense vesicle is much larger than the neural tube and is much more like a vertebrate brain than is that of Amphioxus, since it has connected with it paired *optic vesicles* and otocysts. The tadpole larva therefore makes up for the deficiencies of the adult in having a good notochord and a neural tube. It is, however, embryonic in the pharynx and alimentary tract. At first there is no mouth and no anal opening and the pharynx has only a very few clefts. The atrium and atriopore also are at first absent.

Larval Metamorphosis.—The tadpole larva leads an active free-swimming life for a short time and then settles down upon the bottom or upon some other solid object. Attachment is secured by means of three sucker-like papillæ or "*chin-warts*" and undergoes a very rapid metamorphosis. All of the structures pertaining to a free active life atrophy and the food-concentrating apparatus becomes greatly elaborated. Tail, notochord, and neural tube disappear. The brain degenerates to a ganglion. While the tail is atrophying the chin or ventral side of the pharynx grows out of all proportion to the rest of the body so that the mouth is pushed away from the point of attachment and comes to be at the free end. The dorsal side of the pharyngeal region diminishes rapidly also, so as to cause the stomach and intestine to turn upward into a U-shaped body, the anus opening in the same direction as the mouth. The atrium arises as two ectodermal invaginations on right and left sides. These remain separate for some time, but, after they have grown in so as to surround the whole pharynx, the two cavities fuse and lose the bounding wall in the dorsal side, but remain separate at the median ventral line near the endostyle. There is therefore not much resemblance, so far as mode of origin is concerned, between the atrium of a tunicate and that of Amphioxus. The one point in common is that it is lined with ectoderm.

The tadpole larva is a very good chordate larva, quite comparable in most respects to a vertebrate larva, such as that of a fish or a frog tadpole; but the adult ascidian is a lowly creature with an organization not much higher than that of a cœlenterate or at best, a worm. It is quite evident then that the ascidian presents a typical case of degeneration accompanying the assumption of sessile life. The well-developed brain of the ascidian larva indicates that the ancestral chordate had a better brain and head than has Amphioxus, and had an organization not far from intermediate between that of the adult Amphioxus and that of the ascidian larva.

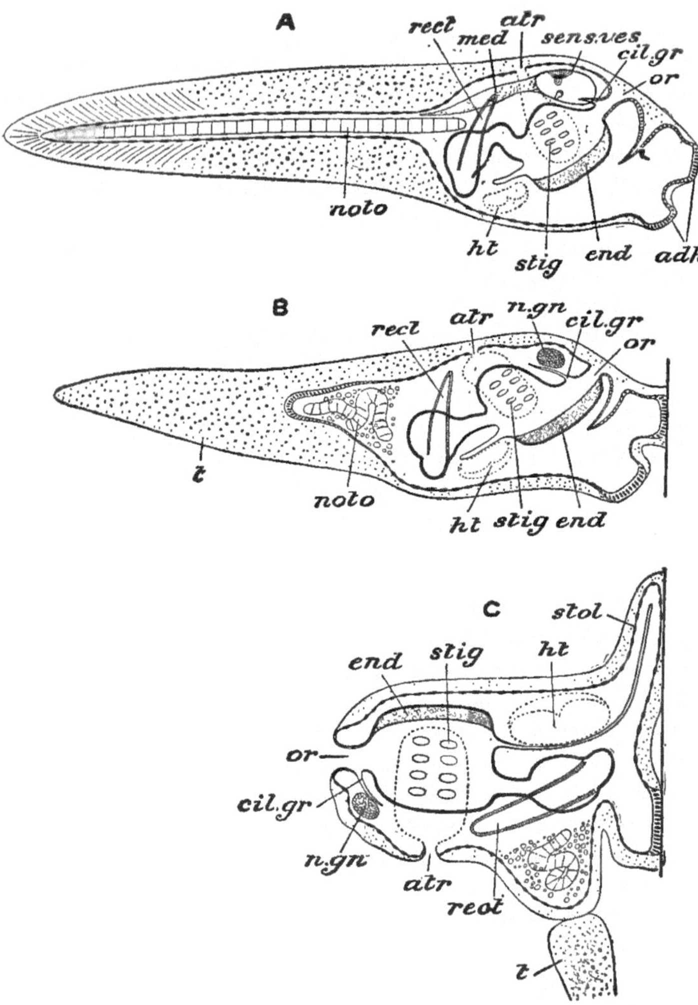

FIG. 25.—Diagram showing the metamorphosis of the free-swimming "tadpole" larva of an Ascidian into the sessile or fixed adult condition. A, stage of free-swimming larva; B, larva recently fixed; C, older fixed stage. *adh*, adhesive papillæ; *atr*, atrial cavity; *cil. gr*, ciliated diverticulum, becoming ciliated funnel; *end*, endostyle; *ht*, heart; *med*, nerve cord of trunk; *n. gn*, nerve ganglion; *noto*, notochord; *or*, oral aperture; *rect*, rectum; *sens. ves*, sense vesicle or brain; *stig*, stigmata; *stol*, stolon; *t*, tail. (From Parker and Haswell, after Seeliger.)

Colonial Ascidians (Ascidiæ Compositæ).—All of the colonial ascidians are produced by budding of a single parent individual. It is usual to find a group of ascidiozoöids grouped together in such a way that they have a common cloacal opening (Figs. 26 and 27). Naturally also the tunics of the various individuals form a common matrix in which they appear to be embedded. The so-called "sea-pork" from our Atlantic coast is a typical example of that fixed type of colonial ascidian, which forms incrusting masses on rocks and piles and gives the appearance of a piece of pinkish fat pork, perforated at intervals by groups of pores, which are the mouths and atrial pores of groups of zoöids.

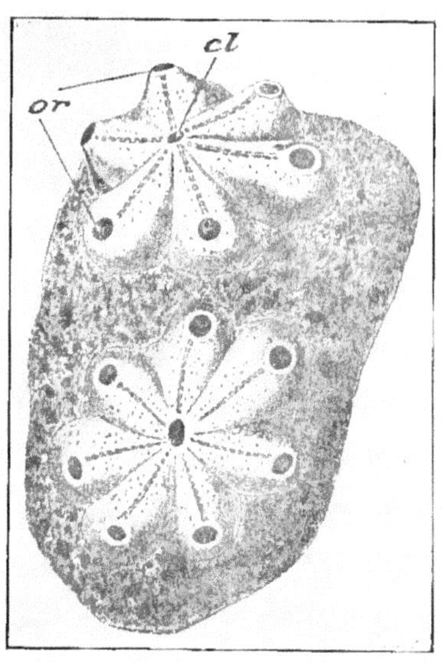

Fig. 26.—Surface view of two systems of colonial ascidians of the species *Botryllus violaceus*. *cl*, cloaca or common atrial funnel; *or*, oral funnels of the individual zooids. (From Herdman, after H. Milne-Edwards.)

Free-Swimming Colonies of Ascidians (Ascidiæ Luciæ).—These are pelagic forms in which the ascidiozoöids are arranged in such a way that the mouth opening is on the outside of a hollow cylinder and the atrial opening on the inside (Fig. 28). Since only one end of the cylinder is open, a steady current of water is forced out at this end

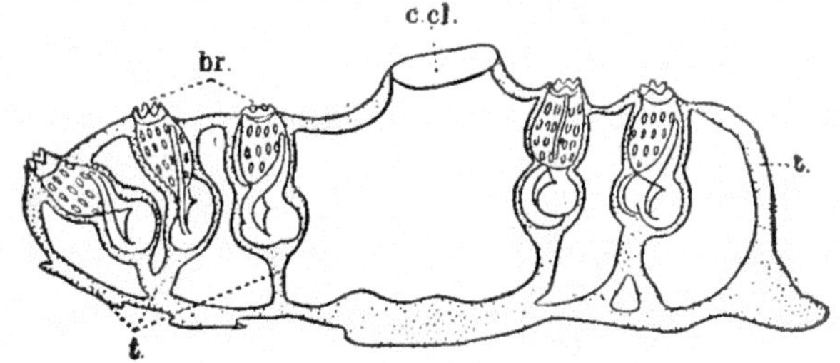

Fig. 27.—Sectional view of a portion of a colony of a colonial ascidian, (*Diplosoma*) showing: *c. cl*, common cloaca into which atrial openings of all zoöids empty; *br*, branchial apertures or oral funnels of individual zoöids; *t*, test or tunic forming the common matrix of the colony. (From Herdman.)

and serves to propel the colony through the water. *Pyrosoma*, a typical member of this group, is found swimming near the surface of

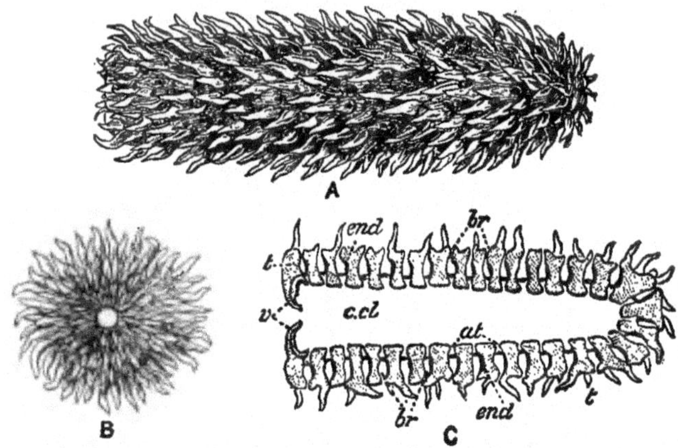

FIG. 28.—The free-swimming colonial ascidian, *Pyrosoma*. A, lateral view (nat. size); B, view of the open end; C, diagram of longitudinal section; *at*, atrial pores opening into the central cavity or cloaca, *ccl*, of the colony; *br*, branchial or oral apertures opening to the outside; *end*, endostyle; *t*, test or tunic; *v*, velum or diaphragm at terminal opening. (From Herdman.)

warm seas and is brilliantly phosphorescent. Colonies vary from an inch to upwards to twelve feet in length.

ORDER II. THALIACEA (SALPIANS)

These are all free-swimming tunicates that live near the sea surface and are doubtless derived from the free-swimming ascidians. They may be either solitary or colonial. In the typical salpians the body is strikingly like a barrel, open at both ends, but with a partition across near the middle. The resemblance to a barrel is enhanced by the presence of a considerable number of muscle bands that encircle the cylinder like barrel-hoops. They are semi-transparent forms, often beautifully colored, and are capable of making very good progress by forcing water out of the broad cloacal opening by contracting the muscles of that region.

The structure is well shown in the classic salpian type *Doliolum* (Fig. 29). At the left there is the widely open *oral funnel* which opens into the broad sac-like *pharynx*. This has a large number of distinct *branchial clefts*, an *endostyle*, and *peripharyngeal grooves*, but no distinct dorsal lamina. The *atrium* has two wings that extend laterally to receive water from the pharyngeal clefts, and opens broadly on the

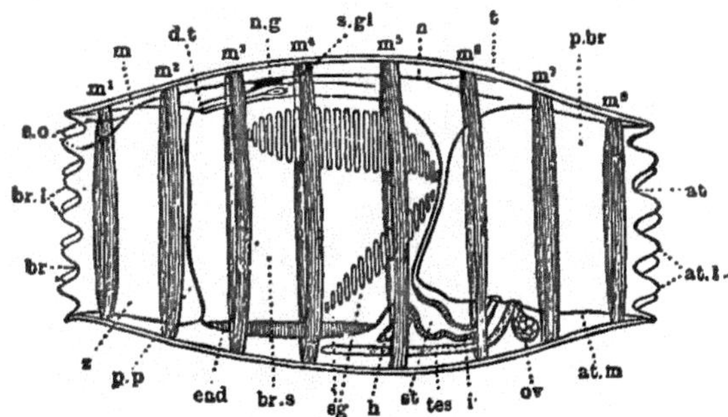

FIG. 29.—Individual of the sexual generation of the Salpian *Doliolum tritonis* × 10. *at*, atrial aperture; *atl*, atrial lobes; *atm*, wall of atrium; *br*, branchial or oral aperture; *brl*, branchial lobes or tentacles; *brs*, branchial sac or pharynx; *dt*, dorsal tubercle; *end*, endostyle; *h*, heart; *m*, mantle; m_1–m_8, circular muscle bands; *n*, nerve; *ng*, nerve ganglion; *ov*, ovary; *pbr*, peribranchial or atrial cavity; *pp*, peripharyngeal bands; *sg*, stigmata; *sgl*, subneural gland; *so*, sense organ; *st*, stomach; *t*, test or tunic; *tes*, testis; *z*, prebranchial zone. (After Herdman.)

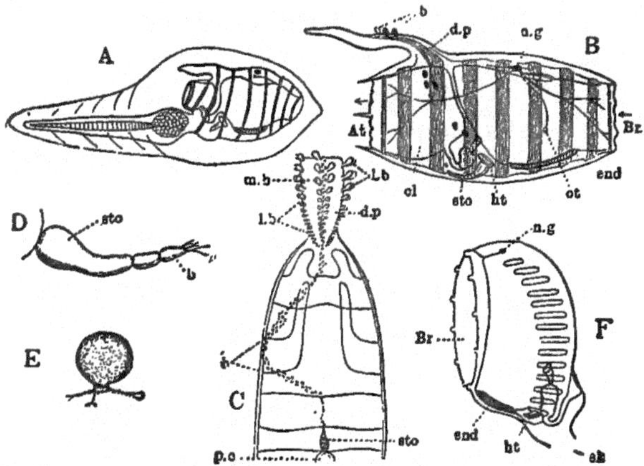

FIG. 30.—Life history of *Doliolum*. A, tailed larval stage; B, "nurse" or oözoiöd, showing buds (blastozoöids) migrating from the ventral stolon to the dorsal process; *c*, posterior part of much later oözoöid to show buds arranged in three rows on the dorsal process; D, stolon segmenting; E, young migrating bud; F, trophpzoöid developed from one of the buds of a lateral row.

At, atrial aperture; *b*, buds; *Br*, branchial or oral aperture; *cl*, cloaca; *dp*, dorsal process; *end*, endostyle; *ht*, heart; *lb*, lateral buds; *mb*, median buds; *ng*, nerve ganglion; *ot*, otocyst; *pc*, pericardium; *sk*, stalk; *sto*, stolon. (From Herdman, after Uljanin and Barrois.)

right as an *atriopore*. The salpian is therefore not twisted into a U-shaped body but is secondarily somewhat straightened out, as compared with an ascidian, since the atrium opens in the opposite direction from the mouth. There is, however, evidence in the position of the small, bent stomach and intestine that these now free-swimming forms have been derived from fixed ancestors.

Life Cycle of a Salpian (Doliolum).—An alternation of generations exists between a sexual generation and a sexless generation, which reproduces only by budding. The eggs produced by the sexual form just described for *Doliolum* go through a typical early development and produce tadpole larvæ with large body and comparatively small tail (Fig. 30, A). The whole animal is embedded in a thick coat of jelly. These "tadpoles" undergo a metamorphosis by resorption of the tail, and produce a type of individual much like the sexual invividuals, called *nurses* (Fig. 30, B). On the ventral side near the heart they have a short finger-like process which produces *primary buds* by a process of constriction. These buds migrate over the surface of the nurse by ameboid movement of the peripheral cells, and take up their positions in three longitudinal rows on the *cadophore*, a dorsal process of the body that comes off near the atrial end. The two lateral rows are specialized as *nutritive individuals* and never leave the nurse. Of those in the median row some become detached early and form a second generation of *foster-parents* or nurses, while the rest grow to be *sexual individuals* and ultimately separate one after the other from the nurse.

Colonial Salpians.—The genus *Salpa* is the only representative of the Thaliacea that is colonial in the sexual condition; chains of individuals that remain as colonies, adhering together by means of their mantle walls, are released by the nurse. The development from the egg to the nurse is direct, without a free-swimming tadpole stage. In other respects they are quite like the solitary salpians.

Order III. Larvacea (Appendicularians)

These are free-swimming pelagic forms with a much simplified body region and a persistent tail like that of the larval ascidian or salpian, supported by a well-developed notochord. The whole body (Fig. 31) is usually inclosed in a voluminous gelatinous envelope or test, called a "house," which is secreted by the mantle ectoderm just as a tunicate secretes its wooden tunic. The "house" is, however, merely a temporary structure that serves for a night's lodging, as it were; for the animal leaves it at intervals and is capable of making another "house" in an hour or so. The tail is rather loosely jointed to the body and serves as a flexible paddle in locomotion. The body is like that of a tunicate, with the whole food-concentrating apparatus reduced and simplified (Fig. 32). There is only one pair of *pharyngeal clefts* that open directly through paired atriopores; a condition suggestive of the larval atrial cavities in the ascidians. The *endostyle* also occupies an anterior position similar to that of a larval Amphioxus.

In almost every respect the Larvacea sug-

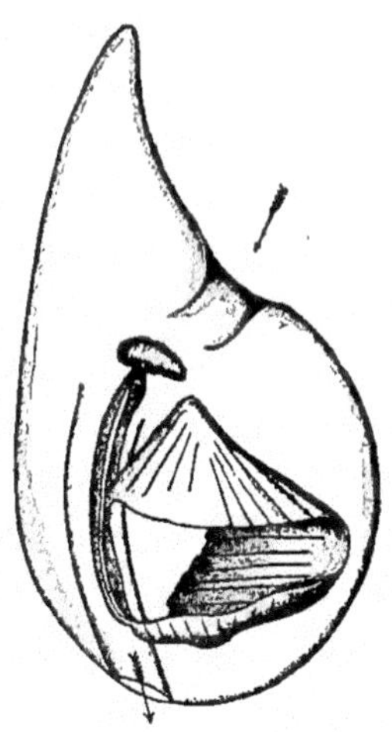

Fig. 31.—An individual belonging to the class Larvacea (*Oikopleura*) in its gelatinous "house." Only the small hammer-shaped object in the main passage-way is the animal itself. The arrows show the current of water through the "house." (From Herdman after Fol.)

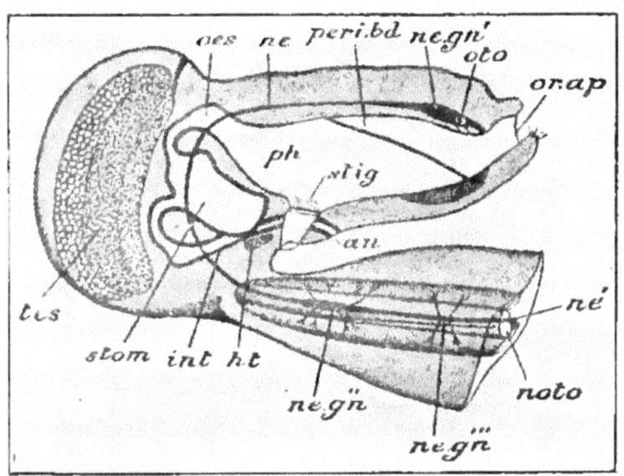

Fig. 32.—Diagram of a larvacean (*Appendicularia*) from the right side, with most of the tail removed. *an*, anus; *ht*, heart; *int*, intestine; *ne*, nerve; *ne'*, caudal portion of nerve; *ne. gn*, principal nerve ganglion; *ne. gn''*, *ne. gn'''*, first two ganglia of nerve of tail; *noto*, notochord; *oes*, œsophagus; *or. ap*, oral aperture; *oto*, otocyst (statocyst); *peri. bd*, peripharyngeal band; *ph*, pharynx; *tes*, testis; *stig*. one of the single pair of pharyngeal clefts, stigmata; *stom*, stomach. (From Parker and Haswell, after Herdman.)

gest a permanent larval condition akin to neotony or pædogenesis. The alternative view, that these forms are prototypic of the ancestral chordate, is, we believe, not well taken; for there are evidences in the U-shaped intestine that the Larvacea have come from a sessile ancestor. It is because we consider the Ascidiacea as most generalized and the Larvacea as degenerate or pædogenetic, that we have arranged the orders of Urochordata as follows:

Order I. Ascidiacea (Ascidians).
  Sub-Order 1. Ascidiæ Simplices.
  Sub-Order 2. Ascidiæ Compositæ.
  Sub-Order 3. Ascidiæ Luciæ.

Order II. Thaliacea (Salpians).
  Sub-Order 1. Cyclomyaria (e.g. Doliolum).
  Sub-Order 2. Hemimyaria (e.g. Salpa).

Order III. Larvacea (Appendicularians).

SUB-PHYLUM III. HEMICHORDATA

The status of this group in the chordate phylum is at best very insecure. While the Enteropneusta, as exemplified by *Balanoglossus*, seem, on account of certain resemblances to Amphioxus, to have some rather well-founded claims to vertebrate relationship, the Pterobranchia and Phoronidia are admitted to the phylum Chordata largely by virtue of the almost unmistakable affinities of *Cephalodiscus* with *Balanoglossus*, and of *Phoronis* with *Cephalodiscus*. Without *Balanoglossus*, it is unlikely that *Cephalodiscus* would be considered as a chordate, and without *Cephaldiscus*, there would be little to connect *Phoronis* with this phylum.

ORDER I. ENTEROPNEUSTA

These are worm-like, burrowing forms with numerous paired gill-slits; intestine running straight from mouth to terminal anus. There are three main body divisions: anterior *proboscis;* ring-shaped *collar;* and segmented *trunk*, resembling the body of an annelid worm. They are of moderate size, ranging from one inch to four feet in length. The burrowing habits of *Balanoglossus* (Fig. 33) reveal the significance of these structural peculiarities. Spengel and Ritter have described these in some detail. The proboscis (Fig. 34), which is capable of becoming swollen or turgid by taking in water through the *collar pore*, is pushed forcibly against the sand much like the pig's snout in "rooting." Sand is loosened and pushed aside until a hole is made deep enough for the proboscis to bury itself. Then the *collar*, by filling itself tightly with water through the *collar pore*, comes to fit itself rigidly against the sides of the hole and to act as a fulcrum for the further operations of the proboscis. Thus progress is accelerated. Sand is quickly loosened and taken up by the *scoop-like mouth* which is situated between the base of the proboscis and the ventral part of the collar. If

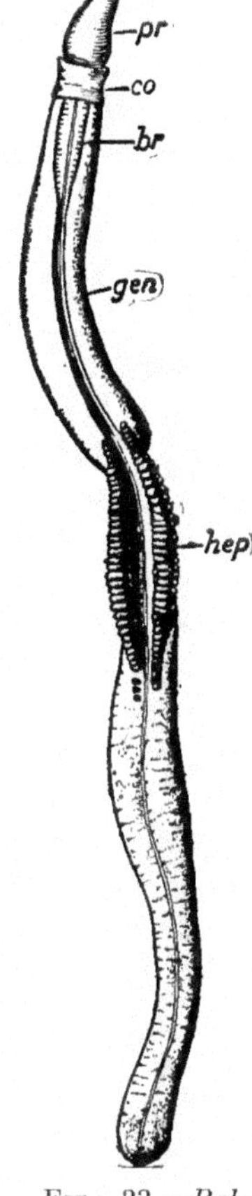

FIG. 33.—*Balanoglossus*. *br*, branchial region; *co*, collar; *gen*, genital ridges; *hep*, hepatic prominences formed by hepatic (liver) cæca; *pr*, proboscis. (From Lull after Spengel.)

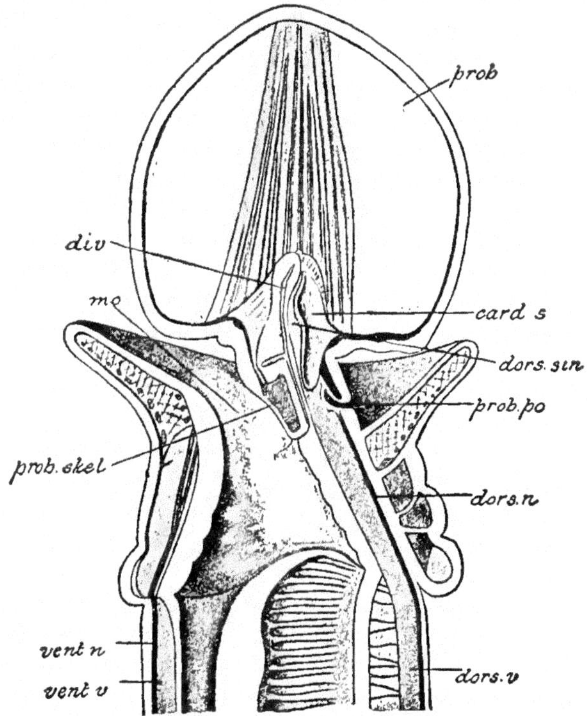

FIG. 34.—*Balanoglossus*. Diagrammatic sagittal section of anterior end. *card. s*, cardiac sac; *div.*, diverticulum, (supposed notochord); *dors. n*, dorsal nerve strand; *dors. sin*, dorsal sinus; *dors. ves*, dorsal vessel; *mo*, mouth; *prob*, proboscis; *prob. po*, proboscis pore; *prob. skel*, proboscis skeleton; *vent. n*, ventral nerve strand; *vent. v*, ventral vessel. (From Parker and Haswell, after Spengel.)

we may be pardoned a somewhat homely analogy, we may compare the collar of *Balanoglossus* to a too-large collar of a man and the base of the proboscis to the too-small neck of the same individual. The space between the collar and the neck in front is like the ventral mouth of *Balanoglossus*. Earth scooped in by this mouth is passed rapidly through the alimentary tract and is deposited on the surface by the anus. Digging in this way the animal quickly disappears from view.

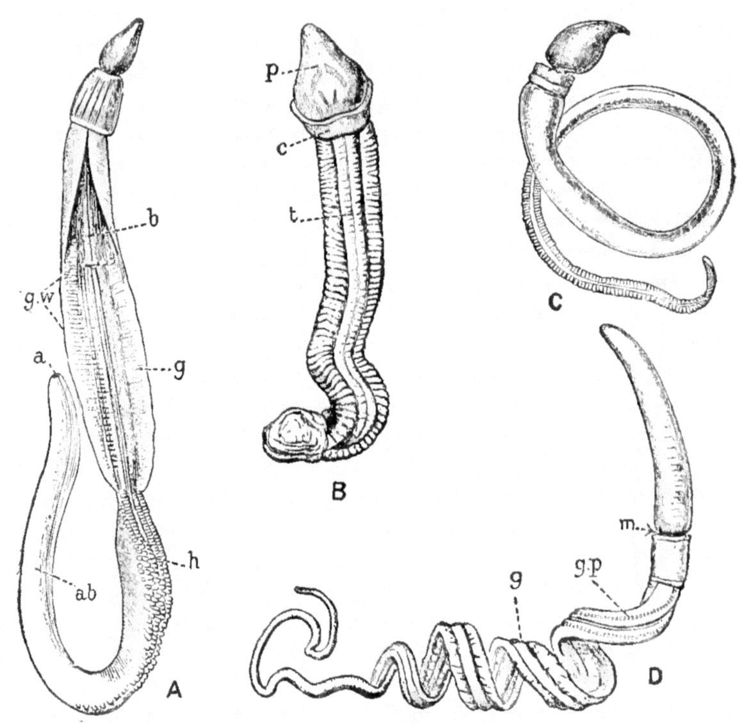

FIG. 35.—Various types of Enteropneusta, relatives of *Balanoglossus*. A. *Balanoglossus clavigerus;* B. *Glandiceps hacksi;* C. *Schizocardium brasiliense;* D. *Dolichoglossus kowalevskii;* *a*, anus; *ab*, abdominal and caudal regions; *b*, branchial region; *c*, collar; *g*, genital region; *gp*, gill pore or branchial cleft; *gw* genital wing; *h*, hepatic region; *m*, position of mouth; *p*, proboscis; *t*, trunk. (From Harmer: A, B, and C, after Spengel; D, after Bateson.)

There are about thirty species of Enteropneusta, grouped into nine genera. Their mode of life is essentially the same as that of the genus *Balanoglossus*. The other genera differ mainly in the relative importance and size of the three body regions, in the shape of these regions, and in their coloration, which is often quite striking, the prevailing colors being brilliant yellows, red-orange tints, and various shades of green. The most essential anatomical differences between the genera are concerned with the degree of development of the so-called notochord.

A description of the more significant **anatomical details of Balanoglossus** will serve as a characterization of the group.

The Notochord.—The structure which is identified as homologous with the true notochord of typical chordates is identified as such largely on account of its relations to other structures. It consists of a short thick-walled *diverticulum* of the mid-dorsal region of the anterior end of the alimentary tract. The diverticulum projects forward as a rod into the cavity of the proboscis and is stiffened by a Y-shaped "*proboscis skeleton*," a chitinous secretion of the sheath of the diverticulum. The histological structure of the notochord is not unlike that of the true notochord in that its cells are vacuolated. The diverticulum is therefore diagnosed as a notochord by virtue of its derivation from a median dorsal portion of the alimentary tract and because it seems to serve some skeletal function. *Schizocardium* (Fig. 36) has a much longer "notochord" due to its extension forward into a long *vermiform process*.

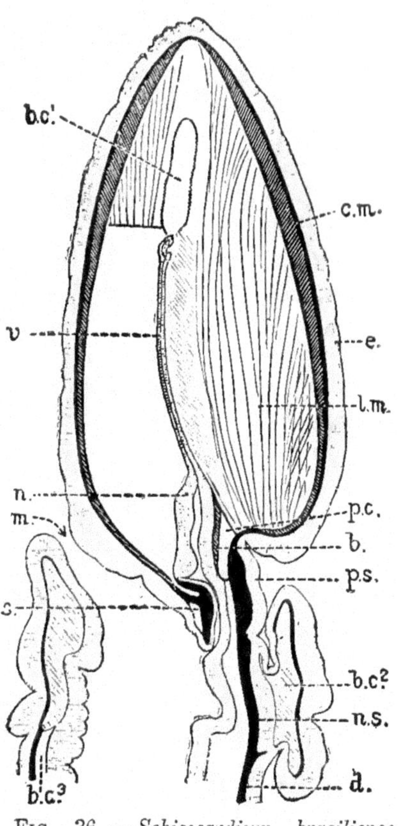

FIG. 36. —*Schizocardium brasiliense*, longitudinal, median section through the anterior end. *b*, blood space; *b. c*¹, *b. c*², *b. c*³, first, second and third body cavities; *cm*, circular muscles of proboscis; *e*, epidermis; *lm*, longitudinal muscles; *m*, mouth; *n*, notochord; *ns*, central nervous system; *d*, dorsal nerve; *pc*, pericardium; *ps*, proboscis stalk; *s*, proboscis skeleton; *v*, vermiform process or extension of notochord. (From Harmer, after Spengel.)

The Pharyngeal Clefts.—These structures take the form of "gill-sacs," each of which opens into the pharnyx by a U-shaped slit, resembling that of Amphioxus, and opens to the exterior by a small pore. These "gill-slit" openings through the pharynx are supported by thin *chitinous bars,* also resembling the gill-bar system of Amphioxus.

The Neural Tube.—The nervous system of *Balanoglossus* is in general quite unlike that of the true chordate. Though it is diffuse and rather poorly centralized, a distinct *ventral as well as a dorsal nerve cord occurs,* the two being connected at the base of the collar by a *commissure.* In the collar region a short posterior part of the dorsal nerve cord seems to be distinctly tubular. In *Glossobalanus* and *Ptychodera,* two other enteropneustan genera, the entire dorsal nerve cord of the collar is said to be tubular. The claim therefore of the Enteropneusta to chordate affinities are on this score rather strong, for no other phylum of animals has a tubular dorsal nervous system.

The most pronounced dissimilarities to the Chordates are seen in connection with the *cœlomic cavities,* for there are five separate cavities: an unpaired proboscis cavity, paired collar cavities, and paired trunk cavities. In this respect there are rather striking suggestions of echinoderm conditions.

The Tornaria Larva.—Perhaps the most marked evidences of echinoderm affinities, however, are seen in the larva of *Balanoglossus,* the *Tornaria larva* (Fig. 46) which is so strikingly like that of the young *Auricularia* larva of a holothurian that it was originally classed as an echinoderm larva by Johannes Müller. The relationship that seems to be indicated is not so much between the Enteropneusta and the modern echinoderms as between a remote bilateral ancestor of the echinoderms and an equally remote pelagic ancestor of the Enteropneusta. It is probable, however, that the rather superficial resemblance between the larvæ of these two groups has been overemphasized. Larval resemblances present at best a very uncertain basis for phylogenetic conclusions, since so many of the structures present are mere provisional larval organs of probably cænogenetic origin.

We must then conclude that the Enteropneusta show evidences of having been derived from a stock that was related to the remotest ancestors of the chordates (the so-called protochordates) and to-day represent a rather unsuccessful lateral offshoot from the base of the chordate branch of the phylogenetic tree. The closest linkage between the Hemichordata and the vertebrates is through Amphioxus, especially in the close resemblance between the gill-slits and gill-bars of the two forms.

ORDER II. PTEROBRANCHIA

These are forms in which sessile life has profoundly modified the primitive structures. There are two genera, *Cephalodiscus* and *Rhabdopleura*. Many species of *Cephalodiscus* (Figs. 37 and 38) are found in both deep and shallow water, mostly oriental in distribution. Though only about two or three millimeters in length, the individual *Cephalodiscus* has certain unmistakable resemblances to *Balanoglossus*. As an adaptation to sessile life the whole body is bent into a U-shape so that the mouth and anus open close together. Instead of a digging proboscis, it has a flattened structure called the "buccal shield" that overhangs and conceals the mouth, and whose cavity opens to the exterior by paired *proboscis pores*. The collar, which has paired collar cavities with pores, is provided with from four to six *plume-like tentacles* on the dorsal side. These tentacles are provided with ciliated grooves which sweep food toward the mouth. There is but one pair of pharyngeal clefts opening in the trunk just back of the collar. The *notochord* is a slender diverticulum of the proboscis region of the dorsal wall of the alimentary canal, practically identical with that of *Balanoglossus*. The nervous system is not tubular but is a mere plexus of nervous tissue on the dorsal epidermis of the collar. The trunk is short and plump, and has paired body cavities in which lie the paired gonads. The ecology and habits of *Cephalodiscus* are similar to those of other colonial sessile forms and especially like those of some of the colonial ascidians. They live in comparatively deep water attached to the bottom, forming large colonies, each individual of which is embedded in a hollow pocket of the common "house."

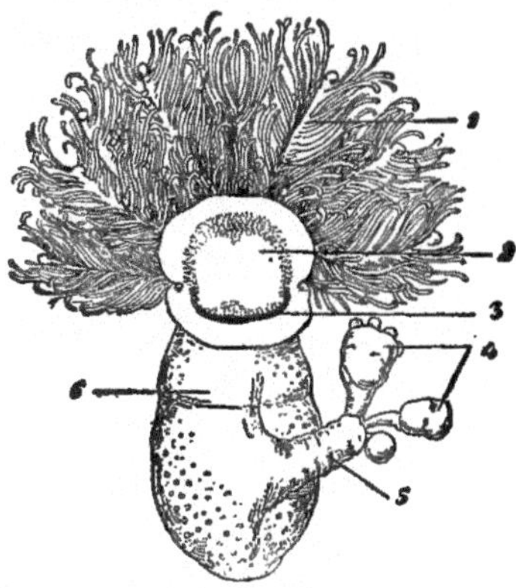

FIG. 37.—*Cephalodiscus dodecalophus*, anterior view. *1*, tentacles; *2*, proboscis (buccal shield); *3*, pigment band on proboscis; *4*, buds; *5*, pedicle; *6*, trunk. (From Hegner, after McIntosh.)

They depend for food upon minute organic particles that come to them in the water and may be swept into the mouth by means of the ciliated grooves of the tentacles. Many other sessile animals have a similar feeding mechanism. In fact there are few fixed forms that do not have feeding adaptations of this sort. Like the colonial ascidians,

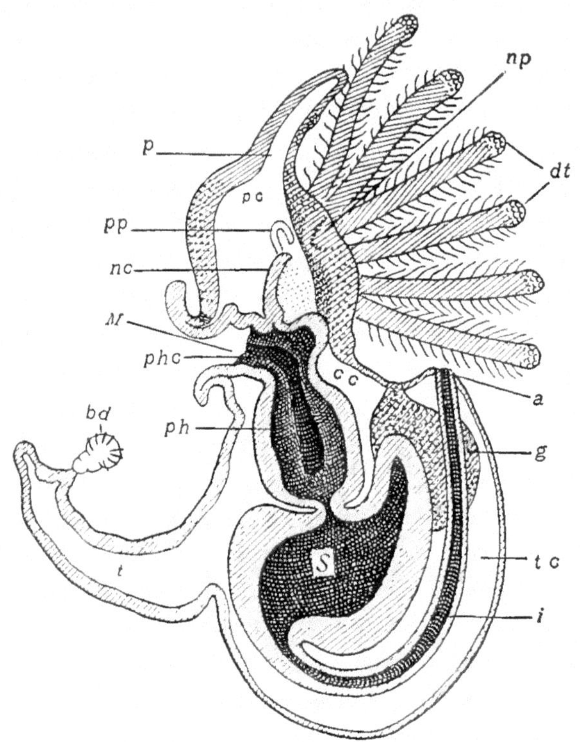

FIG. 38.—Sectional view of *Cephalodiscus*. *a*, anus; *bd*, bud on stolon (*t*); *cc*, collar cavity; *dt*, dorsal tentacles; *g*, gonad; *i*, intestine; *M*, mouth; *nc*, notochord; *np*, neural plate of collar region; *p*, proboscis; *ph*, pharynx; *phc*, pharyngeal cleft or gill slit; *pp*, proboscis pore; *pc*, proboscis cavity; *S*, stomach. (Redrawn after Patten.)

Cephalodiscus reproduces by asexual budding as well as by eggs and sperm.

Rhabdopleura (Figs. 39 and 40), a genus with about four known species of minute tubicolous forms, is obtained by deep-sea dredging. They are of microscopic size, scarcely more than one-tenth of a millimeter in length. Each individual inhabits a delicate flexible tube into which the body may be withdrawn. *Rhabdopleura* in general resembles *Cephalodiscus*, but differs chiefly in the lack of some of the organs possessed by the latter. There are *no gill-slits nor proboscis pores.* The dorsal region of the collar bears but a single large pair of *feathery tentacles* which form the most conspicuous part of the animal.

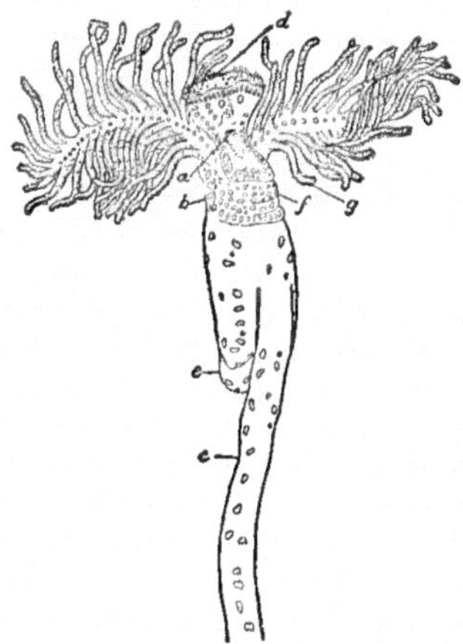

Fig. 39. — *Rhabdopleura.* *a*, mouth; *b*, anus; *c*, stalk; *d*, proboscis; *e*, intestine; *f*, anterior region of trunk; *g*, a tentacle. (From Hegner, after Lankester.)

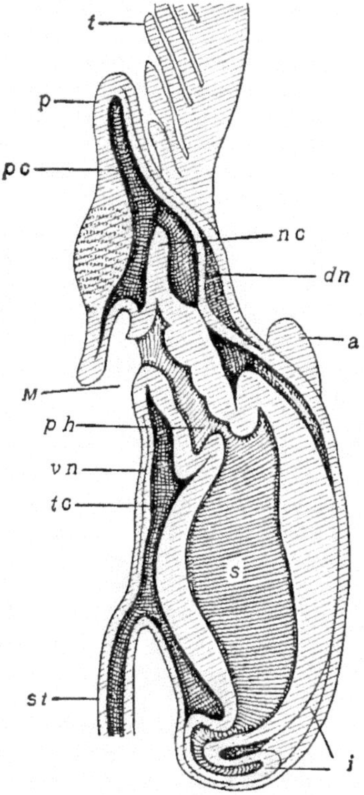

Order III. Phoronidea

Phoronis is a small tubicolous animal with a general resemblance to *Rhabdopleura*. It has been classified as a gephyrean (an aberrant annelid type). The really striking point of contact is between the larva of *Phoronis* and *Balanoglossus*. This larva may be described as a somewhat simplified *Balanoglossus*, since it has the proboscis, the collar fringed with tentacles, and the short trunk terminating in

Fig. 40.—Internal view of *Rhabdopleura*. *a*, anus; *dn*, dorsal nerve; *i*, intestine; *m*, mouth; *nc*, notochord; *p*, proboscis; *pc*, proboscis cavity; *ph*, pharynx; *s*, stomach; *st*, stalk; *t*, tentacles; *tc*, trunk cavity. (Redrawn from Parker and Haswell, after Schepotieff.)

the anus. The "notochord" is a very short rudimentary evagination of the alimentary canal. There are no pharyngeal clefts. In *Phoronis* we approach very close to the outskirts of the phylum Annelida; in fact it is usually classed as an aberrant annelid. In a sense the Phoronidia may be considered as a link between the annelids and the chordates. The propriety of classing them with the chordates is, however, open to serious question.

SUB-PHYLUM IV. CRANIATA (THE VERTEBRATES)

All of the remaining chordates differ chiefly from those of the three lower sub-phyla in possessing a *cranium* of some sort and a vastly *more complex brain* than is to be found elsewhere. Although the craniates are separated by a wide gap from the lower chordates, they form within the group, especially when the extinct forms are considered, a graded series from the lowest to the highest forms, which is taken to indicate approximately the general course of evolution of the sub-phylum. The group has been compared to the "fairly reliable and complete records of a country during the historical period," while the lower sub-phyla are comparable with "the few scattered and scarcely decipherable documents of prehistoric epochs."

The characteristics of the Craniata have already been outlined in the first chapter in connection with the account of vertebrate morphology. The craniates are subdivided into six classes, the various groupings of which are indicated in the following table:—

Div. 1. Ichthyopsida—breathing by means of *gills* at some period in the life history; therefore essentially *aquatic vertebrates.* They are sometimes called *Anamniota* (*Anamnia*) on account of the lack of an *amnion*, and *Anallantoida* on account of the lack of an *allantois.*

Class I. Cyclostomata—round-mouthed fishes; lampreys and hag-fishes.

Class II. Pisces—gnathastome or jaw-mouthed fishes—true fishes.

Class III. Amphibia—newts, salamanders, frogs and toads.

Div. 2. Sauropsida—breathing with *lungs* and never developing functional gills; therefore essentially *terrestrial vertebrates.* They are, together with the *Mammalia*, called *Amniota* because they have an *amnion*, and *Allantoida* because they have an *allantois*.

Class IV. *Reptilia*—reptiles, such as lizards, turtles, crocodiles and snakes.

Class V. *Aves*—birds.

DIV. 3 AND CLASS VI. MAMMALIA—hairy quadrupeds and other mammals.

The most significant general subdivision of the Craniata is one that recognizes what was probably the most important evolutionary crisis encountered by the vertebrates: the transition from an aquatic to a terrestrial mode of life. The lower craniates (cyclostomes and fishes) are all aquatic; the Amphibia constitute a transitional group, but are primarily aquatic, in that some of the degenerate forms are permanently aquatic and all are essentially aquatic during the embryonic and larval periods; all of the higher craniates are terrestrial in the sense that they breathe with lungs instead of with gills. The *Ichthyopsida* are all fish-like creatures having the aquatic mode of locomotion and of respiration and a circulation designed for gill respiration. The fish-like form common to the members of the group is to be considered as a general adaptation to aquatic life. They are *Anamnia* because their eggs develop in water and need no amniotic water-bag to protect the growing embryo. Likewise they are *Anallantoida* because the allantois is essentially an adaptation for late embryonic respiration in the air and is therefore not needed by forms that develop gills early in larval life.

The air-breathing classes, Reptilia, Aves, and Mammalia, are *Amniota* and *Allantoida* for the obvious reason that they are terrestrial in embryonic as well as in adult life. The grouping of the Reptilia and Aves into the *Sauropsida* indicates a conviction prevalent among comparative anatomists that there is an unusually close affinity between these two classes. In fact some go so far as to deny class value to the Aves, on the ground that they are merely specialized flying reptiles. The validity of the subdivision into *Agnathostomata* and *Gnathostomata* is questioned by some authorities on the ground that the jawless condition of the cyclostomes is said to be due to degeneration, owing to the highly specialized condition of the "rasping tongue" and its accessories. The claim is made that some of the cranial cartilages are the rudiments of a former lower jaw. If the jawless condition is secondary there is little reason for separating the cyclostomes from the Pisces. A more nearly acceptable theory is that the cyclostomes represent an early side line of vertebrate evolution quite distinct from the true fishes.

CHAPTER III

THE ORIGIN AND EVOLUTION OF THE VERTEBRATES

The problem of vertebrate ancestry is an old one and one that evades a direct solution. Only through circumstantial evidence are we at present likely to reach even a reasonably satisfactory answer. Certain postulates must be made, however, upon which may be built up a theory. In the first place it may be assumed that the first vertebrates were aquatic. Almost equally warranted is the assumption that they were free-swimming, active creatures. They were also evidently axiate animals with fairly advanced cephalization, *i. e.*, with well-differentiated heads and sense organs. They were metameric, with well-developed segmental musculature and with a large open cœlom. They also had a dorsal tubular central nervous system, a notochord and pharyngeal clefts—three fundamental chordate characters.

An animal with these characters could not be very different from a fish. Let us assume, then, with Osborn, that the first true vertebrate was a "free-swimming, quickly darting type, with double pointed, fusiform body in which the segmented propelling muscles are external and a stiffening notochord is central." We have found no fish either past or present that appears to be primitive enough to be considered as the first fish. The earliest fossil fishes are armored types that have evidently become specialized secondarily for bottom feeding habits. Some of the most primitive sharks, especially the Devonian shark *Cladoselache* (Fig. 65), almost meet our expectations of what a primordial fish may have been, but even these fishes bear evidences of having evolved from much simpler fish-like ancestors. Where then can we turn for a true fish prototype? The existing lancelets (*Branchiostoma* or *Amphioxus*) with their fusiform bodies, notochord, and segmental musculature, appear to be the closest approximation we can find of what the ancestral fish creature may have been, and one of the prevailing theories of vertebrate ancestry is the so-called "Amphioxus theory," a theory that appears to the writer to be more satisfactory than any other and which we shall herewith present in a new form.

The Amphioxus Theory of Vertebrate Origin

Amphioxus, called by someone the "chordate Adam and Eve," bears many evidences of being a very antique type. Its cosmopolitan distribution, the essential uniformity of all the different species, and its almost diagrammatic simplicity argue for its status as a truly primitive creature. Even though it appears to be somewhat degenerate as to its head parts and though it is suspected by some authors of being pædogenetic, there is much to show that its fundamental structures are nearly prototypic.

The habits of Amphioxus show an interesting combination of two types of life, the sedentary and the free-swimming. While they are capable of rapid and vigorous darting movements in the water, they spend much of the time as sedentary creatures, living in burrows in the sand with only the oral end protruding (Fig. 9). This life along the sandy shores requires just exactly this double adaptation, since it is in this region that we would find the most effective tidal currents. The lancelets are evidently able to swim rapidly against the tidal currents and to change their locations as often as the state of the tide demands. They also burrow into the sand by vigorous undulating movements and, when once buried, they remain for long periods quietly feeding, as shown in Fig. 9, by means of their unique food-concentrating mechanism, a complex of structures that is highly adapted for sedentary life and ill adapted for active free-swimming life. The lancelet then has a combination of two opposed tendencies, that of the quiet sedentary animal, and that of the free-living rapid-swimming animal. Specialization might readily be carried out in either direction.

If specialization followed the lines of an increasing fixity of position, we might easily conceive of a type of lancelet that became fixed while still a larva. The functional stimulus for further development of the locomotor musculature and the notchord would be wanting and the resultant creature would consist largely of a food-concentrating and digestive body without any of the organs essential for active life. This we believe is just what happened. Some of the ancestral lancelets migrating into deeper waters, where the lack of tidal currents would no longer stimulate the swimming about of the larvæ and young, would remain stationary throughout life, would not even be able to make burrows, but would fix themselves to rocks, etc., at the bottom. The secretion of a protective coat or tunic would be a logical consequence of such a step and we would have the typical *tunicates*. The salpians probably were an offshoot of the primitive tunicates that migrated away from the shores, but instead of following the bottom they acquired the floating or pelagic habit.

A totally different situation, however, would meet those primitive lancelets that migrated from the shores up the mouths of rivers. The water currents would become more constant and there would be less opportunity for sedentary life. In river currents the microscopic organic particles suspended in the water would be much less abundant and hence it would be more difficult to subsist by the old lancelet method of food concentration. This whole apparatus as a feeding apparatus would therefore lose its significance and would be used largely as a respiratory mechanism. Perhaps the first part of the mechanism to be lost would be the one last formed in ontogeny—the atrium. Possibly the remains of the atrium would persist as paired lateral ridges, or metapleural folds, that might act as lateral flanges or balancing organs in swimming, and may have been the primordia of the paired fins of fishes. New feeding methods had to be acquired and at least two schemes were adopted, one the jaw apparatus involving the opening of a new ventral mouth, and the second the oral funnel apparatus with its accessory, the rasping tongue. The jaw apparatus characterizes the true fishes (Pisces), while the oral funnel and rasping "tongue" apparatus characterizes the Cyclostomata. Both of these groups appear to have originated at about the same time and independently of each other. In many ways the cyclostomes have been much the less successful type and comparatively little evolutionary progress has resulted. The true fishes, on the other hand, appear to have furnished the basis of all future vertebrate specialization. That the rapid stream environment was the stimulating factor in the production of the first true vertebrates has been ably upheld by Prof. T. C. Chamberlin, on purely theoretic grounds. He contends that the sea does not furnish the dynamic stimuli necessary to bring about the evolution of vertebrate characters. The rivers, however, furnish just the needed conditions to bring forth the complex of energy elements that we associate with a vertebrate.

> "There is only one conspicuous type that is facilely suited to free life, independent of the bottom, in swift streams, and that is the fish-form. The form and the motion of the typical fish are

a close imitation of the form and motion of wisps of water-grass passively shaped and gracefully waved by the pulsations of the current. The rhythmical undulations of the lamprey, which perhaps best illustrates the primitive vertebrate form, and is itself archaic in structure, are an almost perfect embodiment in the active voice of the passive undulations of ropes of river confervæ. The movement of the fish is produced by alternate rhythmical contractions of the side muscles, by which the pressure of the fish's body is brought to bear in successive waves against the water of the incurved sections. In the movement of a rope of vegetation in the pulsating current, it is the pressure of the pulses of the water against the sides of the rope that give the incurvations. The two phenomena are natural reciprocals in the active and passive voices.

"The development in the fish of a rhythmical system of motion responsive to the rhythm impressed upon it by its persistent environment and duly adjusted to it in pulse and force, is a natural mode of neutralizing the current force and securing stability of position or motion against the current, as desired. Beyond question the form and movement of the typical fish are admirably adapted to motion in static water and that has been thought a sufficient reason for the evolution of the form, and so possibly it may be, but fishes in static water have not as uniformly retained the attenuated spindle-like form and the extreme lateral flexibility as have those of running water. Among these latter it is rare that any great departure from typical lines and from ample flexibility has taken place, while it is common in sea fishes. Among the latter not a few have lost both the typical form and the flexibility. The porcupine fish, the sea-horse, the flounders, and many others are examples of such retrogressive evolution, which is doubtless advantageous to them within their special spheres in quiet waters, but would quite unfit them for life in a swift stream. And if the view be extended to include the low degenerate forms, like the Ascidians (tunicates), that are by some authors classed as chordates, the statement finds further emphasis.

"It is not difficult for the imagination to picture a lowly aggregate of animal cells, still plastic and undeterminate in organization, brought under the influence of a persistent current and caused to develop into a determinate organization under its control, and hence to acquire, as its essential features, a spindle-like form, a lateral flexibility, and a set of longitudinal side muscles adapted to rhythmical contractions, since these are but expressions of conformity and responsiveness to the shape and movement normally impressed by the controlling environment upon plastic bodies immersed in it. The necessity for a stiffened axial tract to resist the longitudinal contractions of the side muscles and thus to prevent shortening without seriously interfering with lateral flexibility, is obvious, and is supplied by the notochord. Thus by hypothesis, the primitive chordate form may be regarded as a specific response to the special environment that dominated the evolution of a previously indeterminate ancestral form."

It will be seen that the Amphioxus theory of vertebrate descent carries with it certain implications:

(1) That the earliest pro-fishes were of a grade of organization not very different from Amphioxus.

(2) That the first fishes developed directly from these pro-fishes under the influence of a rapid stream environment.

(3) That the ostracoderms, the oldest fossil fishes found in the middle Ordovician, are specialized bottom-feeding types and are not ancestral to any of the modern types.

(4) That the true ancestral fishes were, according to Osborn, "active, free-swimming, double-pointed types of fusiform shape, adapted to rapid motion through the water and to predaceous habits in pursuit of swift-moving prey."

(5) These primitive fishes must have existed before the Ordovician, *i. e.*, in Cambrian times, which is the earliest period of which we have positive fossil records. This makes the chordates as old as any of the invertebrate phyla and tends to detract from the validity of any theory involving the idea that the chordates have been derived from any of the higher invertebrate phyla.

(6) That the tunicates, instead of being ancestral to Amphioxus, have been derived from an early Amphioxus-like stock.

Other Theories of Vertebrate Ancestry

The traditional method of phylogenetic research has led to the belief that each higher group of animals has been derived from one of the known lower groups. It is generally believed, for example, that the Arthropoda are descended from the Annelida, the Annelida from the Platyhelminthes, the Platyhelminthes from the Trochelminthes, etc. It is likely to be forgotten that these groups as we know them, both recent and fossil, are highly specialized, and that a highly specialized group is not likely to retain sufficient plasticity to give origin to any radically new departures that might lead to new phyla. The modern phylogenist has come to believe that the early ancestors of such large fundamentally isolated groups as the phyla all originated far back in pre-Cambrian times and are therefore forever lost to us as actual relics. Nevertheless there are still extant some theories that derive the vertebrates from various contemporaneous invertebrate phyla. If the vertebrates did come off from any known invertebrate types, what more natural than to look to the other great metameric phyla for the ancestral conditions? The only truly metameric phyla among the invertebrates are the Annelida and the Arthropoda and there are two rival theories, one claiming that the annelids and the other that the arthropods are the ancestors of the vertebrates.

Both theories base their argument on fundamental morphological resemblances between the vertebrate and the annelid, or the arthropod, as the case may be. The annelid is unquestionably more generalized in its organization than the vertebrate, and it is true that the vertebrate embryo much more closely resembles the annelid than does the vertebrate adult. There are many striking homologies between all three groups (annelids, arthropods, and vertebrates) and nothing could be further from the thought of the writer than to deny that any phylogenetic relationship exists between them. Such a denial would be equivalent to a denial of the validity of the whole principle of homologies, upon which the science of morphology rests. An admission of relationship between annelid and vertebrate or between arthropod and vertebrate is, however, quite different from an admission that the vertebrate is descended from either annelid or arthropod. Is it not much more reasonable to suppose that all three of these highly specialized groups, that have so much in common, have been derived from a primitive ancestor characterized by the features that all three groups have in common? Such an ancestor would be metameric, cœlomate, with antero-posterior, dorso-ventral, and bilateral axes, with probably ciliary bands as a mode of locomotion, with tubular nephridia, with well-defined double nerve-cord and paired ganglia, and possibly some sort of primitive paired appendages. That such an ancestral form will ever actually be discovered is highly improbable, for the annelids and arthropods, and probably also the vertebrates, were old at the dawn of Cambrian times, and the pre-Cambrian creatures are not preserved for our edification. It would, however, be unfair to leave the discussion of the ancestry of the vertebrates from annelids or arthropods without presenting some of the evidences that have been advanced in support of these theories.

The Annelid Theory of Vertebrate Ancestry

The chief basis for this hypothesis lies in the unmistakably close resemblance between the embryonic characters of the vertebrates and some of the structural peculiarities of the annelid worms. It has already been shown that the vertebrate embryo is distinctly metameric, especially in the cœlom and its derivatives. The nephridia of the lower vertebrates are cœlomoducts with the same relations as those of the annelids. The vascular system, the nervous chain of paired ganglia, the relations of the intestine, nervous system, and circulatory system, are so much alike that a diagram (Fig. 41) of

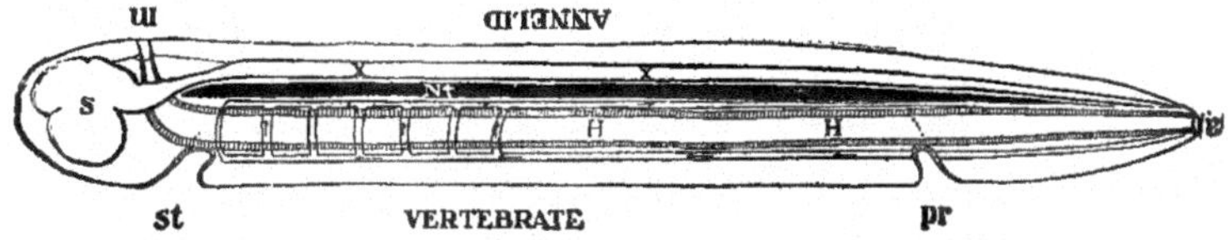

Fig. 41.—Reversible diagram illustrating the Annelid Theory. Reversible designations applying to both forms: *S*, brain; *x*, nerve cord; *H*, alimentary canal. Designations applying to annelid only: *m*, mouth; *a*, anus. Designations applying to vertebrate only: *st*, stomodæum; *pr*, proctodæum; *nt*, notochord. (From Wilder.)

these structures for the annelid serves, *if inverted*, as a diagram of a vertebrate. A vertebrate is said to be merely an annelid turned over on its back and with certain minor changes to adjust the animal to the new position, such as the development of a new mouth and a new anus. Such positional reversals are not without parallel, for the squid is reversed as compared with other mollusks, and a king-crab (Limulus) swims upside down. The chief stumbling-block of the annelid theory is the notochord of the vertebrate; even this is not unsurmountable for there has been found in annelids a *Faserstrang*, which is described by Wilder as "a bundle of fibres running along the nerve chain and serving as a support. This and the notochord lie in precisely similar positions in relation to the other organs, and in both cases they are inclosed with the nerve córd in a common sheath of connective tissue."

Coming back to the idea that a vertebrate is an annelid turned over, let us follow up with the aid of the diagram (Fig. 41) the results of such a reversal. The annelid mouth is ventral and the œsophagus passes through the brain in such a way that one pair of ganglia (*supraœsophageal ganglia*) are dorsal to the œsophagus while the rest of the ganglionic chain is ventral to the alimentary tract. The change from annelid to vertebrate involves doing away with the annelid or primitive mouth and the opening up of a secondary mouth on the new ventral side, which does not pass through the nervous system. The entire nervous system, including the supraœsophageal ganglia (the vertebrate brain) is thus left on the new dorsal side. Two pieces of embryological evidence are offered in support of this contention. First, the vertebrate mouth is quite late in breaking through and this has been taken as a sign that it is an afterthought. Second, there are vestiges of the primitive or ancestral mouth in the neuropore of vertebrate embryos and in the adult Amphioxus, as well as in the hypophysis, a structure located where the old mouth might have been, and apparently without any other significance unless it does stand for the point of closure of the primitive mouth. In the annelid the blood flows forward in the dorsal vessel and passes across through segmental arches to a ventral vessel in which it flows backward. Reverse this condition and we have the primitive vertebrate condition with the blood flowing forward on the ventral side, crossing in certain special arches (branchial arches) to the dorsal side and from there flowing backward. Some of the annelids even have specialized branchial tissues developed segmentally near the anterior end. In the annelid the anus opens at the posterior extremity, but in the vertebrate a new anus is formed on the ventral side, leaving in the embryo a blind hindgut, which subsequently disappears, leaving a part of the trunk back of the anus which has no alimentary tract. This is the vertebrate tail. Both the new mouth (stomodæum) and the new anus (proctodæum) are lined with ectoderm.

"Convincing as these comparisons seem," says Wilder, "when taken by themselves, the influence of later investigation has tended rather away from the annelid hypothesis, and at present, although there are many investigators who seek the ancestor of vertebrates in some worm-like form, there are few who wish to definitely assert that this ancestor was an annelid."

The writer sees nothing in the annelid theory seriously out of harmony with the Amphioxus theory, for Amphioxus may have been derived from some early metameric, cœlomate type that would be as much like an annelid as anything else. It is much more likely, however, that the resemblances between annelids and vertebrates are merely the necessary similarities that result from the fact that they are constructed upon the same fundamental lines. Given two groups having in common the axis of polarity, the dorso-ventral, the bilateral axes, together with a metameric arrangement of organs, and the expressions in structural differentiation must of necessity be fundamentally similar. Their differences are more puzzling than their resemblances, and therein lies the real problem.

The Theory of Arthropod Ancestry of the Vertebrates

This theory is presented compactly by Lull, as follows:

"In addition to the annelid theory, recent authorities have tried to prove vertebrate descent from the Arthropoda, especially from the more primitive arachnoids such as today are represented by the scorpion and the horseshoe crab (*Limulus*) and formerly by the extinct Merostomata. By this hypothesis we must set aside as primitive such forms as Amphioxus and the cyclostomes and start with the highly specialized ostracoderms which lived in Ordovician and Devonian times and thus were contemporaneous with and in general appearance and probable habits quite similar to the Merostomata. The soft parts of the Merostomata are of course unknown, but it is reasonable to suppose that they were not unlike those of the related scorpions and *Limulus*, and, as Patten has shown, especially in the brain and cranial nerves of vertebrates and the fused cephalothoracic ganglionic mass found in such arachnoids, there are many points of resemblance (Figs. 43 and 44). Then, too, the sense organs, especially the eyes, are more or less comparable, and there is in *Limulus* an internal skeletal piece known as the 'endocranium' or sternum which serves to protect the central nerve complex, and which in general form and in its relation to other parts resembles the primordial vertebrate skull. Similarities also exist between the heart and arterial systems of each group, and the appendages may be compared. There are, again, the very arthropod-like jaws which Patten has demonstrated for the ostracoderm *Bothriolepis*, a type which, on the other hand, shows many vertebrate-like characteristics; and the general arrangement of the plates by

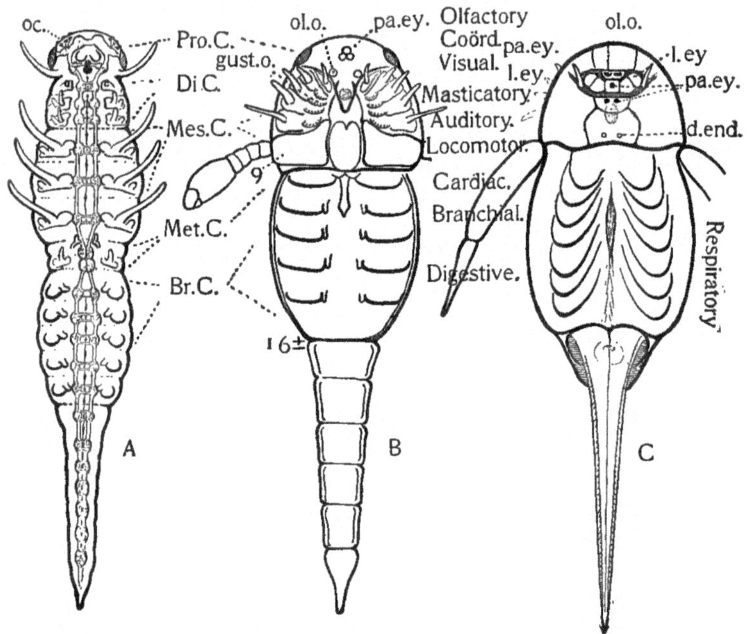

FIG. 42.—Diagram showing the supposed homologies between: A, insect; B, arachnid (merostome) and C, ostracoderm (*Bothriolepis*). *Pro. C*, procephalon or primitive head; *Di. C*, and *Mes. C*, dicephalon and mesocephalon, usually spoken of as thorax; *Met. C*, metacephalon or vagus region; *Br. C*, Branchiocephalon, or respiratory region. *d. end*, ductus endolymphaticus; *gust. o*, gustatory organ; *l. ey*, lateral eye; *pa. ey*, parietal eye. (After Patten's "Evolution of the Vertebrates and their Kin" [P. Blakiston's Sons & Co.].)

which the cephalothorax is covered is also very similar in the ostracoderms and in contemporary arachnoids, but unfortunately for the argument *Bothriolepis* is a highly specialized end-form from the Upper Devonian. Nevertheless, while the arachnoid theory has been set forth by Gaskell (*The Origin of Vertebrates*, 1908) and by Patten (*The Evolution of the Vertebrates and their Kin*, 1912) the main thesis has received thus far but little

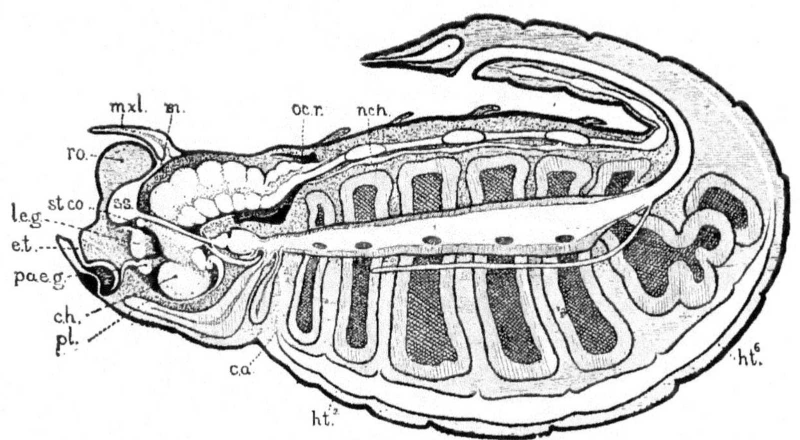

FIG. 43.—Sagittal section of a young scorpion (arachnid) to be compared with fig. 44. *c. a*, carotid artery; *ch*, cerebral hemispheres; *e. t*, parietal eye tube; ht^2-ht^3, heart; *l. e. g*, lateral eye ganglion; *mxl*, maxillaria; *m*, mouth; *nch*, notochord; *oc. r*, occipital region; *pae. g*, parietal eye ganglion; *ro*, rostrum; *st. co*, stomodæal commissure. (After Patten's "Evolution of the Vertebrates and their Kin" [P. Blakiston's Sons and Co.].)

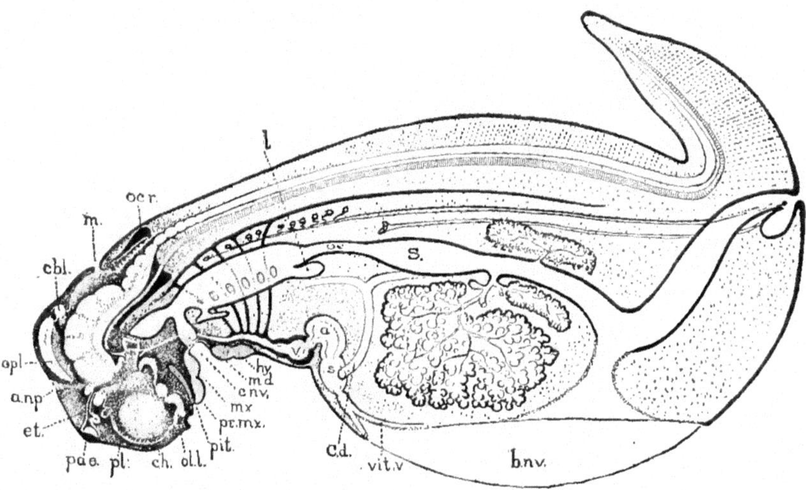

FIG. 44.—Sagittal section of a primitive vertebrate embryo, showing the relation of its principal organs to those in the arachnids (cf. fig. 43); schematic. *a. np*, anterior neuropore; *b. nv*, belly navel; *cbl*, cerebellum; *C. d*, duct of Cuvier; *ch*, cerebral hemispheres; *cnv*, new ventral mouth; *et*, eye tube; *hy*, hyoid; *l*, lung; *m*, mouth; *oc*, œsophagus; *ocr*, occipital region; *ol. l*, olfactory lobe; *op. l*, optic lobe; *pa. e*, parietal eye; *pit*, pituitary body; *pr. mx*, premaxilla; *md*, mandible; *mx*, maxilla; *S*, stomach; *s. v*, sinus venosus; *v*, ventricle. (From Patten's "Evolution of the Vertebrates and their Kin" [P. Blakiston's Sons and Co.].)

recognition; although the evidence, especially in Professor Patten's book, is based upon an admirably executed piece of research."

As Lull has intimated, it is highly improbable (a) that the ostrocoderms, especially those of the order Antiarchi to which *Bothriolepis* belongs, are primitive vertebrates, and (b) that the Merostomata were sufficiently generalized or plastic to have given origin to a group radically different from that to which they belong. The theory requires the origin of one highly specialized group of one phylum from an equally highly specialized group of another phylum. It is much more likely that the superficial resemblances of the two groups are both adaptations for bottom-feeding, and that the heavily armored condition is evidence in both groups of senility. In general, groups of animals are thought of as becoming armored as the result of racial old age and a slowing down of the developmental vigor of the constituent protoplasmic materials. The plastic young races retain their flexibility and are as a rule without heavy integumentary deposits. From this point of view the ostracoderms fit very poorly into the rôle of primitive or ancestral vertebrates. If the ostracoderms cannot be used as the link between the fishes and the Merostomata, the whole fabric of vetebrate phylogeny, as erected by Patten, breaks up and falls apart. There is no question, however, as to the value of Patten's careful and exhaustive studies of comparative anatomy and embryology of vertebrates and arachnoids, and especially valuable is his contribution to our knowledge of the anatomy of the ostracoderms, a group hitherto all too imperfectly known. The chief fault that one finds with Patten's method of argument and exposition concerns his skillful illustrations, which contain an ingenious intermingling of fact and interpretation that is insidiously convincing unless one be on his guard. Figures 42, 43, and 44 are among the most characteristic of Patten's illustrations; and they speak for themselves.

Minor Theories of Vertebrate Ancestry

There are several other invertebrate groups that might conceivably have given rise to the vertebrates. One of these is the vermian group Nemertea, a group of uncertain affinities but related to the flat worms. The *Nemertean Theory of the Origin of the Vertebrates* has been advocated by Hubrecht, and the argument is based largely on the nervous system. The nemertean (Fig. 45) has two lateral nerve cords and a more slender dorsal nerve cord. The three cords are connected with one another by cross commissures. It is thought that in the evolution of the Nemertea into vertebrates the dorsal nerve cord becomes the central nervous system and is then enlarged at the anterior end into a brain; while the two lateral nerve cords become compatively small

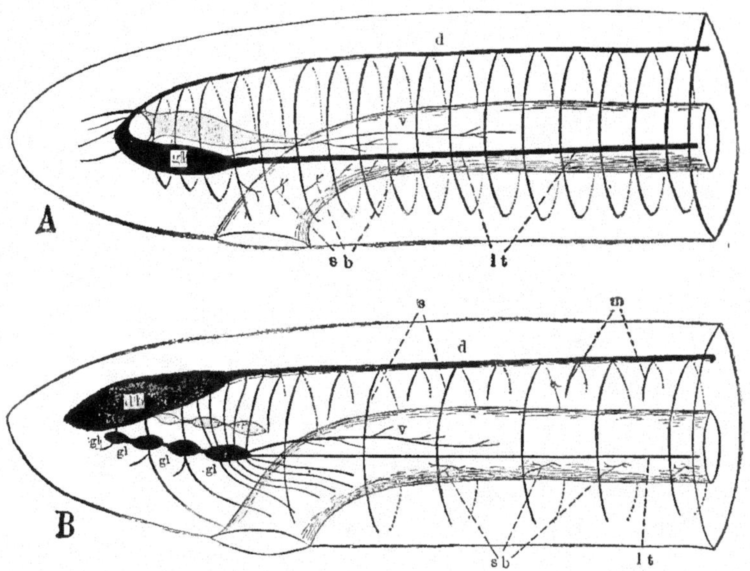

FIG. 45.—To illustrate the Nemertean Theory of the origin of the Vertebrates. A. A typical diagram of nemertean. *d*, dorsal nerve cord; *gl*, ganglion; *lt*, lateral nerve cord; *v*, intestinal nerve; *sb*, small intestinal branches.

B. Typical diagram of vertebrate. *db*, dorsal brain; *d*, dorsal nerve cord; *s*, sensory, and *m*, motor spinal nerves; *gl*, sympathetic ganglia; *v*, ramus intestinalis vagi; *sb*, sympathetic branches. (From Wilder, after Hubrecht.)

and of secondary importance, persisting however in the so-called "Vagus system, rami lateralis X" in the lower vertebrates. It can scarcely be said that this theory carries force in view of the fact that the other structures of the body show little resemblance to vertebrate conditions.

Another theory of vertebrate ancestry associates the latter with the Echinodermata through the connecting link Enteropneusta (*Balanoglossus*). The Balanoglossus situation is indeed a puzzling one. It does not seem to fit in with any of the other theories of vertebrate ancestry. The adult *Balanoglossus,* and even more so some of its relatives, such as *Harrimania,* shows certain striking chordate resemblances. The branchial orifices are similar in form and relations and remind one of the pharyngeal clefts of Amphioxus; the notochord is somewhat doubtful in *Balanoglossus,* but in *Harrimania* it appears in a much more obvious form and has an origin quite like that in Amphioxus; the nervous system is highly generalized, consisting of four longitudinal cords, with the dorsal somewhat more strongly developed than the rest. In *Glossobalanus* and *Ptychordera,* according to Harmer, "a central canal, opening in front and behind, exists throughout the entire length of the central nervous system. . . . *Balanoglossus* is thus typically provided with a dorsal, tubular, central nervous system, and although it does not extend beyond the limits of the collar, it shows noteworthy resemblances to Vertebrate Animals."

It is the opinion of many students of comparative anatomy that *Balanoglossus* is a chordate or pro-vertebrate more primitive than Amphioxus. According to Wilder, *Balanoglossus* and its relatives the Enteropneusta "lie nearly in the line of vertebrate descent, and represent an earlier stage than that of the tunicates. But here the chain seems to end, for *Balanoglossus* is itself unusually isolated and shows no close affinity to any other invertebrate types."

Although we may accept as unquestioned the above view of the chordate affinities of *Balanoglossus* there are also very striking resemblances between the latter and certain echinoderms which seem to the writer to be as weighty as are those relating it to the chordates. Foremost of these echinoderm resemblances of *Balanoglossus* is that involved in the structure of the larvæ of the two groups. The *Tornaria* larva of *Balanoglossus* (Fig. 46) is compared with the *Auricularia* larva of the holothurian and the *Bipinnaria* larva of the starfish. The resemblance is obvious. Moreover, there is in *Balanoglossus* a system quite like the water vascular system of the echinoderms, which is unique for that group. It will be recalled that there is a water-pore communicating with the proboscis cœlom and a pair of water-pores communicating with the paired collar cœloms. "Recent studies on the development of Echinoderms," says Wilder, ' have made it probable that the five body cavities of *Balanoglossus* are represented in the larvæ of these animals; and this materially strengthens the probability of the view that the respective adults are also allied. It may be added that the relationship which appears to be indicated is between

Balanoglossus and the bilateral ancestors from which the radially-symmetrical Echinoderms are probably descended."

The view is taken by several authors that the Enteropneusta and the echinoderms were derived from a common ancestral stock, which is now copied in a simplified form by the larvæ of both groups. "The common characters of all the larvæ are," says Wilder, "bilaterality,

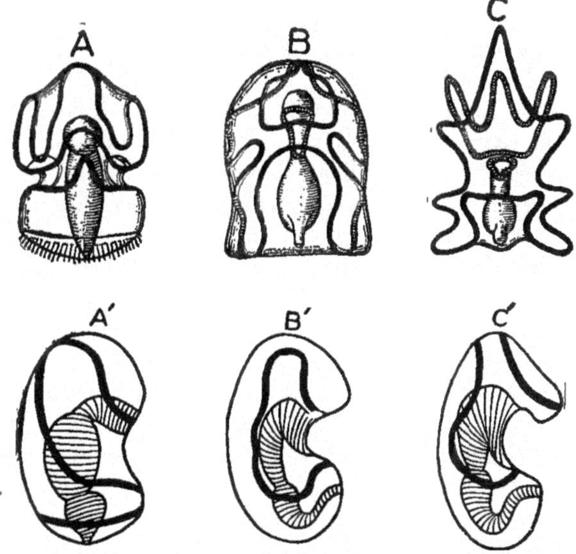

FIG. 46.—Comparison of Tornaria larva with larval echinoderms. Main ciliated bands in black, lesser systems cross-lined. Upper row ventral aspect; lower row right lateral aspect. A. A ′, *Tornaria;* B. B ′, *Auricularia* (sea cucumber); C. C ′, Bipennaria (star-fish). (From Lull, after Wilder.)

transparency, locomotion by bands of cilia, and pelagic life." A common ancestor having these characters may well have existed. "The lineal descendents of this hypothetical ancestor chose two paths, the one leading to the Echinodermata, the other to *Balanoglossus,* the Tunicata, Amphioxus, and eventually the Vertebrate." This view makes the Enteropneusta ancestral to the tunicates and the tunicates ancestral to Amphioxus. In a previous discussion of the Amphioxus theory we have dealt with the tunicates as degenerate derivatives of Amphioxus-like ancestors; just the reverse of the opinion expressed by Wilder. It should also be said that, although Amphioxus is obviously not at the very bottom of the vertebrate ancestral trunk, there is little evidence of its having descended from any form at all like *Balanoglossus*. Rather would we believe that *Balanoglossus* is an early lateral offshoot from the main line of chordate stock that leads more directly to Amphioxus and the vertebrates.

Summary of the Theories of Vertebrate Ancestry

After a review of these various and contradictory theories as to the origin of the vertebrates, are we any nearer a solution of this perplexing problem than when we started? Certainly it cannot be claimed that the problem is solved, but at least we have examined the question and have considered the various possibilities. Of all the alternatives, the one that makes Amphioxus a central main-line ancestral form, perhaps slightly degenerate, but only a little specialized, seems best supported by evidence. Amphioxus, on the one hand, is so much like the vertebrates that it has been classed as an acraniate vertebrate by so able a writer as Osborn. It is also, on the other hand, unquestionably related to the tunicates. The assumption of an Amphioxus-like creature as a common ancestor of vertebrate and tunicate, and the diagnosis of the Hemichordata as an early lateral offshoot of a still more primitive chordate trunk, seems much more logical than alternative assumptions. It also rids us of the necessity of deriving one phylum from the highly differentiated members of another phylum, and is therefore more acceptable than annelid, arthropod, or nemertean theories.

CHAPTER IV

CLASS I. CYCLOSTOMATA

The name "round-mouth eels" is often applied to these fish-like forms to distinguish them from the "jaw-mouth" fishes (gnathotomes). The group consists of two distinct types, the "hag-fishes" and the lampreys. Both of these have a superficial resemblance to the eels, but they differ from the group to which the eels belong (Pisces) in many important respects. Some of the differences are to be interpreted as evidences of primitiveness and others of specialization and degeneration. Since there are no certain fossil remains of the cyclostomes, it is an exceedingly difficult matter to decide as to which of their characters are palingenetic (truly primitive) and which are cænogenetic (due to specialization or degeneration). Evidently, however, the characters that are of universal occurrence in the group are more likely to be primitive than are those in which the "hags" differ from the lampreys.

The cyclostomes are a minor class of vertebrates as compared with the other five classes, since there are only a few genera and species. They are also of secondary significance phylogenetically; for they are believed to represent a comparatively unsuccessful lateral branch of the vertebrate ancestral tree, that came off probably from an Amphioxus-like stock prior to the origin of the fishes proper. If this view is valid, the true fishes and all of the other vertebrate classes represent an evolutionary series totally independent from the cyclostomes. Any attempts, therefore, to establish detailed homologies between the two groups must be viewed with suspicion. On the whole, the cyclostomes have departed less widely from the Amphioxus-like prototype than have the fishes, and in that sense they represent a lower grade of vertebrate organization.

The characters in which the cyclostomes in general differ from the fishes are as follows:

External features:

1. *No jaws.* Attempts have been made to homologize the so-called "tongue cartilages" with the first visceral or mandibular arch, but the comparison is far-fetched.

2. The *mouth* is round and is closed only by the end of the "tongue."
3. There is a median unpaired nostril, and the nasal passage in some (hag-fishes) opens into the mouth; in others (lampreys) it ends blindly beneath the brain.
4. The *branchial clefts* are in the form of pouches or pockets, a character that has given to the group the name "Marsipobranchii." The number of clefts is, in general, greater than in fishes.
5. There are *no paired appendages.*
6. There are *no scales.*
7. The *lateral line organs* lie in open grooves.

Internal features:

1. There is, in lieu of a masticatory apparatus, a complex rasping apparatus, called a "*tongue*", that is armed at the mouth end with chitinous teeth, is supported by several cartilages, and is worked by a specialized set of muscles. Since both hags and lampreys have this apparatus, in spite of their very different modes of life, it is likely that this is a very old character and represents an evolutionary experiment even more ancient than does the jaw apparatus of fishes.

2. The *notochord* is a persistent unbroken rod much like that of Amphioxus, but extends only to the hind-brain and is covered with an extra sheath of connective tissue.

3. The *vertebræ* are represented by primitive neural arches quite separate from the notochord.

4. The *brain* is small but typically vertebrate in structure, with the vagus nerve not included in the cranial region.

5. The *cranium* is entirely beneath the brain and forms neither sides nor roof for the latter. Attempts have been made to homologize the cartilages of this cranium with those of the embryonic fish skull.

6. The *gonads* give off their products into the cœlom. Eggs and sperm make their exit directly to the exterior through genital pores situated near the urinary opening.

7. The *inner ear* is more primitive than in fishes, having only one or two semicircular canals.

8. The *afferent branchial arteries* go directly to the gill pouches instead of into the arch between two gill clefts.

9. A *branchial basket* composed of cartilaginous bars and rods forms a support to the branchial part of the pharynx. This cannot successfully be compared with the branchial arches of fishes.

The *myotomic musculature*, the *median fins*, the *digestive, circulatory*, and *excretory* system are much as in the fishes.

A more special account of the characters of cyclostomes will best be presented when the two sub-classes Myxinoidea and Petromyzontia are compared.

Sub-Class I. Myxinoidea

These "hag-fishes" or "borers" are called myxinoids because of their habit of producing, when captured, great quantities of slimy

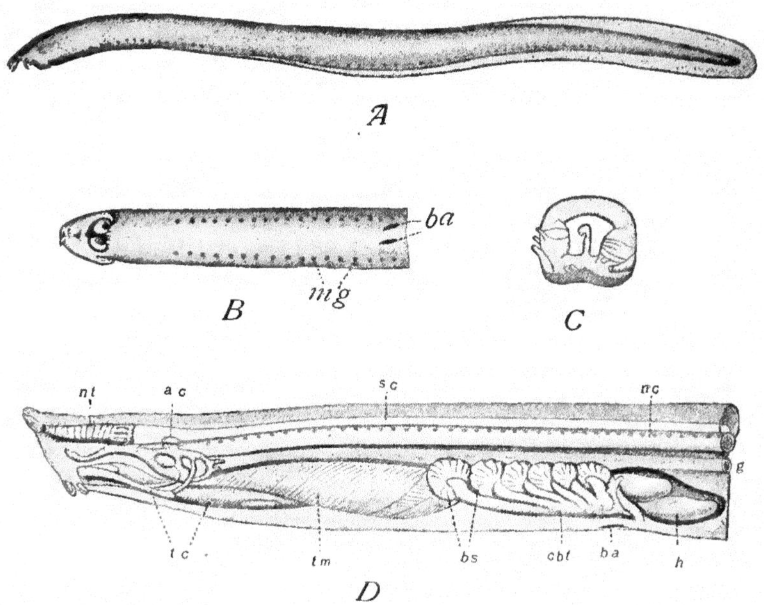

Fig. 47.—*Myxine*. A. External view of entire animal; the rows of pores are openings of mucous glands; no eyes. B. Ventral view of anterior end, showing terminal nostril, oral hood with buccal tentacles; *ba*, single pair of branchial apertures (*cf.* D); *mg*, openings of mucous glands. C. inner ear showing single semicircular canal. D. Internal anatomy. *ac*, auditory capsule; *bs*, branchial sacs, opening by means of a common branchial tube (*cbt*) into the common branchial aperture (*ba*); *g*, gut; *h*, heart; *nc*, notochord; *nt*, nasal tube; *sc*, spinal chord; *tc*, tongue cartilages; *tm*, tongue musculature. (Redrawn after Parker and Haswell.)

mucous jelly. It is said that one large specimen will make a bucketfull of jelly. Hags lead a quasi-parasitic life, in that they commonly enter the gills or mouths of dead or disabled fishes and remain inside till they have so completely gutted the quasi-host that only the shell remains. It seems unlikely that they attack living active prey, as do the lampreys, but they undoubtedly do prey upon fishes caught in gill-nets or upon trot-lines. On account of these habits the hags have acquired a well-earned unpopularity among the fishermen of the North Sea and elsewhere. The flesh of the prey is shredded by means of the strong rasping apparatus, and the digestive tract is so capacious that one good meal lasts some time. All active hunting is done at night, for the hags are blind; the day is spent buried in the mud of the sea bottom at depths around three hundred fathoms, where they are themselves safe from enemies. They swim swiftly with an eel-like, undulatory motion. When caught they secrete from the skin glands a great quantity of gelatinous mucus. Hags are said to be the only vertebrates that are specifically hermaphroditic. This character may be associated with their solitary life and with the fact that they are blind.

Some of the anatomical characters of the myxinoids that differ from those of the petromyzonts are herewith listed. These characters are illustrated in Figs. 47 and 48.

1. The mouth is terminal and there is no real buccal funnel.
2. The naso-pituitary sac (nasal passage and infundibulum) opens into the pharynx and through it water is drawn into the gills while the mouth is engaged in feeding.
3. There are four pairs of tentacles surrounding the mouth and the terminal nasal opening, which have been compared with the oral tentacles of Amphioxus.
4. The branchial skeleton (basket) is poorly developed, since it does not have to withstand strong suction.
5. Dorsal arcualia (vertebral arches) are confined to the tail region or extend only slightly forward from the tail.
6. No spiral valve in the intestine.
7. There is a row of mucous sacs on each side.
8. The brain has no distinct cerebrum or cerebellum.
9. The eyes are degenerate and without muscles or nerves.
10. There is no regional specialization of the median fin.
11. There is only one semicircular canal in the inner ear.
12. The tongue apparatus is larger and more elaborate than in the lampreys and this pushes the gill-slits further back.
13. Some of the myxinoids have a larger number of gill-slits than the lampreys, which is considered a primitive character.
14. The pronephros (larval kidney) functions in the adult.

15. The eggs are large and rich in yolk. It naturally follows that the cleavage is meroblastic and the development is direct without larval metamorphosis.

It is probable that numbers 3, 4, 5, 6, 8, 10, 11, 13, and 14 represent more primitive conditions than those found in the petromyzonts;

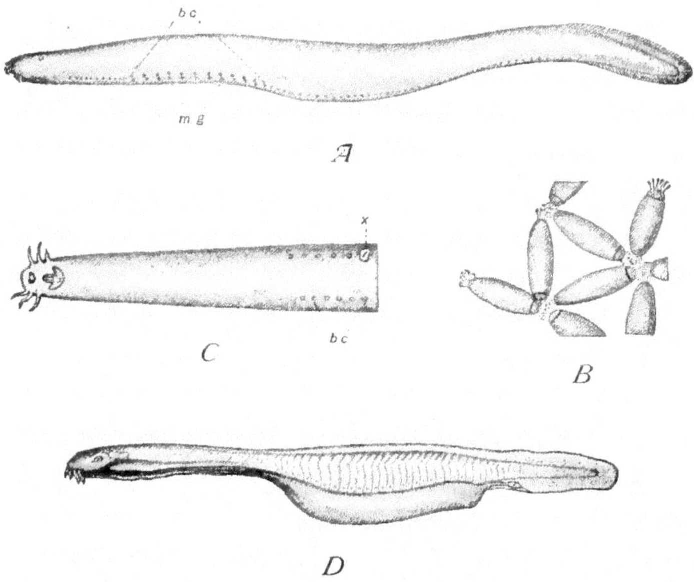

FIG. 48.—*Bdellostoma*. A. External view of whole animal, showing: *bc*, branchial clefts; *mg*, mucous glands. B. group of eggs adhering by anchor-like hooks. C. ventral view of anterior end, showing somewhat ventral nostril, ventral mouth, and oral tentacles; *bc*, branchial clefts; *x*, œsophageo-cutaneous duct. D. Larva, showing functional eye. (Redrawn mainly after Dean.)

while numbers 1, 2, 7, 9, 12, and 15 represent cænogenetic characters resulting from specialization or degeneration.

SUB-CLASS II. PETROMYZONTIA

The lampreys live in both fresh and salt water, the marine species being larger than those inhabiting streams. They live an active predaceous life, attacking frequently fishes much larger than themselves, such as sharks. Their mode of attack is to give chase and to attach themselves to the body of the prey by means of the sucker-mouth or oral funnel, which is lined with chitinous teeth to help in securing a firm hold. The flesh of the fish is then lacerated by the rasping tongue and swallowed. Ultimately, of course, the prey is killed and then feeding may be carried on at leisure. While they are feeding, the mouth is closed and respiration is carried on by incurrent and excurrent streams of water through the branchial clefts. The branchial part of the pharynx is cut off from the alimentary part and ends blindly, so that respiration is independent of the mouth.

The fresh-water lampreys are lovers of rapid waters and are often seen in rocky streams attached by the mouth to stones.

The Petromyzontia represent an evolutionary stage a step or two beyond that occupied by the Myxinoidea and a formal list of their characters furnishes an interesting comparison with those given above for the latter. The characters of the adult lamprey are shown in Fig. 49.

1. The mouth is provided with a perfect oral funnel, used as a vacuum cup, and armed with chitinous teeth.

2. The naso-pituitary sac ends blindly beneath the brain.

3. No buccal tentacles.

4. Branchial skeleton a stiff intricate basket of cartilaginous bands and rods.

5. Vertebral arches extend well forward from the tail and are of a more advanced structure than in the myxinoids.

6. A rudimentary spiral valve is present in the intestine, probably indicating a specialization of the intestine paralleling that seen in the sharks and certain other fishes.

7. No distinct mucous sacs.

8. The cerebral hemispheres are distinct and a band-like cerebellum is recognizable.

9. Eyes well developed, with good muscles and nerves.

10. The median fin system is specialized into two dorsal fins and a caudal fin.

11. There are two semicircular canals in the inner ear, a condition intermediate between that seen in the myxinoids and that in the true fishes, where three canals are always present.

12. "Tongue" apparatus less elaborate.

13. Number of gill-slits uniformly 7, more primitive than fishes, but more specialized than in some of the myxenoids.

14. Pronephros non-functional in the adult.

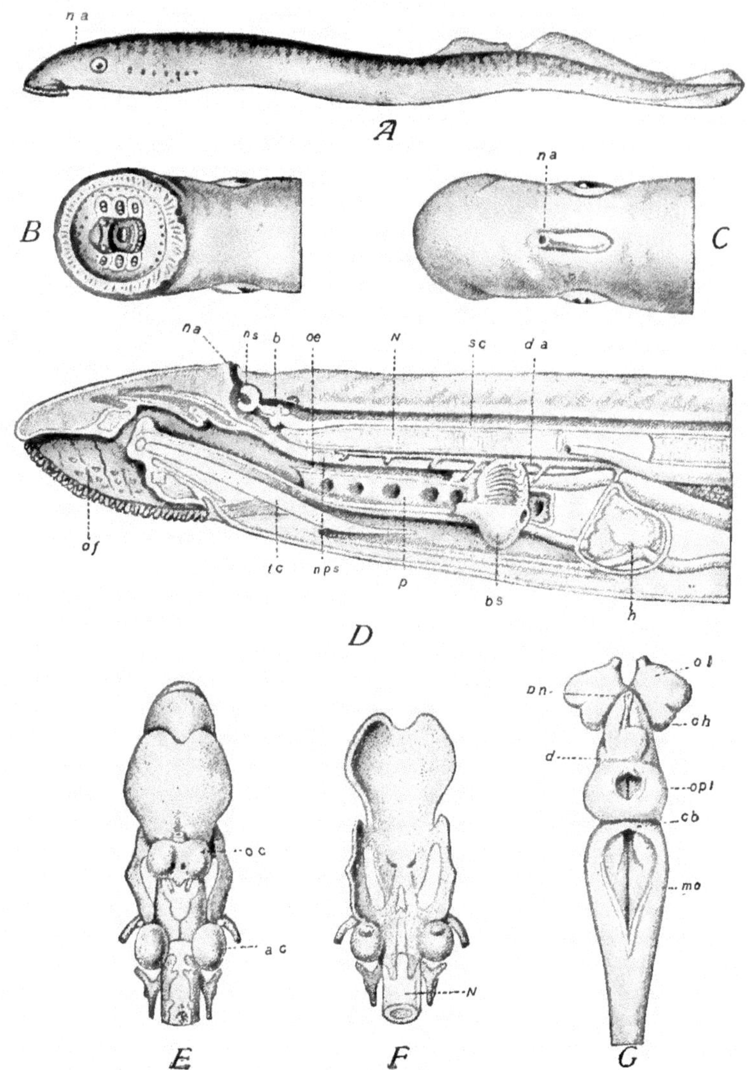

Fig. 49.—*Petromyzon*. A. External view of entire animal from specimen; *na*, median nasal aperture. B. Ventral view of head showing funnel-like oral sucker armed with chitinous teeth; end of "tongue" in mouth opening. C. Dorsal view of head, showing median nostril. D. Internal anatomy: *bs*, branchial sac cut open to show gill tissue; *b*, brain; *da*, dorsal aorta; *h*, heart; *N*, notochord; *na*, nasal aperture; leading into naso-pituitary sac (*nps*), *ns*, nasal sac; *oe*, œsophagus quite independent of pharyngeal diverticulum (*p*); *of*, oral funnel; *sc*, spinal cord; *tc*, tongue cartilage. E. Ventral view of cranium: *ac*, auditory capsule; *oc*, olfactory capsule. F. Dorsal view of cranium: *N*, notochord. G. Dorsal view of brain: *ch*, cerebral hemispheres; *cb*, cerebellum; *d*, diencephalon; *mo*, medulla oblongata; *ol*. olfactory lobe; *opl*, optic lobe; *pn*, pineal body. (Redrawn after Parker and Haswell.)

15. The eggs are comparatively small and with little yolk. Cleavage is holoblastic. The larva has a long life and undergoes a radical metamorphosis into the adult state.

Thus the lampreys may be considered as a step in advance of the hags in numbers 1, 2, 3, 4, 5, 6, 8, 10, 11, 13, 14. In other respects

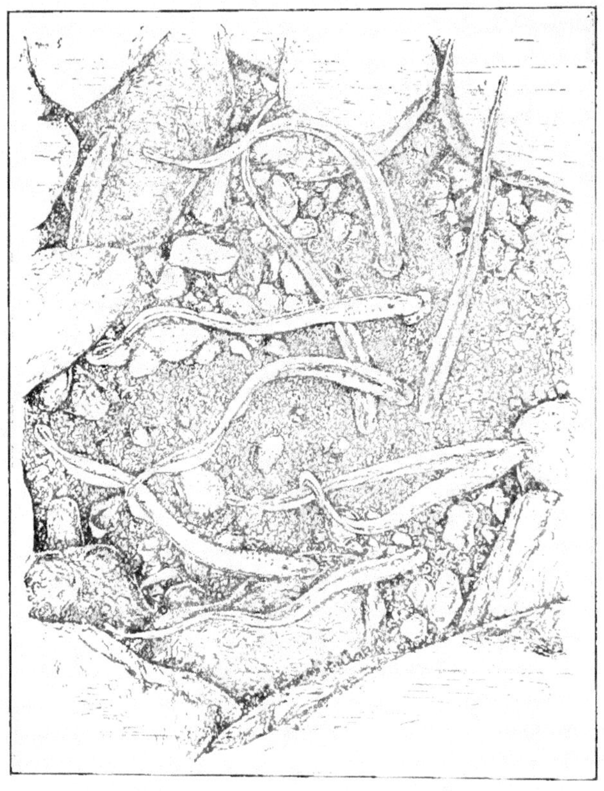

Fig. 50.—Spawning of the Brook-Lamprey (*Petromyzon wilderi*). On the right of the figure a male is attached to the head of a female. (From Cambridge Nat. Hist., after Dean and Summer.)

they show either more primitive conditions or merely different adaptations for their totally different life.

Life Cycle of the Lamprey. The lamprey breeds in fresh water, even in the case of the marine forms. The male coils the tail about the body of the female in the spawning act (Fig. 50) and,during a vigorous vibration of the two bodies, eggs and sperm are extruded in close contact. The eggs sink to the gravelly bottom in a place that has been cleaned up for a "nest." The nest has been prepared by moving many stones, both males and females using the buccal funnel for this purpose.

The egg, which measures about a millimeter in diameter, goes rapidly through cleavage, blastula, and gastrula stages and forms a tiny larva which has come to be known as "*Ammocœtes*," because when first discovered it was believed to represent a separate genus of lowly chordates.

The most significant characteristics of the Ammocœtes larva (Fig. 51) are those in which it strikingly resembles Amphioxus: (1) a hood-like upper lip resembling the oral hood of Amphioxus; (2) a well-defined parietal or median eye; (3) a food-concentrating apparatus consisting of an endostyle and dorsal mucous groove; this implies a similar mode of feeding; (4) median fins continuous and unspecialized.

Ammocœtes is more advanced than Amphioxus in other respects, as for example:—the paired eyes that lie deep in the head, a small cranium, a much more advanced brain, a reduced number of branchial clefts, a concentrated kidney (pronephros), a distinct ventral heart, a liver with a gall bladder.

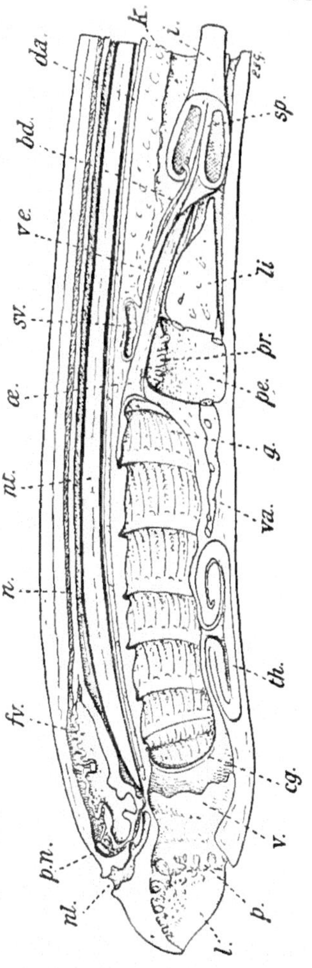

Fig. 51.—Ammocœtes larva of *Petromyzon*, enlarged sagittal section. *bd*, bile duct; *cg*, ciliated groove; *da*, dorsal aorta; *fv*, cavity of brain; *g*, gills; *i*, intestine; *k*, kidney (mesonephros); *l*, lip; *li*, liver; *nl*, nasal aperture; *nt*, notochord; *n*, neural tube; *oe*, œsophagus; *p*, oral papillæ; *pe*, pericardium with heart removed; *pn*, pineal eye; *pr*, pronephros showing nephric funnels; *sp*, spiral valve; *th*, thyroid forming from endostyle; *v*, velum; *va*, ventral aorta; *ve*, intestinal vein. (From Lankester's "Treatise on Zoölogy," Vol. IX, Goodrich.)

After living in the larval state for from three to four years, during a few weeks in the winter it undergoes a profound metamorphosis of structure and habits and emerges as a small juvenile lamprey. During the metamorphosis the buccal funnel is developed; the eyes come to the surface and begin to function; the median fin becomes specialized; the skull and branchial basket become more elaborate, the gall bladder is lost except for a few vestiges. The most remarkable transformation, however, concerns the pharyngeal region and the food-concentrating mechanism. The dorsal part, which represents the site of the dorsal groove, becomes pinched off from the old pharynx and forms a new œsophagus that breaks into the mouth. The *endostyle* becomes pinched off from the ventral floor of the pharynx and coiled up into a *thyroid gland.* The middle part of the pharynx remains as the blind, lung-like branchial sac which operates independently of the mouth.

All of these facts seem to point unmistakably to an affinity between the cyclostomes and the cephalochordates. This helps to bridge the gap between the vertebrates and the lower chordates. It has been suggested that Amphioxus is simply a case of pædogenesis, or extreme racial senescence involving development arrested in a larval condition. According to this view, which is not acceptable to biologists in general, Amphioxus is a permanent larva of some species of cyclostome. The fact that Ammocœtes spends as much as four years in the larval stage lends color to this contention, for it indicates a developmental retardation, that, if carried a step farther, might result in a permanent pædogenetic condition akin to that which occurs in the Amphibia (Axolotl larva of Amblystoma). There are, however, arguments against this view which would be out of place here. We have already put ourselves on record as supporting the theory that the ancestor of the vertebrates was Amphioxus-like. Possibly, however, such an ancestor was more like Ammocœtes than like Amphioxus.

CHAPTER V

CLASS II. PISCES (TRUE FISHES)

The fishes are at present, and have been since Silurian and Devonian times, the dominant creatures of the waters, both fresh and salt. The chronological history of the fishes is well shown in Fig. 52. Although the environment has been practically constant both as to temperature and chemical constitution the evolution of form and function has gone on rapidly. "This indicates," says Osborn, "that a changing physicochemical environment, although important, is not an essential cause of the evolution of form."

Although the aquatic environment may in a sense be thought of as constant it is not a uniform or homogeneous medium, for, within aquatic confines, there are several life zones that differ radically from one another. There are: (a) the region of river or tidal currents in which the fish must be active, swift-moving, and predaceous; (b) the surface strata of still bodies of water, where life may be comparatively passive and where only a moderate speed is necessary; (c) the region at the bottom, which may be either at moderate depths or abysmal, where life may be sluggish. These types and derivatives from them are concisely shown in the accompanying pictorial table (Fig. 53).

The body form of the **type living in currents** and depending for food and safety on swiftness is naturally the double-pointed, elongated, submarine-shaped animal (Fig. 53, *a* to *c*), illustrated by some sharks, the pickerel, the trout, the salmon, and many of the minnows. Few of these swift-moving fishes have heavy armor, nor have they any excessive development of fins, spines, or other projecting structures that might interfere with swift progress through the water. The dog-shark or common spiny dog-fish (Fig. 62) may be taken as an excellent example of this type of fish, a type that is believed to copy, perhaps more nearly than any other, the ancestral fish form.

The fishes of the open seas or lakes living at moderate depths or near the surface, where they are little affected by currents, are often of the deep-bodied, laterally compressed type (Fig. 53 *h*, *i*). A good example of this group is the sun-fish, representing that vast assemblage of modern fishes, the Acanthopterygii, which includes more species than all other groups of fishes combined. Extreme cases of the laterally compressed type are seen in the "Head-Fishes" and in *Zanclus* (Fig. 86), which are about as high as they are long and are very much compressed. This foreshortening of the long axis accompanied

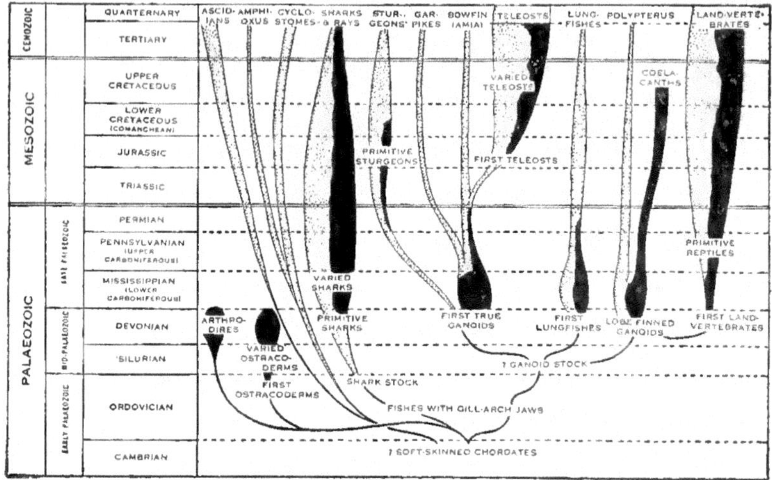

Fig. 52.—Origin and Adaptive Radiation of the Fishes. Dotted areas represent groups still existing; black areas represent extinct groups. (After Osborn and Gregory.)

by corresponding increase in height may be taken as a symptom of racial senescence; for, according to this view, there has been a retardation in growth vigor down the principal axis accompanied by a marked acceleration of growth in the secondary (dorso-ventral) axis.

The bottom-feeding type (Fig. 53, *j*, *k*, *l*) is one that involves many grades and types of specialization. In general, bottom feeders are depressed dorso-ventrally and are broad bilaterally. Common representatives of this adaptive assemblage are the skates and rays, the extinct ostracoderms, and several types of bottom-feeding teleosts. They are essentially the oldest or most senescent of the fish types, as evidenced by the comparative suppression of the primary or longitudinal axis and the secondary or dorso-ventral axis by the tertiary or bilateral axis.

Thus it will be seen that the fishes, although in a less varied habitat than that of terrestrial animals, tend to radiate adaptively in several main and many minor directions. This has occurred not once in the

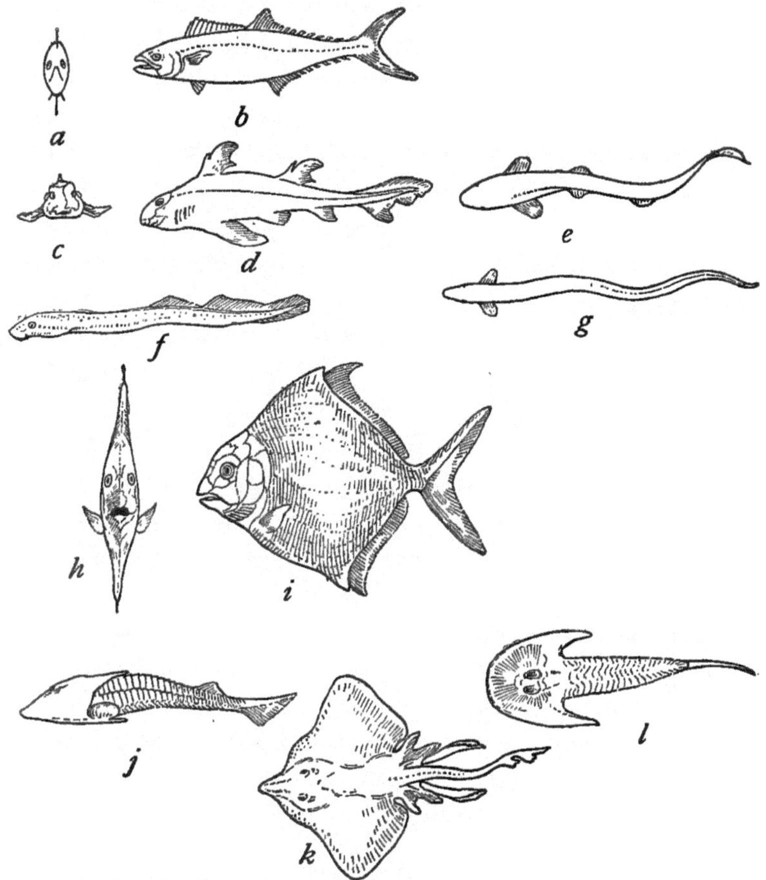

Fig. 53.—The principal types of body form in fishes. *a*, *b*, swift moving compressed, fusiform; *c*, *d*, *e*, elongated, swift, fusiform types, grading into—*f*, *g*, elongated eel-like forms; *h*, *i*, laterally compressed, slow moving, deep-bodied; *j*, *k*, *l*, laterally depressed, flat, bottom-feeding. (After Osborn's "Origin and Evolution of Life." [Charles Scribner's Sons].)

evolution of fishes, but many times. Every large group of fishes exhibits "adaptive radiation" in Osborn's sense; for we find in nearly every order of fishes the swift fusiform types, the short laterally compressed types, the broad, shallow, bottom-feeding types, and many minor types.

Structural Features of the Fishes

The fish in its typical form is essentially an aquatic mechanism—a submarine automaton. Like a submarine vessel it has a fusiform shape; steering and balancing devices; hydrostatic appliances, such as air reservoirs; a mechanism for extracting oxygen from the water; optical devices adapted for aquatic vision; and instruments for detecting vibrations in water, warning of the approach of the enemy or of the nearness of prey.

The following *formal characterization of the Class Pisces* will serve to distinguish them from other classes:

1. **Jaws**: fishes proper are all Gnathostomata (hinged-mouthed as distinguished from the Cyclostomata or round-mouthed eels.
2. **Gills or Branchiæ.**—Their method of respiration is distinctly aquatic throughout life, though accessory organs for air-breathing occur in several distinct orders of fishes. The gills are vascular processes of the walls of the branchial clefts.
3. The **Circulation** is built about the gill system. The blood is pumped forward from the ventral heart through the gills and is thence, as arterial blood, carried backward in the dorsal aorta. This scheme of circulation wherever found will be interpreted as primarily aquatic.
4. The **Heart** is a single S-shaped muscular tube with but one auricle, one ventricle, and a *bulbus arteriosus,* and receives only venous blood.
5. **Tail.**—The principal organ of locomotion is the tail, terminated by a paddle-like expansion, the *caudal fin,* and sculled by means of the powerful segmental muscles.
6. **Fins.**—These are of two sorts, the *paired* and the *median* fins. The paired fins, *pectoral* in front and *pelvic* behind, are homologous to the fore and hind limbs of terrestrial vertebrates and are supported by bony or cartilaginous rays articulated with a simple pectoral or pelvic girdle, which may be either bony or cartilaginous. These appendages are essentially balancing organs though they may be modified for various purposes, or even lost. The median fins occur in both ven-

tral and dorsal positions and may be greatly modified or wanting in places. They are supported by bony or cartilaginous rays.

7. **Exoskeleton.**—The integumentary units are *scales* of various sorts, and are found in every degree of exaggeration or degeneration. In general they may be said to be *placoid, ganoid, cycloid*, or *ctenoid* in form and structure. Sometimes the scales fuse together to form a coherent armor.
8. **Endoskeleton.**—The endoskeleton consists of appendicular and axial elements. The *appendicular skeleton* is mentioned under "Fins." The *axial skeleton* consists of a bony or cartilaginous cranium and a vertebral column more or less completely organized, consisting of bone or cartilage. A *notochord* persists in some of the more primitive orders.
9. **Lateral line of sensory organs.**—These strictly aquatic sense organs are found arranged in linear tracts along the sides and over the head. Their exact function is not definitely known, but they are probably associated with the perception of vibrations in the water.
10. **Olfactory organs.**—These are paired and end blindly, not communicating with the pharnyx as in terrestrial animals and in the hag-fishes.
11. **Auditory Organs.**—These are entirely internal and have no communication with the exterior. They serve largely the function of equilibration, though they also perceive vibrations.
12. **Eyes.**—The eyes are much like those of other vertebrates, except that they are lidless, and have spherical lenses for short-range vision in the water.
13. **Brain.**—The brain is small and shows no flexures. It nevertheless has all of the characteristic features of the vertebrate brain, though there are but ten cranial nerves.
14. **Spinal Cord:** like that in other vertebrates.
15. **Alimentary Tract.**—The pharynx is extensive and perforated by branchial clefts. The œsophagus is simple and short, for in the fish there is no neck. The stomach is little differentiated. The intestine is short and, in order to increase the digestive surface, a *spiral valve* is often present, especially in all of the more primitive orders.

16. A **Swim-Bladder** occurs in all but the Elasmobranchii, the Holocephali, and in a few degenerate teleosts. It is a gas-filled bladder, derived from, and frequently connected with, the pharynx. In some fishes it is used as an accessory lung, but it is usually for hydrostatic purposes.
17. **Kidneys.**—The nephridial system consists of elongated bodies situated in the median dorsal part of the cœlom. The units of the system are *nephric tubules* that have *nephrostomes*, funnel-like openings into the cœlom. The functional kidney is a *mesonephros*.
18. **Gonads.**—The *ovaries* and *testes* are simple sac-like structures that have ducts, *oviducts* and *vasa deferentia*, developed in connection with the primitive nephridial ducts, as in other groups.
19. **Eggs.**—The eggs of different fishes range from large, heavily-yolked eggs with chitinous shells, as in the modern elasmobranchs, to small pelagic eggs of many modern teleosts. The eggs are for the most part fertilized in the open water, but many fishes of various orders practice internal impregnation and are viviparous.

An understanding of the majority of these characters will doubtless be acquired in connection with the laboratory exercises that accompany courses in vertebrate zoölogy, but further comment on three of the most significant characteristics of fishes, the fins, the respiratory organs, and the integument, seems to be necessary.

The Fins of Fishes

Of all characters of fishes the fins are, perhaps, the most distinctive since they are adaptations for aquatic life. Analogous structures have been secondarily developed by reptiles and mammals, such as the extinct ichthyosaurs and the porpoise. (Fig. 4).

The median fin system appears primitively as a continuous median fold supported by cartilaginous or bony rays, running from just back of the head on the dorsal side, round the tail and ending behind the vent on the ventral side. According to one view the median fin system bears no relation to the paired fin system, which is thought of as being derived independently from gill septa. The prevailing view or "**continuous fin-fold theory,**" however, holds that originally the median fin system, instead of terminating back of the vent, bifurcated around the latter and continued forward as two lateral fin-folds that reached almost to the head. Two pieces of evidence support this theory. The first is, that in Amphioxus the median fin-fold actually does fork and continue forward on the ventral side as the *metapleural folds.* The second is that in the primitive shark *Cladoselache* (Fig. 65) the paired fins have rays parallel to one another much as in the median fin system, and give the impression that they are merely parts of a system continuous with the latter. Two diagrams are here presented that illustrate the origin of the paired fins from

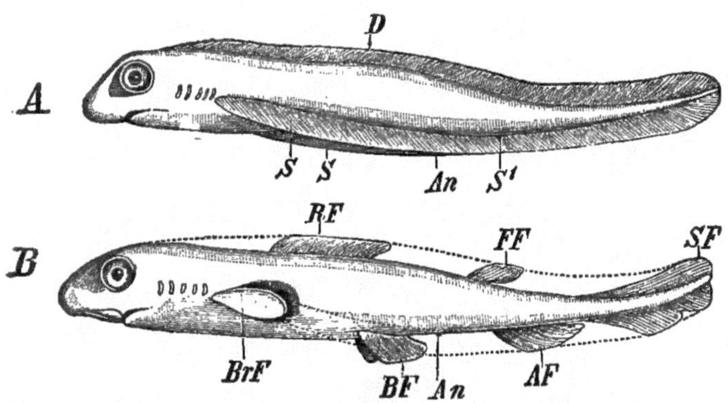

FIG. 54.—Fin-fold origin of paired fins. A. The hypothetical undifferentiated condition. B. The manner in which it is thought that the permanent fins were derived from the continuous fin-folds. *AF*, Anal fin; *An*, anus; *BF*, pelvic fins; *BrF*, pectoral fins; *D*, continuous dorsal fin-fold; *FF*, posterior dorsal fin; *S* and *S*, paired ventral lateral fin-folds; S^1, median ventral fin-fold continuous with *S and S; SF*, superior or dorsal lobe of caudal fin. (From Wiedersheim.)

lateral fin-folds. The first (Fig. 54) is that of Wiedersheim, and the second is adapted from that of Kingsley (Fig. 55), involving double dorsal as well as double ventral fin-folds.

The most primitive types of fishes or fish-like creatures have the median fin system unbroken and regionally unspecialized, but most of the fishes proper have the continuous median fin subdivided into one, two or three dorsal fins, a caudal fin, and one or two anal or ventral fins. The degree of development of these various regional specializations of the median fin system varies greatly according to the habit and degree of specialization of the group. In the more generalized types the size and degree of elaboration of these fins remain well within the bounds of bodily symmetry, but in some of the highly specialized groups these fins may become greatly exaggerated as in *Zanclus* (Fig. 86) and in *Dendrochirus* (Fig. 90).

The most significant evolutionary processes concern the caudal fin and in this region of the fish's body we have changes that parallel those that have occurred in the caudal regions of various land animals, involving, on the one hand, more or less pronounced foreshortening or atrophy, and, on the other hand, excessive prolongation or hypertrophy of the tail region. Beginning with what appears to be the

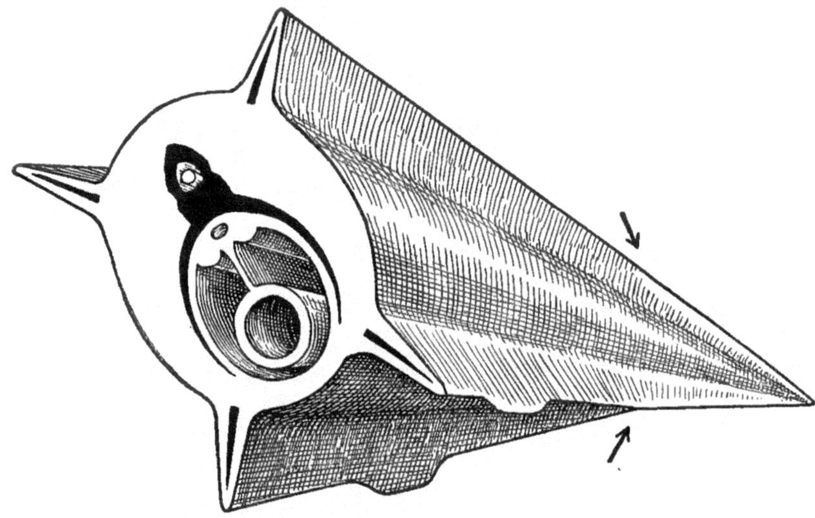

Fig. 55.—Diagram of the origin of the median and paired appendages from lateral fin-folds. The arrows indicate the points of junction, dorsal and ventral, of the paired fin-folds with the median fin-folds. (Modified after Kingsley.)

ancestral condition, we have a type of caudal fin called *diphycercal* (Fig. 56, A), which is evenly developed both above and below the notochord of the tail. The supporting rays above the notochord (*epichordal rays*) are as well developed as those below the notochord (*hypochordal rays*). A type of caudal fin that appears to be next in order of antiquity is the *heterocercal* type (Fig. 56, B), in which the epichordal rays, and consequently the upper lobe of the fin, are much less strongly developed than the hypochordals. This type of fin is found in most elasmobranchs, in the Holocephali, in the Chondrostei and, in a modified form, in the Holostei. A third type, or *homocercal* caudal fin (Fig. 56, E), is the result of a foreshortening of the terminal portion of the caudal axis and results in the typically "fish-tailed" type of fin. In a teleost fish such as the salmon the young fish has first a diphycercal, then a heterocercal, and finally a homocercal type of fin. Various modifications of these three main types are found and will be commented upon in appropriate places.

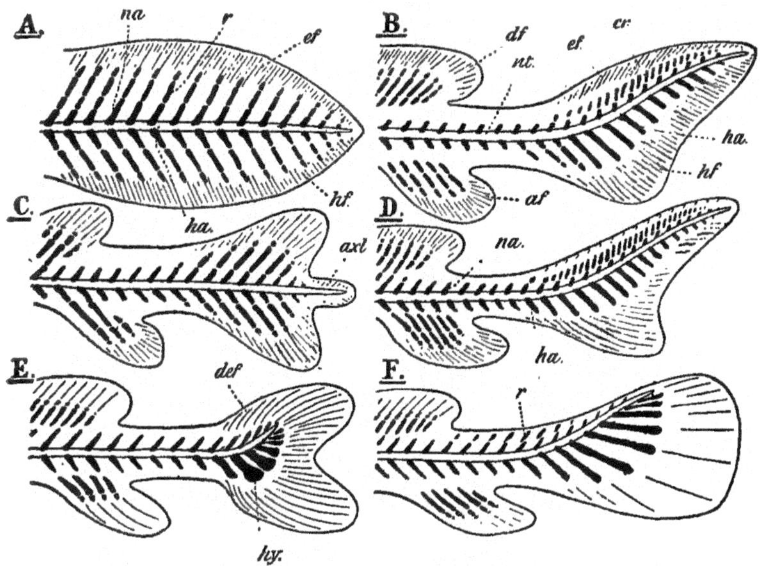

Fig. 56.—Types of Caudal Fins. A, Diphycercal, with equal dorsal and ventral lobes; B, Heterocercal (Selachii); C, Modified diphycercal (some teleosts); D, Heterocercal (Chondrostei); E, Homocercal (teleosts); F, Abbreviated heterocercal (some Holostei). *af*, anal fin; *axl*, axillary process; *cr*, caudal fin rays; *def*, dorsal lobe of caudal fin; *df*, dorsal fin; *ef*, epicaudal lobe of caudal fin; *ha*, hæmal arches; *hf*, hypocaudal lobe of caudal fin; *na*, neural arches; *nt*, notochord; *r*, dermal fin rays. (From Lankester's "Treatise on Zoölogy," Vol. IX. [A & C. Black].)

The most primitive type of paired fin is believed to be that seen in *Cladoselache* (Fig. 65). The "lobe-fins" of the Crossopterygii are next in primitiveness, while the fins of other groups are specialized types derived from these two primitive types. As will appear later the "lobe-fin" is the most nearly hand-like in architecture and is believed to have given rise to the hand-type of paired appendage seen in primitive land vertebrates.

THE RESPIRATORY ORGANS OF FISHES

The characteristic respiratory organs of aquatic vertebrates are gills or branchiæ. These structures are finely divided outgrowths of the ectodermal or endodermal epithelium lining the branchial clefts. The number of clefts or gill-slits varies from five to seven in number, each cleft being separated from its neighbors by branchial septa. The most primitive fishes have the larger number of branchial clefts and the more modern types have regularly five. *Heptanchus*, sometimes mentioned as the most primitive living species of shark, has seven clefts, *Hexanchus*, another primitive shark, has six, while the elasmobranchs in general have five fully developed clefts and a vestigial anterior first cleft called a spiracle. The spiracle or rudimentary first cleft is also found among the most primitive Teleostomi (Crossopterygii and Chondrostei), and is present in the embryos of Teleostei and Holostei, but is closed before hatching. In the Holocephali, an aberrant group of elasmobranch fishes, the fifth cleft is closed in the adult, thus reducing the number of functional clefts to four. It will be recalled that the pro-vertebrates have much larger numbers of pharyngeal clefts, over fifty pairs in Amphioxus and even larger numbers in some tunicates. The cyclostomes have on the whole larger numbers of clefts than the true fishes. Though the hag-fishes of the family *Myxinidæ* have no more than six pairs, those of the family *Bdellostomidæ* have as many as fourteen pairs, while the lampreys all have seven pairs. The direction of evolution appears to be one of reduction in number of clefts from around fifty in the Amphioxus-like ancestor, fourteen to six in the cyclostomes, seven to five in the true fishes, and four in the Holocephali.

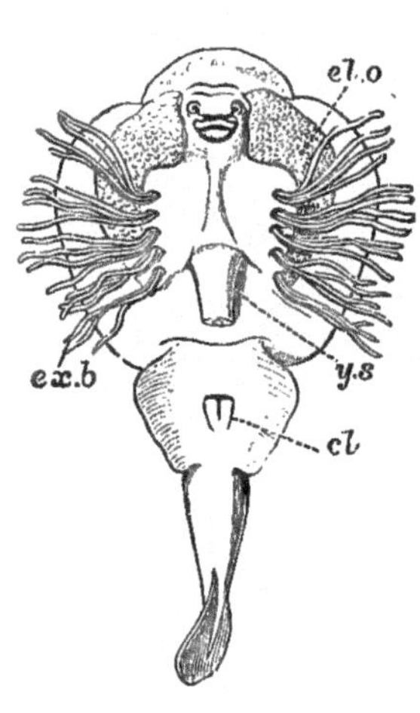

FIG. 57.—External gills in embryo torpedo. *cl*, cloaca; *ex. b*, external gills; *el. o*, electric organ; *ys*, yolk stalk. (From Bridge, Cambridge Nat. Hist., Vol. VII.)

The openings of the clefts to the exterior differ in different groups of fishes. Among the elasmobranchs the usual situation is that each cleft opens separately and is not covered by any flap or operculum; though in *Chlameidoselachus* the primitive "frilled shark" (Fig. 67, A) each cleft has a backwardly directed flap or gill-cover. In the Holocephali the first three clefts are covered by an operculum and only the fourth, or last functional cleft, opens freely to the outside. In the great majority of Teleostomi and in the Dipneusti the five clefts are covered with a flap-like operculum, capable of opening and closing and effectively protecting the branchial filaments from injury. In some of the eels and in other specialized types of teleosts the gills are

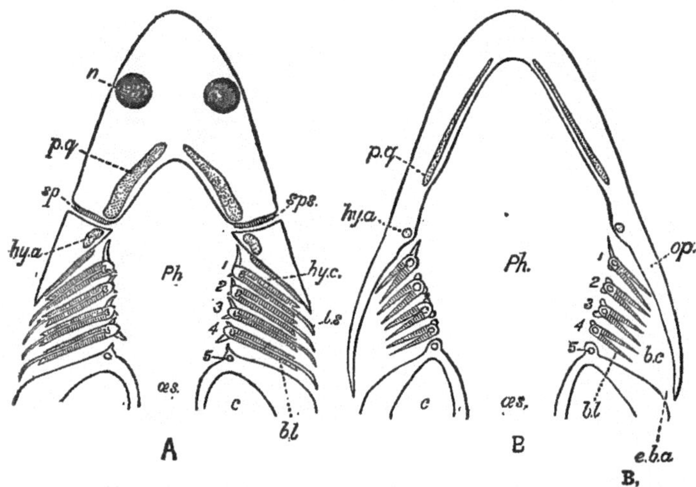

Fig. 58.—Diagram of gills of fishes. A, Horizontal section through the head of an Elasmobranch; B, Similar section of a Teleost. *bc*, branchial cavity; *bl*, branchial lamellæ; *c*, cœlom; *eba*, external branchial aperture; *hy. a*, hyoid arch; *hy. c*, hyo-branchial cleft; *ls*, interbranchial septum; *n*, nasal organ; *oes*, œsophagus; *op*, operculum; *pq*, palatoquadrate cartilage; *Ph*, pharynx; *sp*, spiracle; *s. ps*, spiracular pseudobranch; *1–5*, 1st to 5th branchial arches. (From Bridge, after Boas.)

completely covered with a fold of skin and the only exit is through one or a pair of small water-pores.

Two quite different and distinct kinds of gills are found among fishes: external and internal gills.

External gills are purely larval or embryonic organs and are not functional in any adult fish; though their homologues are found in the perennibranchiate Amphibia, believed to be pædogenetic or permanent larval types. External gills are finely branched processes of the ectodermal epithelium of the branchial clefts. They are found in the embryos of many elasmobranchs (Fig. 57) and in some teleosts. A notable case of larval gills is seen in the advanced larva of *Polypterus* (Fig. 70, C).

Internal gills are true functional gills of adult fishes. They are finely divided diverticula of the endodermal epithelium of the branchial clefts. Their location is well shown in the diagrams of elasmobranch and teleost heads (Fig. 58).

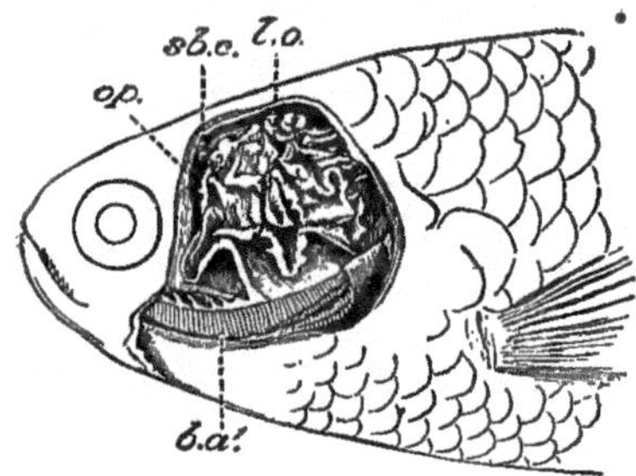

FIG. 59.—Respiratory labyrinth of the Climbing Perch (*Anabas scandens*) exposed by removal of part of operculum. *ba'*, first branchial arch; *lo*, labyrinthiform organ; *op*, operculum. *sbc*, suprabranchial cavity. (From Bridge).

THE AIR-BLADDER AND ACCESSORY ORGANS OF RESPIRATION

In all of the groups of fishes above the elasmobranchs there is a single or paired air-bladder, a sac-like diverticulum of the pharynx derived from either dorsal or ventral sides of the alimentary tract. It is in all cases supplied with blood from the "pulmonary artery" and, primitively at least, subserves two functions: that of a hydrostatic or buoyancy organ and that of an accessory respiratory organ or primitive lung. In

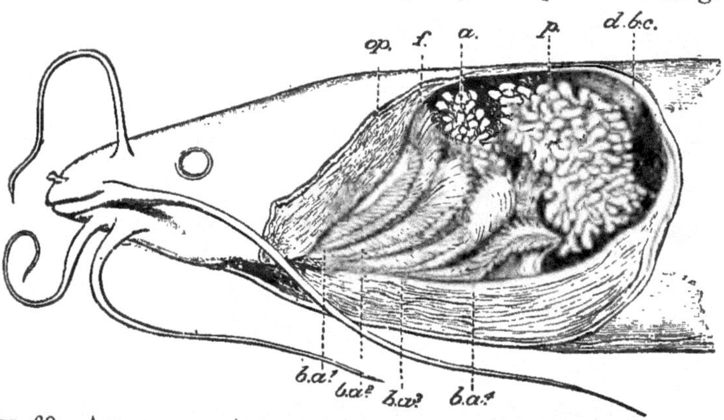

FIG. 60.—Accessory respiratory organs of the cat-fish, *Clarias*, as seen after removal of operculum. *a*, anterior arborescent organ; *b. a* $^{1-4}$, first four branchial arches; *d. b. c*, dorsal extension of left branchial cavity; *f*, modified gill-filaments; *op*, base of operculum; *p*, posterior arborescent organ. (From Bridge.)

the most primitive teleostome fishes, the Crossopterygii, it is used as a lung when the water is foul; in *Amia* it is constantly functional as an air-breathing apparatus; while in the Dipneusti (lung-fishes) it is an elaborately pouched lung, used to tide the fish over a period of drought.

In certain other fishes that have acquired terrestrial habits, such as the Climbing Perch, *Anabas* (Fig. 59), and in the air-breathing eel, *Clarias* (Fig. 60), there is an extensive post-branchial chamber provided with labyrinthine or arborescent elaborations of the epithelium that are highly vascular and play a pulmonary rôle.

The Integument of Fishes

The integument of fishes differs from that of land vertebrates in being soft and slimy on the surface, though usually well protected beneath the soft epidermis by means of scales or plates composed of hard materials such as bone and ganoin. The slimy condition of the epidermis is not due to a deposit from the water, as is commonly believed, but is a mucous secretion from numerous cutaneous glands.

The *scales* of fishes differ materially from those of reptiles, birds, and mammals in that they are composed exclusively of dermal elements. In the land vertebrates the scales are mainly cornifications of the epidermis. The great majority of fishes are more or less completely covered with scales of moderate size, but some fishes such as the eels, cat-fishes, etc., have secondarily lost their scales or have them in a degenerate condition, imbedded deeply in the skin; other fishes have the scales replaced by bony plates, which may form a more or less solid armor, as in the trunk fishes (Fig. 93) and the porcupine fishes in which the plates are provided with spines. The possession of a coating of small, equally distributed scales is considered as the primitive or generalized condition for fishes, and the loss of scales or the development of heavy plates, as specialized or senescent conditions.

Four main types of scales are distinguished: *placoid*, *ganoid*, *cycloid* and *ctenoid*. The *placoid* scale, characteristic of elasmobranch fishes, is believed to be the original type of scale. It consists of a basal plate of bony substance derived from the dermis and a spine-like external protuberance covered with an enamel-like substance derived from the epidermis. In the sharks and their kin it is clear that from this type of scale are developed the *true teeth*—the latter being merely enlarged and flattened placoid scales derived from the oral integument. Thus the placoid scale is the ancestor of all true vertebrate teeth.

The ganoid scale is an archaic type of integumentary unit, found only in the so-called ganoid orders of Teleostomi and in a few primitive teleosts. The ganoid scale is believed to be derived from the placoid condition by the loss of the spike-like protuberance and by the addition of a hard external coating of glistening substance called *ganoin* which is secreted by the dermis and is not homologous with enamel. Usually ganoid scales are rhombic in form and are laid like tiles, but in some ganoid fishes the scales overlap like shingles. Cycloid and ctenoid scales are of similar structure to ganoid scales except that they have lost the ganoin covering and are thinner and less protective. Cycloid scales have a smooth circular margin and are characteristic of the older groups of fishes, while ctenoid scales have comb-like edges and are present mainly in the higher orders of teleosts.

The **coloration** of fishes is due to the presence of dermal pigment-cells or chromatophores, which carry variously colored pigments and are under the control of the central nervous system. The more primitive groups of fishes are colored in rather neutral fashion and the more highly specialized types are highly colored. The colors of tropical fishes, especially those of the coral reefs, run riot and rival those of the birds in elaborateness and brilliancy. Most of these highly colored fishes belong to the climax order of modern fishes, the *Acanthopterygii*, in which the presence of high coloration is taken as evidence of the onset of racial senescence. The flounders are the "chameleons" among fishes. They are perhaps among all animals the most efficient in their ability to modify their color patterns in response to varied backgrounds.

CLASSIFICATION OF PISCES

SUB-CLASS I. ELASMOBRANCHII.

Order I. *Pleuropterygii* (Family. Cladoselachidæ)—extinct.
Order II. *Ichthyotomi* (Family. Pleuracanthidæ)—extinct.
Order III. *Acanthodei* (Family 1. Diplacanthidae)—extinct.
(Family 2. Acanthodidæ)—extinct.

Order IV. *Plagiostomi*

Sub-Order 1. *Selachii* (12 living and 3 extinct families of sharks and dog-fishes).

Sub-Order 2. *Batoidei* (Saw-fishes, skates, rays and torpedoes, 7 families).

Order V. *Holocephali* (Chimæras, 3 extinct and 1 living family).

SUB-CLASS II. TELEOSTOMI

Order I. *Crossopterygii*

Sub-Order 1. *Osteolepida* (4 extinct families).

Sub-Order 2. *Cladista* (recent genera, *Polypterus* and *Calamichthys*).

Order II. *Chondrostei* (5 extinct and two living families, including paddle-fishes and sturgeons.)

Order III. *Holostei* (6 extinct and 2 living families, including bow-fins and gar-pikes).

Order IV. *Teleostei.*

Sub-Order 1. *Malacopterygii* (21 families, including tarpons, herrings, salmon, etc.).

Sub-Order 2. *Ostariophysi* (6 families, including carp, tench, cat-fishes, etc).

Sub-Order 3. *Symbranchii* (2 families).

Sub-Order 4. *Apodes* (5 families of eels).

Sub-Order 5. *Haplomi* (14 families, including pickerel and killifishes, etc.).

Sub-Order 6. *Heteromi* (5 families, *Fierasfer*, etc.).

Sub-Order 7. *Catosteomi* (11 families, including stickle-backs, pipe-fishes, sea-horses, etc).

Sub-Order 8. *Percesoces* (12 families, including *Belone*, sand-eels, rag-fishes, etc.).

Sub-Order 9. *Anacanthini* (3 families, including cod, etc.).

Sub-Order 10. *Acanthopterygii* (78 families, including a large proportion of our commonest fishes—perch, bass, mackerel, flounders, gobies, shark-suckers, etc.).

Sub-Order 11. *Opisthomi* (1 family—eel-like fishes).

Sub-Order 12. *Pediculati* (5 families, including the Anglers, Bathymal Sea-Devils, etc.).

Sub-Order 13. *Plectognathi* (7 families, including file-fishes, trunk-fishes, puffers, porcupine fish, and sun-fish).

SUB-CLASS III. DIPNEUSTI (Dipnoi) Lung-Fishes.
(2 extinct and 2 living families, including Neoceratodus, Protoperus, and Lepidosiren).

APPENDIX (TO TRUE FISHES)

I. PALÆOSPONDYLIDÆ (1 family, between cyclostomes and fishes).
II. OSTRACODERMI (3 orders of 8 families, mostly armored fishes).
III. ANTIARCHI (1 family of mailed fishes).
IV. ARTHRODIRA (1 family of mailed fishes).

It will be of interest to note that of the 226 families of true fishes listed in the above classification, 172 belong to the order Teleostei. There are 32 families of Elasmobranchii, 9 of which are extinct. The remaining 22 families are divided among the ganoids and dipnoans,

Although the Teleostei are unquestionably the characteristic fishes of the present and may be considered the dominant creatures of the waters of modern times, the elasmobranchs, represented most typically by the sharks, are still, as they have been since Devonian times, a powerful, predaceous tribe, that has ruled the waters through strength and savagery; though not characterized by such sheer numbers nor by so many specializations and adaptations as are the teleosts.

SUB-CLASS I. ELASMOBRANCHII

The present day sharks, though not as primitive as the extinct types of elasmobranchs, are relatively a conservative group. Though they have come through millions of years of evolution some of the sharks (Fig. 67) notably the Notidanidæ and the dog-sharks and relatives, are still very generalized aquatic vertebrates and serve well to illustrate primitive vertebrate morphology. On that account they are used extensively as a main-line vertebrate type (stem-type) in courses in comparative anatomy. Adaptive radiation has taken place among the elasmobranchs less than in other groups. For the most part they have retained the elongated spindle-shaped body, the lightly armored integument, and active predaceous habits, characteristic of plastic or ever-juvenile races. The broad, flat types such as skates, rays and torpedoes, illustrate the bottom-feeding, comparatively sluggish adaptive complex, but this has not reached its extreme; for though these forms are more heavily armed with spines and small dermal plates than are the sharks, none of them have acquired really heavy armor, nor a definitely sedentary habit.

A Typical Elasmobranch

The dog-fish *Squalus acanthias* (Fig. 61) may be taken as a typical elasmobranch in order to introduce the group. These rather small predaceous sharks are called by the natives of our Atlantic coast "Horned Dogs" or "Spiny Dogs" to distinguish them from the similar but smoother dog-fish, *Mustelus*, which belongs to a different family. In Buzzards Bay and Vineyard Sound the species was a few years ago so abundant as to be a real pest and they were caught in large quantities and used only for fertilizer. They have recently been discovered to be an excellent food fish and are now being put on the

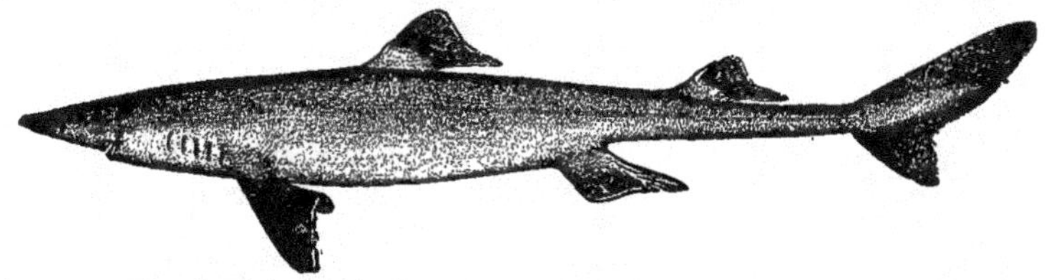

Fig. 61.—The dogfish shark, *Squalus acanthias*. (From Hegner, after Dean.)

market in canned form under the name of "gray fish." They rove the coastal waters in schools and destroy large numbers of smaller fishes, squids, ctenophores, and many worms. They are caught in fish traps where with their sharp teeth they do a good deal of damage to the mesh. *Squalus* is viviparous, giving birth to young "pups" upwards of six inches in length. The figures used to illustrate the anatomy of the dog-fish, are for good reasons, not of the species likely to be used in the laboratory, but correspond sufficiently closely with descriptions.

External Characters (Fig. 61).—The body is submarine-shaped, sharp at both ends. The steering and balancing organs consist of two *median dorsal fins*, two pairs of lateral fins, *pectoral* and *pelvic*, the latter of which are in the male provided with stiff specialized portions known as *claspers*, used for holding the female during copulation and for introducing sperm into the latter's oviducts. The eggs are internally fertilized and embryonic development takes place in the uterus of the mother. This is a highly specialized character, as will be clear when the more primitive sharks are described. The mouth is ventrally situated and is armed with a number of rows of *teeth* on both upper and lower jaws. The teeth are obviously modified placoid scales. The body is covered all over with small scales (*placoid scales*), with a bony plate-like base of dermal origin and a sharp protruding spine covered with hard enamel of ectodermal origin. Between the mouth and the pectoral fins are the *gill-slits* (pharyngeal clefts) each of which opens separately to the exterior. The anterior gill-slit on each side is small and modified into a *spiracle*, situated just back of the eye. Several external features discussed in other connections are *nasal apertures, cloaca, lateral-line* organs, and lastly, the tail, which is provided with a typical heterocercal caudal fin.

The Skeletal System.—The entire skeleton is cartilaginous with only a slight impregnation of *calcareous* matter. The *cranium* is a *chondrocranium*, a solid, one-piece capsule completely inclosing the brain and the principal sense organs. The cranium proper is fused with paired nasal capsules, and paired auditory capsules. The vertebral column consists of a series of hour-glass-shaped vertebræ, with lens-shaped pieces of the original notochord between adjacent vertebræ, and connected with each other by a strand of notochordal tissue perforating the entire set of vertebræ like the string through a chain of beads. Closely associated with the skull but not fused with it, is the *mandibular skeleton*, consisting of an upper jaw (*palatoquadrate cartilage*) and a lower jaw (*Meckel's cartilage*). Back of the jaws are the *visceral arches*, that are composed of upper and lower parts like the jaws; the first pair being specialized as the *hyoid arch*, the five others being the more generalized *branchial arches* that afford support for the gills. The fins all have cartilaginous *ray-like supports*, and the pectoral and pelvic limb skeletons are supported upon simple horseshoe-shaped girdles (*pectoral* and *pelvic girdles*) each composed of but one piece of cartilage.

The Alimentary System.—The mouth opens directly into the capacious *pharnyx*, which is perforated by five *gill-clefts* and the paired *spiracles*. A short *œsophagus* of large caliber leads into a U-shaped *stomach*, which in turn communicates through a valvular opening, controlled by a sphincter muscle, with the *intestine* (Fig. 62). The latter

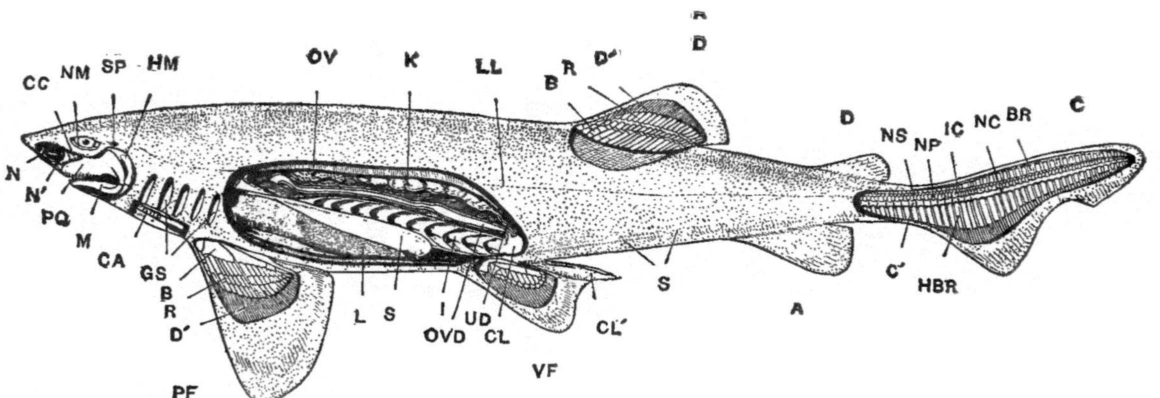

FIG. 62.—General anatomy of a shark. *A*, anal fin; *B*, basal cartilaginous fin supports; *B*, *R*, basals and radials;
C, caudal fin; *C'*, centrum; *CA*, conus arteriosus; *CC*, cartilaginous cranium; *CL*, cloaca; *CL'*, clasping appendage of
ventral fin; *D*, dorsal fin; *D'*, dermal fin-rays; *GS*, gill-slits; *HBR*, hæmal, basal, and radial supports; *HM*, hyomand-
ibular; *I*, intestine, showing spiral valve; *IC*, intercalary plate; *K*, kidney; *L*, liver; *LL*, lateral line; *M*, mandible
(Meckel's cartilage); *N*, position of anterior, and, *N'*, of posterior, nasal aperture; *NC*, notochord; *NM*, nictitating
membrane of eye; *NP*, neural process; *NS*, neural spine; *OV*, ovary; *OVD*, oviduct (Müllerian duct); *PF* pectoral
fin; *PQ*, palatoquadrate cartilage; *R*, radials; *S*, stomach; *SP*, spiracle; *UD*, urinary duct; *VF*, ventral fin. (From
Hegner, after Dean.)

is short but of large diameter and has a secreting surface greatly enlarged by a fold, in the shape of a spiral staircase, called the *spiral valve* (a primitive fish character). Into the intestine empties the large bilobed *liver* which is provided with a *gall-bladder* and a *bile-duct*. A diffuse *pancreas* also pours its secretion into the intestine.

The Respiratory System.—Branchial respiration is carried on in the six pairs of branchial clefts. These branchiæ are primitive respiratory organs consisting of mere diverticula of mucous membrane, richly vascular and supported by cartilaginous processes, called *gill-rays*. The water enters the mouth and is forced out through the gill-slits. In doing so, it aërates the gill-filaments and provides oxygen for the blood that circulates rapidly through them.

The Circulatory System (Fig. 63).—The architecture of this system is in the main laid out in accord with the branchial system. The *heart* receives only venous blood from a single *precaval vein*, and pumps it forward in a *common ventral aorta*, which divides into five pairs of *afferent branchial arteries*, each of which carries blood to one set of branchiæ. A corresponding *efferent branchial vessel* picks up the aërated blood from each gill and carries it to a *dorsal aorta* which in turn distributes the blood to all parts of the body both anteriorly and posteriorly. A complex system of veins consisting of a general *systemic* part, a *hepatic portal*, and a *renal portal* system, returns the blood to the heart along paired channels called *anterior* and *posterior cardinal veins*.

Urogenital System.—The *nephridia* (kidneys) consist of paired strap-like organs lying side by side along the roof of the body cavity. Microscopic examination shows that these long organs are composed of paired nephric tubules each of which opens at one end into the cœlom and at the other into a nephric duct that leads to the cloaca. The functional adult kidney is a *mesonephros* or "mid-kidney," the *pronephros* being reduced to a mere vestige, though functional in the larva. The *testes* are paired whitish bodies of rather flat, long and narrow shape, that communicate with the *cloaca* by paired ducts, the *vasa deferentia*. The *ovaries* are, when mature, large lobulated bodies attached to the dorsal anterior part of the body cavity and communicating with the exterior by large paired *oviducts*, which unite anteriorly and have a single funnel-like opening for receiving the large ova. Each oviduct is regionally modified into a *shell gland* and a *uterus*.

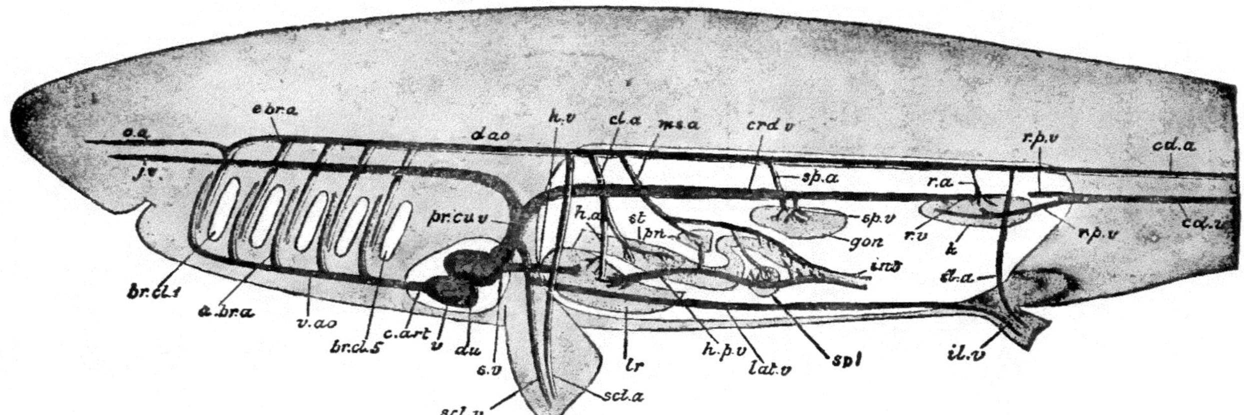

FIG. 63.—Semi-diagrammatic side view of the vascular system of a dogfish-shark. *a. br. a*, afferent branchial artery; *au*, auricle; *br. cl. 1–5*, branchial clefts; *c. a*, carotid artery; *c. art*, conus arteriosus; *cd. a*, caudal artery; *cd. v*, caudal vein; *cl. a*, cœliac artery; *crd. v*, cardinal vein; *d. ao*, dorsal aorta; *e. br. a*, efferent branchial artery; *gon*, gonad; *h. a*, hepatic artery; *h. p. v*, hepatic portal vein; *h. v*, hepatic vein; *il. a*, iliac artery; *il. v*, iliac vein; *int*, intestine; *j. v*, jugular vein; *k*, kidney; *lat. v*, lateral vein; *lr*, liver; *ms. a*, mesenteric artery; *pn*, pancreas; *pr. cv. v*, precaval vein; *r. a*, renal artery; *r. p. v*, renal portal vein; *r. v*, renal vein; *scl. a*, subclavian artery; *scl. v*, subclavian vein; *sp. a*, spermatic artery; *spl*, spleen; *sp. v*, spermatic vein; *st*, stomach; *s. v*, sinus venosus; *v*, ventricle; *v. ao*, ventral aorta. (From Hegner, after Parker.)

Nervous System.—Though the brain (Fig. 64) is very small, it is larger in proportion to body size than that of the Cyclostomata. The most striking feature is the large size of the *olfactory bulbs. Cerebral hemispheres* are well defined, *cerebellum* is large and overlaps anteriorly a part of the *optic lobes* and posteriorly a part of the *medulla oblongata.* The region of the *thalamencephalon* from which come the optic nerves is comparatively small and slender. The *spinal cord* is typical, and inclosed within cartilaginous *neural arches.*

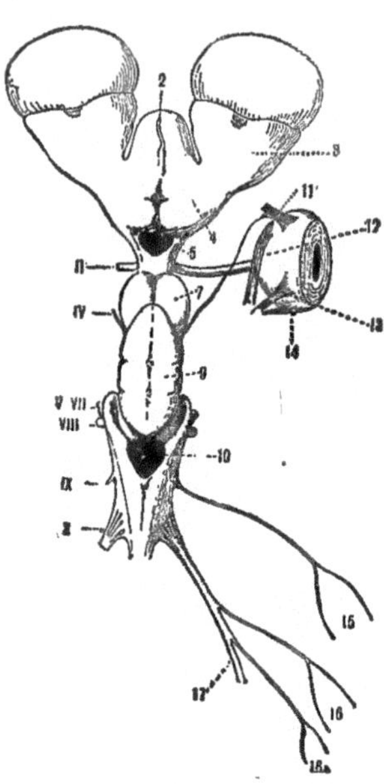

Fig. 64.—Brain of a dogfish shark, *Scyllium catulus,* dorsal view. *2,* pineal stalk; *3,* olfactory lobe; *4,* cerebral hemisphere; *5,* thalamencephalon; *7,* optic lobes; *9,* cerebellum; *10,* roof of hind-brain; *11, 12, 13, 14,* muscles that move the eyeball; *15,* ninth nerve; *16, 16a,* branches of vagus nerve; *17,* main trunk of vagus nerve; II–X, roots of the cranial nerves. (From Hegner, after Shipley and MacBride.)

Sense Organs.—The dominant sense of the elasmobranch is *olfactory;* the sense organ consisting of large convoluted invaginations in close contact with the olfactory bulbs of the brain. The *eyes,* though small and probably not especially keen-sighted, are well developed and connected with the brain by rather slender optic nerves. The *auditory organs* are inclosed in cartilaginous capsules and consist of three *semicircular canals,* a *utriculus,* and a small simple *sacculus.* The *lateral line sense organs* are in grooves of the skin not completely closed; they divide into several branches in the head region, one above and one below the eye and some in the hyo-mandibular region.

The dog-fish represents neither extreme of elasmobranch evolution, but is nearly midway between the most primitive extinct sharks and the most specialized modern skates and rays. It will be instructive to consider some of the most primitive elasmobranchs in order to be able to judge more certainly which of the characters of our favorite laboratory fish are specialized and which are still primitive. A very primitive form of living shark is the Frilled Shark, *Chlameidoselachus* (Fig.

67, A). Its terminal mouth and almost diphycercal caudal fin are decidedly archaic. Whether its frilled branchial clefts are primitive or specialized it is impossible to say.

Extinct Elasmobranchs

Some of the extinct families of sharks: Cladoselachidæ, Pleuracanthidæ, Diplacanthidæ, and Acanthodidæ, bring to light certain conditions that are obviously more primitive than those of any of the modern elasmobranchs. They all have the mouth terminal instead of ventral, indicating that the terminal mouth which characterizes most of the Teleostomi is more primitive than the ventral mouth

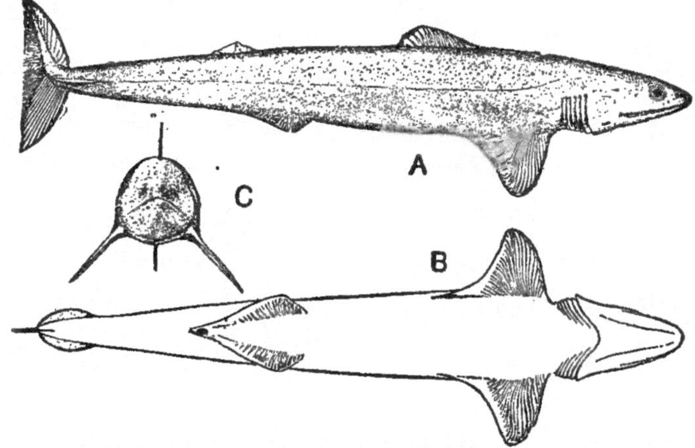

Fig. 65.—*Cladoselache fyleri;* Upper Devonian, Ohio. A, right side view; B, ventral view; C, front view; restored. (From Lankester, after Woodward.)

of the modern elasmobranchs. There are no claspers on the ventral (pelvic) fins, indicating that the habit of copulation, with internal incubation of eggs and consequent viviparity, so common among modern elasmobranchs, is a cænogenetic specialization. The notochord persists as an unbroken elastic rod and the neural and hæmal arches are developed only about as far as they are in the Cyclostomata. The exoskeletal elements appear to be wanting in *Pleuracanthidæ* and, in the other families, are smaller and less developed than in modern sharks.

Each of these families shows some one or two features more primitive than the others. One of the most striking features shown by

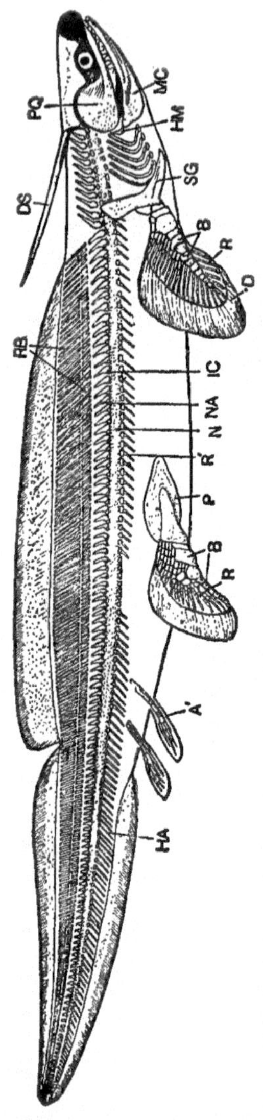

these primitive sharks is the paired fin system of *Cladoselache*, a fish that on the whole is generally admitted to be the most primitive of all true fishes, though its remains were found in the Devonian rocks millions of years later than the earliest ostracoderms. The paired fins of *Cladoselache* (Fig. 65) are so-called "lappet-fins," broad based and closely resembling in bony framework the median fins, such as the dorsals. The skeleton of these fins consists of two sets of elements. The slender, nearly parallel, unjointed cartilages occupy a distal position, and proximal to these is a less numerous set of shorter and stouter cartilages imbedded in the body wall and corresponding to the large cartilages (propteyrgium, mesopterygium, and metapterygium) of the modern shark. The pectoral fin of *Cladoselache* is not quite so primitive as the pelvic and furnishes a transition between the latter and the fins of modern elasmobranchs. The location and general arrangement of the paired fins of *Cladoselache* have given rise to the theory that "the fins of fishes arise from lateral skin folds of the body, into which are extended internal stiffening rods." These folds are supposed to be essentially like those composing the median fin system and are believed to have been at one time continuous with them, as in the hypothetical case described by Dean. This continuous fold is supposed to have been specialized in two regions to form the pelvic and pectoral enlargements and to have disappeared in between.

FIG. 66.—*Pleuracanthus ducheni*, restored. *A'*, ventral fin; *B*, basal fin-cartilages; *D*, dermal margin of fin; *D S.*, dermal fin-spine; *H. A*, hæmal arches; *HM*, hyo-mandibular; *I. C*, inter-neural plates; *M. C*, Meckel's cartilage; *N*, notochord; *NA*, neural process and spine; *P*, supposed pelvic cartilage; *PQ*, palatoquadrate; *R*, radial fin-cartilages; *R'*, ribs; *S. G*, shoulder girdle. (From Parker and Haswell, after Dean.)

Pleuracanthus (Fig. 66) also contributes a very primitive fin character but in another system, the caudal. Instead of having the heterocercal type of tail fin which is characteristic of modern elasmobranchs, it has a still more primitive one, a continuous fin-fold of the diphycercal type, running smoothly about the tail. This is the type that one finds in *Amphioxus*, in the hag-fishes, and, in a slightly modified form, in the lampreys. It is also the earliest embryonic stage in the median fin development of teleosts. There seems little question, then, that the ancestral elasmobranchs had this kind of caudal fin.

Putting together the most primitive characters of all of these extinct elasmobranchs we are able to describe a hypothetical ancestral shark which is also probably the prototype of the earliest real vertebrate.

The Hypothetical Ancestor of the Elasmobranchs, and of Fishes in General

This creature must have had an elongated, spindle-shaped, fusiform body with terminal mouth, armed with dermal teeth, with probably more than seven gill-slits, with small lozenge-shaped dermal denticles scattered over the skin, with lappet-like paired fins, and a diphycercal tail-fin with low dorsal specializations of this fin-fold. Internally, it probably had a persistent notochord with the merest vestiges of vertebral arches. It also doubtless had lateral line organs in open grooves, and, having no claspers, laid small eggs in the open sea. The intestine probably had at least a primitive spiral valve.

Some of the Specialized Modern Elosmobranchs of the Order Plagiostomi

Sub-order 1. Selachii (*true sharks*) have remained on the whole comparatively unspecialized. For the most part they are active, free-swimming, predaceous creatures such as the ancestral sharks must have been. Among the more striking members of the Selachii are the Hammer-Heads, the Whale-Sharks, and the Angel-Sharks.

The Hammer-Heads (*Sphyrnidæ*) are characterized by the lateral protrusion of the eyes on large flat stalks (Fig. 67, D) supported by cartilaginous extensions of the cranium. There are all gradations between the only slightly extended eyes to those in which the eyes extend so far as to give a width five times that of the normal head. It is not known that these greatly projecting eyes are of any especial use to their possessors. It would be very difficult therefore to explain their origin on a natural selection basis. Those who have worked experimentally with fish embryos have often observed in abnormal embryos tendencies for the eyes to project on stalks, and it may well

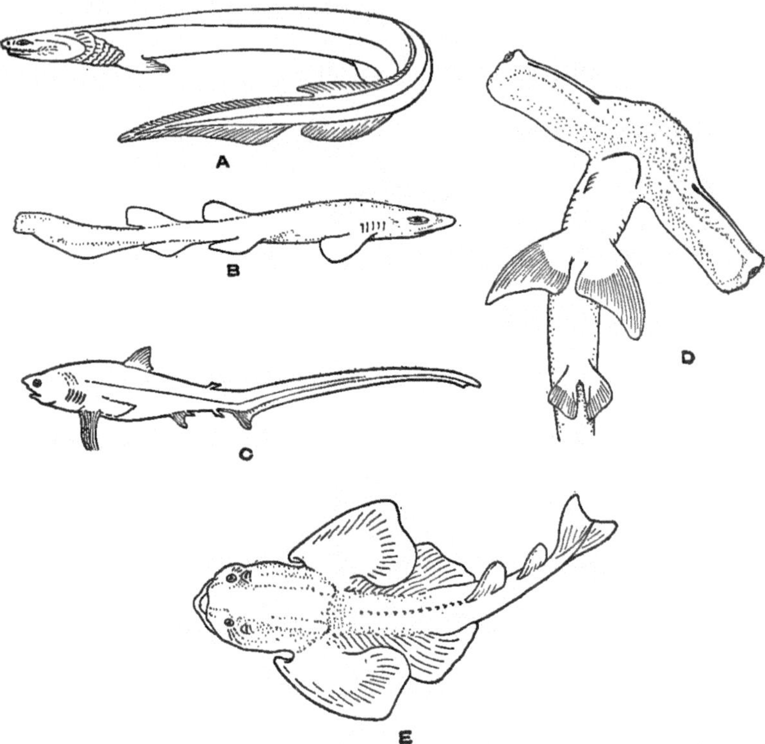

Fig. 67.—Group of Sharks (Selachii). A, Frilled Shark (*Chlameidoselachus anguineus*) after Günther. B, Female Dog-Fish (*Scyllium canescens*), after Günther. C, Thresher Shark (*Alopecias vulpes*), after Jordan and Evermann. D, Hammer-head shark (*Sphyrna zygæna*), male, after Bridge. E, Angel Shark (*Rhina squatina*), after Bridge.

be that some similar explanation would account for this specific character in the Hammer-Heads.

The Whale-Sharks (*Rhinodontidæ*) are of interest because they are much the largest true fishes that have ever lived. They are said sometimes to exceed fifty feet in length and to be of proportionate bulk. Such a shark would be able easily to swallow a man, but it never does. Instead, it feeds only upon small pelagic animals, including fishes, squids, and other small forms, which it strains out of the water by means of the fringes on its long, slender gill-rakers.

The Thresher Shark (Fig. 67, C) is remarkable chiefly for the great length of the upper lobe of the caudal fin, which equals the rest of the body in length.

Angel-Sharks (*Rhinidæ*) constitute an interesting transition between the Selachii and the Batoidei (skates, rays, etc), in that they have a short broad form (Fig. 67, E) with marked lateral expansions of the pectoral and pelvic fins that look like wings and give the group its name. In habits they are more like the rays than the sharks, in that they frequent the bottom and do not follow the free-roving life of the typical sharks.

Sub-order 2. Batoidei (skates, rays, torpedoes), are all, with the single exception of the Saw-Fishes (*Pristidæ*), rhombic or discoidal in form, due to the dorso-ventral flattening of the body and the excessive growth of the pectoral fins. For the most part they are sluggish bottom-feeders that swim slowly over the sea-bottom at various depths, using the pectoral fins as propellers, waves of propulsion passing from in front backward. They are protectively colored on top so as closely to resemble the sea-bottom, but are usually white or uniformly pinkish below. The tail is weak and of little use in locomotion, being used merely as a rudder or, in the sting rays, as a weapon of defense. The typical members of the Batoidei are the common *skates* (Fig. 68, A). These fishes have an almost perfect rhomboidal outline, resembling a broad kite with a short tail. They catch their prey (fishes, crustaceans, etc.) by dropping suddenly upon it and covering it with the broad body and fins. The food is ingested by means of the ventrally placed mouth armed with rasping file-like teeth. Some of the largest species reach a diameter of as much as seven or eight feet.

The Electric Rays (*Torpedinidæ*) are more nearly circular in body outline (Fig. 68, C) than the skates. They are especially noteworthy on account of the presence of paired electric organs, developed from two pillars of muscle situated between the pectoral fins. They are capable to giving at will quite a heavy electric shock. This mode of defense is in accord with the entire absence of dermal spines, for a fish capable of giving a shock needs no armor.

Sting Rays or Whip-Tailed Rays (Fig. 68, D).—These tropical Rays are especially noted for the long flexible tail armed with one or more serrated spines in the position of a dorsal fin. These spines which may be eight or nine inches long are capable of inflicting very severe wounds, which become infected or poisoned by having introduced into them the mucous secretions that bathe the cutting spines.

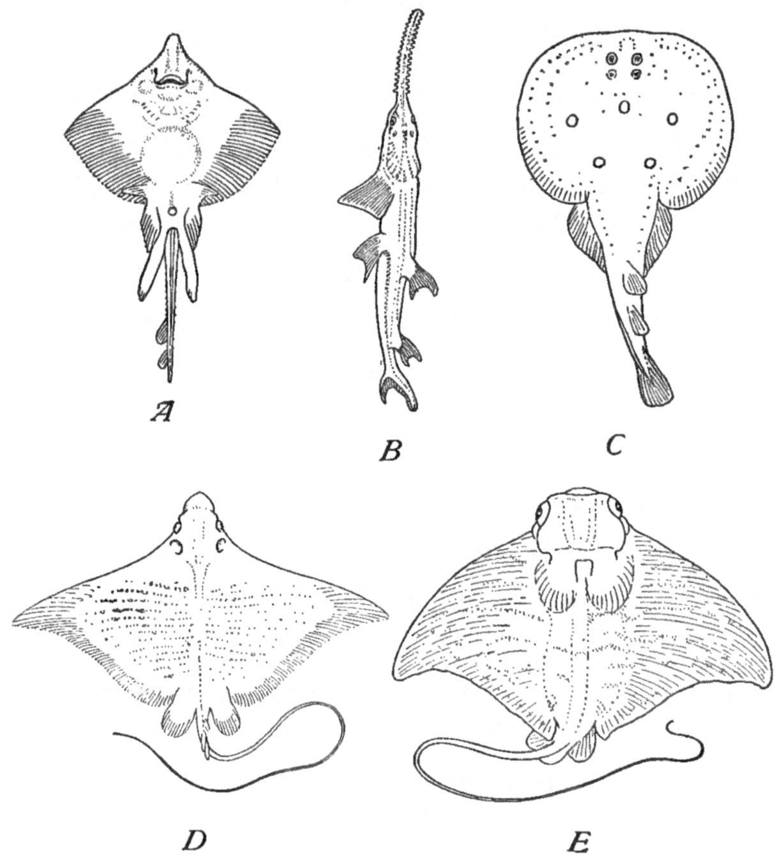

FIG. 68.—Group of Batoidei. A, skate, *Raia batis*, male, ventral view (after Hertwig). B, Saw-Fish, *Pristis antiquorum*, (after Cuvier). C, Electric Ray, *Torpedo ocellata* (after Bridge). D, Sting-Ray, *Stoasodon narinari*, (after Jordan and Evermann). E, Eagle Ray, *Myliobatis aquila* (after Bridge).

Eagle Rays (*Myliobatidæ*) show extremely pronounced specialization of the pectorial fins (Fig. 68, E), giving the body a considerably greater breadth than length, the width being sometimes as great as twenty feet. They catch their prey by enveloping it in their great "wings." They are true "sea-vampires," dreaded by pearl divers near Panama, who are said to have been caught and drowned by these great "winged" creatures. They are sometimes known as "devil fishes."

Saw-Fishes (*Pristiidæ*) exhibit one of the most striking specializations seen among the Batoidei. In them the body (Fig. 68, B) is only slightly broadened laterally, but the rostrum is prolonged until it reaches a length half as great as the rest of the body. The rostrum is armed with two lateral rows of knife-like teeth which enable the fish to deal vicious slashing blows at its enemies. It is said that they attack whales in the soft parts behind the flippers, tearing off and devouring pieces of flesh.

ORDER HOLOCEPHALI (CHIMÆRAS)

Chimæras (Fig. 69) are by some considered as a divergent offshoot of the Elasmobranchii; by others they are placed in a distinct sub-class of coördinate value with the whole sub-class Elasmobranchii. It is difficult to decide between these two alternatives. There are undoubtedly some characters that relate the Holocephali to the Elasmobranchii, but there are also some very fundamental differences. *They agree with the elasmobranchs in the following ways:* a wholly cartilaginous endoskeleton; no cartilage or membrane bones; the limb girdles and the limb skeletons essentially elasmobranch in structure; the dermal denticles, present locally in some modern forms and more generally in extinct forms, agree with those of elasmobranchs; the brain is very similar; the reproductive system, including clasping organs in the male, and large chitinous-shelled eggs, remind one strongly of the elasmobranchs; there is no air-bladder; there is a spiral valve; there is a conus arteriosus; the nostrils are connected with mouth by oro-nasal grooves. The group is undoubtedly one of great antiquity and probably branched off from the elasmobranchs of early times. *The specialized features are:* The skull is *autostylic,* the jaw cartilage (palatoquadrate) being firmly fused with the base of the cranium, a character which accounts for the name (*holos*—whole or undivided; *cephalos*—head). The teeth are modified into large crushing *dental plates.* The *claspers,* instead of consisting merely of one pair derived from the pelvic fins, are five in number. One pair is like that of the elasmobranchs; a second pair occurs in pockets of the skin in front of the pelvic fins, and a median finger-like process is hinged to the forehead between the eyes. Just how these claspers are used is not known. They also show certain tendencies in the

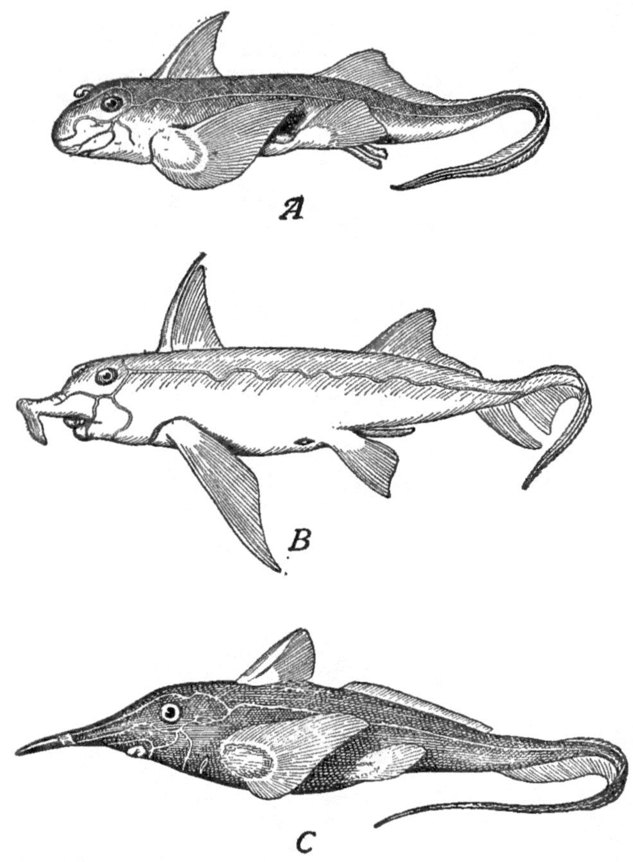

Fig. 69.—Group of Holocephali (Chimæras). A, *Chimæra monstrosa* (after Bridge). B, *Callorhynchus antarcticus*, male (after Parker and Haswell). C, *Harriotta raleighana* (after Goode and Bean).

direction of the Teleostomi, such as the crowding together of the branchial arches underneath the head, the development of an operculum covering all but the last gill-slit, the suppression of the spiracles, and the absence of a cloaca.

These curious fishes are of moderate size, one to three feet in length,

They inhabit the comparatively deep seas, ranging from 200 to 1,200 fathoms, though one species, *Chimæra colliei*, lives at or near the surface. In all of them the median fin system has two peculiarities; a strong spine anterior to the front dorsal fin and long whip-like tail with either a diphycercal or a slightly developed heterocercal fin. The Holocephali may be dismissed then as an interesting but not especially significant side-line of elasmobranch evolution.

SUB-CLASS II. TELEOSTOMI

This great group of fishes to which belong nearly all of the modern species except the Elasmobranchii, consists of a few fragmentary vestiges of formerly abundant fish faunas and a vast assemblage of modern bony fishes. The lowest of the vestigial orders, the Crossopterygii and the Chondrostei, show many evidences of relationship with the primitive sharks and very probably descended from some primitive shark type at about the same time that the sharks gave off the Batoidei and the Holocephali. It may be said in introducing the Teleostomi that the most primitive order, *Crossopterygii*, is from the standpoint of vertebrate evolution of very especial importance; for the group seems to belong to one of the main-trunk-lines of evolution and to be an important junction point for several branch-lines of vertebrate phylogeny. It is now believed that the Crossopterygii gave rise not only to the other orders of Teleostomi, including the modern teleosts, but to the Dipneusti and to the first Amphibia.

The principal morphological characters of the Teleostomi as a whole are:

1. The process of ossification of a primitively cartilaginous endoskeleton has resulted in the appearance of a number of separate bones in the skull, and the jaws have also ossified more or less completely. The roof of the chondrocranium has been covered over by a set of dermal investing bones without teeth or denticles.

2. The **primary jaws** (both upper and lower) have been covered over and reinforced by tooth-bearing investing bones, developed in the dermis.

3. The **operculum** over the gills is supported by a more or less elaborate dermal skeleton that becomes rather intimately associated with the skull and lends a false appearance of complexity to the latter.

4. The **pectoral girdle**, which is often attached to the skull, is also made more complex by the addition of investment bones from the dermis. The pelvic girdle is usually absent or vestigial.

5. The **vertebral column** is not very compact, the vertebræ being often without a centrum; if the latter is present it is an arch-centrum.

6. The **fins** are supported by bony dermal rays.

7. The **integument** is characterized by scales without the denticle or spike seen in the placoid type of scale. They are ganoid, cyloid, or ctenoid in form and may be either tessellated (laid like tiles) or imbricated (overlapping like shingles), the former being the more primitive condition.

8. **No claspers** are known in the group, though in some groups the pelvic fin may be modified as an intromittent organ used in sexual copulation.

9. Most members of the sub-class have an **air-bladder,** which serves to compensate for the additional weight caused by the ossified skeleton.

10. The **gill filaments** project beyond the edges of the inter-branchial septa.

11. The **nasal sacs** have no naso-oral grooves and they open by separate nares.

12. The **brain** has a much reduced cerebrum, with small olfactory lobes.

13. **No cloaca.**

14. The **ova** are usually small and numerous and range, with respect to the cleavage, from holoblastic to meroblastic types, the more primitive types resembling the eggs of Amphioxus in that they have holoblastic cleavage.

Practically all of these characters show an evolution away from the elasmobranch condition. In some cases the evolution is progressive and in others regressive.

Order I. Crossopterygii (Lobe-Finned Ganoids)

This very important group which was abundant in Devonian times and is represented to-day by *Polypterus* and *Calamychthys*, is the most primitive division of the Teleostomi, and, from the phylogenetic standpoint, much more significant than either of the other ganoid orders or the teleosts. There are evidences that this order of fishes is the ancestral group not only of all the higher fishes, but of the terrestrial vertebrates.

There are in this order four extinct families belonging to the sub-order *Osteolepida*, and one family, represented by two genera of present-day African "lobe-fins" of the sub-order *Cladista*. Both of these types are essentially antique, true "living fossils," although no real fossils of these genera have as yet been discovered.

A somewhat detailed description of *Polypterus bichir* will serve to introduce the reader to the characters of the Crossopterygii that are of most importance.

According to Harrington, *Polypterus bichir* inhabits the deeper waters of the Nile but does not bury itself in the mud like a true mud-fish. The large lobate pectoral fins are used as balancers and to some extent as paddles in swimming, but the most significant function of these fins is their use as supports; for when resting on the bottom, the head is held up from the mud by the tips of the fins, much as a mud-puppy (*Necturus*) is supported by its fore legs. The excellent figure (Fig. 102, A), after Osborn, shows these fins used as supporting limbs. Such a habit suggests the mode of origin of the terrestrial limb from the aquatic limb. The air-bladder in this primitive fish is not merely or principally a hydrostatic organ, but is an accessory respiratory organ. It is connected by a primitive trachea with the pharynx and is used as a lung. The supporting character of the pectoral fin and the primitive lung are considered of especial importance as furnishing the beginnings of adaptations for terrestrial life.

The larva of *Polypterus* is interesting on account of its similarity to amphibian larvæ. Budget has described a larva of *P. senegalus* (Fig. 70, C) as quite a striking object, beautiful in color and markings. Its most remarkable feature is the single pair of external gills, which are pinnate in structure and are evidently homologous with the external gills of amphibian larvæ and those of the perennibranchiate urodeles. The larva, like the adult, uses the pectoral fins as supporting appendages. The median fin system is of the primitive diphycercal type like that of a tadpole, but differs from the latter in having cartilaginous supports or rays. The species of *Polypterus* figured (Fig. 70, A) is *P. senegalus*.

The other living genus of Crossopterygii, *Calamichthys* (Fig. 70, B) is much like *Polypterus*, but is a greatly elongated eel-like edition of the latter. It is confined to middle western Africa, living in the small muddy rivers of that region, swimming about with an undulating, snaky motion and feeding mainly on crustaceans.

The more important *anatomical characters of Polypterus* and its allies are as follows:

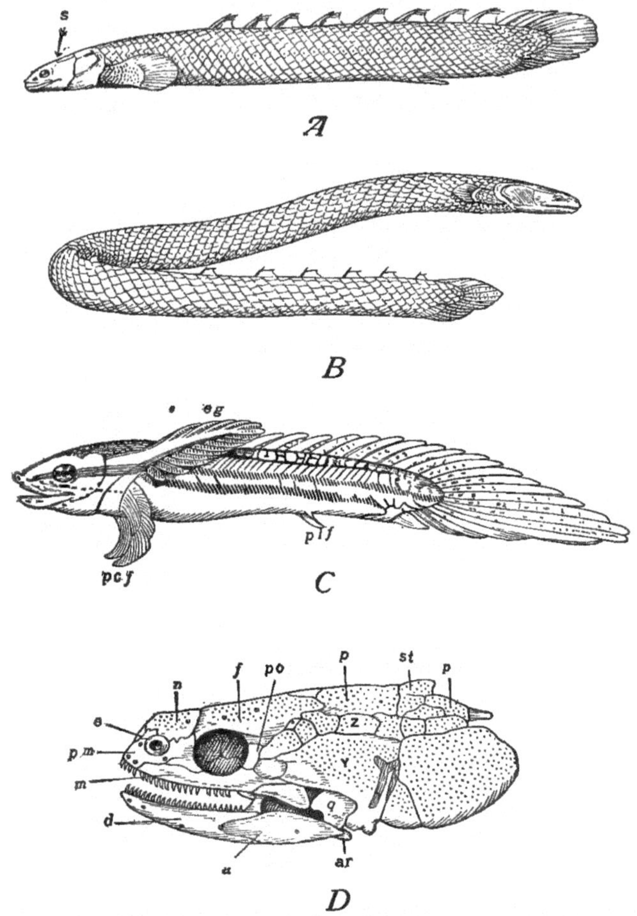

Fig. 70.—Crossopterygii.—A, *Polypterus senegalus; s*, position of spiracle (after Bridge). B, *Calamichthys calabaricus* (after Bridge). Larva of *Polypterus senegalus*, showing characteristic attitude when resting on the bottom, and the large external or cutaneous gills (after Budgett). D, Lateral view of cranium of *Polypterus*. *a*, angulare, *ar*, articulare; *d*, dentary; *e*, ethmoid; *f*, frontal; *m*, maxillary; *n*, nasal; *pm*, premaxillary; *po*, post-orbital; *q*, quadrate; *st*, supra-temporal; *y*, cheek plate; *z*, row of small spiracular ossicles. (After Traquair.)

1. **Endoskeleton.**—The cartilaginous skeleton is largely ossified, the chondrocranium being divided into a number of distinct bones.

The notochord is replaced by bony vertebræ which are hour-glass-shaped.

2. The **skull** (Fig. 70, D) is covered above and below by numerous dermal investment bones that are much like those of the primitive extinct Amphibia (*Stegocephali*).

3. The **integument** is covered by heavy rhomboid scales that are plated externally with a sheet of ganoin.

4. The **pectoral fins** are lobose in outline, but in their skeletal parts show evidences of relationship to the pentadactyl limb. The basal bones are homologous with the finger, wrist, and arm bones of Amphibia, and the fringe of dermal rays is the aquatic part of the appendage.

5. There is a persistent **spiracle,** like that of the elasmobranchs.

6. The intestine has a **spiral valve.**

7. The **conus arteriosus** has several rows of valves.

8. The **median fin system** is essentially continuous from dorsal to ventral side, though a large part of the anterior dorsal end of it is broken into separate spines each with a flap of skin back of it. The *caudal fin* is *diphycercal.*

9. The **pelvic fins** are much reduced.

10. The **air-bladder** is double and opens on the ventral side of the pharynx.

11. The **lower jaw** is sheathed in dermal investment bones, the *dentary* and *angulare.*

Some of the *extinct Crossopterygii* were more highly specialized in certain respects than *Polypterus.* They have the median and paired fins more like those of modern teleosts, overlapping scales, and either heterocercal, or a modified type of the latter called *gephyrocercal,* caudal fins. They are, however, more primitive than *Polypterus* in having persistent notochord and acentrous vertebræ.

On the whole then *Polypterus* may be said to be the most generalized teleostome fish living and may well be considered as the prototype of that group.

Order II. Chondrostei (Cartilaginous Ganoids)

The Chondrostei and Holostei of the ganoid orders, and the Teleostei, together constitute a division known as *Actinopterygii,* in which the paired fins are sharp instead of lobate as in the Crossopterygii. The modern representatives of the Chondrostei are the Paddle-Fish or Spoon-Bill (*Polyodon*), and the sturgeon (*Acipenser*). These are considered to represent the culmination of degenerative evolutionary processes, combined with certain marked specializations.

The **Spoon-Bill** (Fig. 71, B) inhabits the Mississippi River and its tributaries. It is a sluggish creature, feeding chiefly on mud that it shovels up with its spade-like snout. The mud is strained through gills provided with unusually long gill-rakers, which serve to catch the food particles and let the mud go through with the water current. The paddle-like rostrum is richly supplied with tactile end-organs,

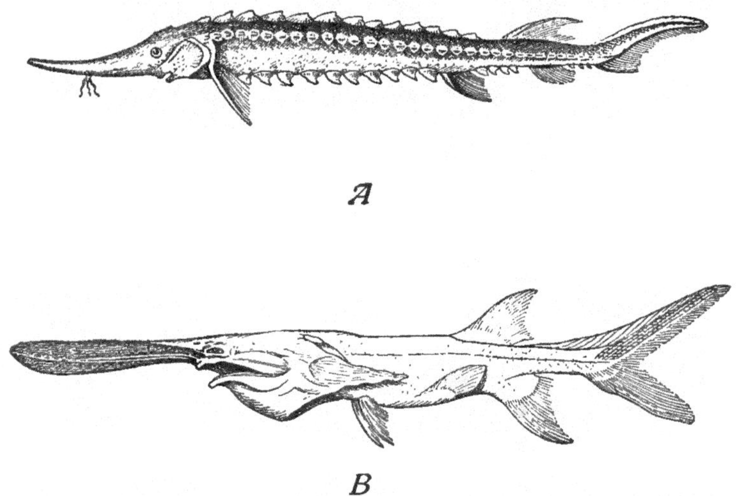

FIG. 71.—Chondrostei. A. The Sturgeon, *Acipenser ruthenus* (after Cuvier). B. The Spoon-Bill or Paddle-Fish, *Polyodon folium* (after Bridge).

enabling the fish to detect the presence of food in the mud. An allied genus of Spoon-Bill (*Psephurus*) inhabits some of the principal rivers of China and evidently lives the same type of life as Polyodon. The shape of the body is decidedly selachoid or shark-like, and the skin is apparently scaleless, though scattered vestigial scales are found, and a patch of rhombic scales occurs on the upper half of the caudal lobe.

The **Sturgeon** family (Fig. 71, A) is a widely distributed group, superficially decidedly selachoid in appearance. The rostrum is prolonged into a snout with a transverse row of barbels depending from its ventral surface. The mouth is small and protrusible, but is without teeth in the adult. The scales are remarkable in that they are arranged in five widely separated longitudinal rows of keeled bony elements. The dermal bones of the roof of the skull are fused into a solid shield. The sturgeons inhabit the inland lakes and seas, being found in our own Great Lakes, in the Black Sea, the Caspian Sea, and the tributaries of these lakes and seas. They feed upon mollusks, worms, small fishes, and vegetation, the mouth being protruded as a cylindrical spout and thrust into the mud in search of food. They sometimes reach a very large size—individuals having been taken that weighed 2,760 and 3,200 pounds. The Russian delicacy, *caviar*, is made of the eggs of sturgeons. The flesh of the sturgeon is also an excellent food and is largely used.

Many extinct Chondrostei are known, and in every case they are nearer the type exemplified by the Crossopterygii than they are like the modern Spoon-Bill and sturgeon. It is therefore highly probable that the group was derived from the Crossopterygii.

Order III. Holostei

This rather large and varied assemblage of fishes is, with the exception of two genera, extinct. It is probably from this order that the malacopterygian Teleostei arose, since there is a gradual transition between the two groups. As the name indicates, these fishes have a completely ossified skeleton, much like that of the Teleostei. The scales are either rhombic or cycloid but are covered externally with a hard coating of ganoin. If one begins with the oldest fossil Holostei and proceeds through a series up to more recent forms, it is possible to trace the development of many of the features that characterize the present teleosts and to note the elimination of many of the more primitive teleostome characters. The Holostei are represented to-day by two families, the *Lepidosteidæ* (Gar-pikes) and the *Amiidæ*) Bow-fins. Only two species of Gar-pike and one species of Bow-fin (*Amia calva*) exist to-day.

The **Gar-pikes,** *Lepidosteus* (Fig. 72, A), are fresh-water fishes of the United States and Canadian waters. They are elongated creatures with long slender snouts heavily armed with teeth. They are predaceous and kill great numbers of other fishes. In some regions a bounty is placed on their heads. The great *Alligator Gar* reaches a length of ten feet or more and with its huge jaws is a really formidable creature. The fins are shaped much as in teleosts, resembling those of the pickerel. The scales are rhombic and heavily coated with ganoin, so as to make a complete coat of mail.

The "**Bow-fin,**" *Amia calva* (Fig. 72, B), or "fresh-water dogfish," is quite like a typical teleost in form, and would readily pass for a member of that group, except for certain minor ganoid features. *Amia* is a voracious, carnivorous fish, living in central and southern North America. Its most salient characters are: its continuous dor-

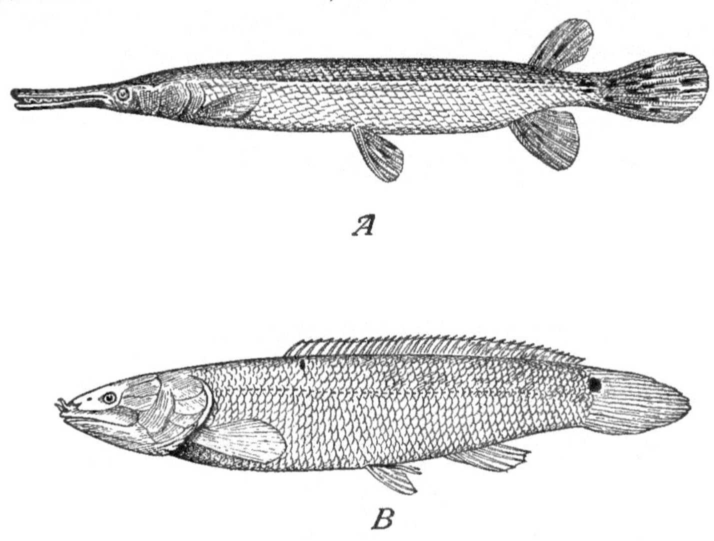

FIG. 72.—Holostei. A, Short-nosed Gar Pike, *Lepidosteus platystomus* (after Goode). B, The Bow-Fin, *Amia calva* (after Bridge).

sal fin (from which its name, "bow-fin" is derived); its heavy ganoin-covered, imbricating scales; its homocercal (really modified heterocercal) tail-fin. It has a cellular air-bladder which it uses as a lung, coming frequently to the surface to gulp air. *Amia* breeds in May and June, building a nest in water weeds, in which it lays its eggs and which the male guards until the eggs hatch. Even after hatching the young remain in a school about the male, who appears to exercise some degree of parental care over them. The egg of *Amia* is a primitive one and the cleavage is holoblastic or nearly so, the whole development being more like that of an amphibian than like that of a teleost.

Order IV. Teleostei

The teleosts as a group appeared first in *Jurassic* times, evidently as a derivative from the holostean ganoids. Some of the more primitive members of the teleostean sub-order *Malacopterygii*, have certain structures that are reminiscent of the Holostei, *e. g.*, ganoid scales, multivalvular conus arteriosus, fulcra in connection with the fin bases, and spiral-valve in the intestine. In the higher teleosts

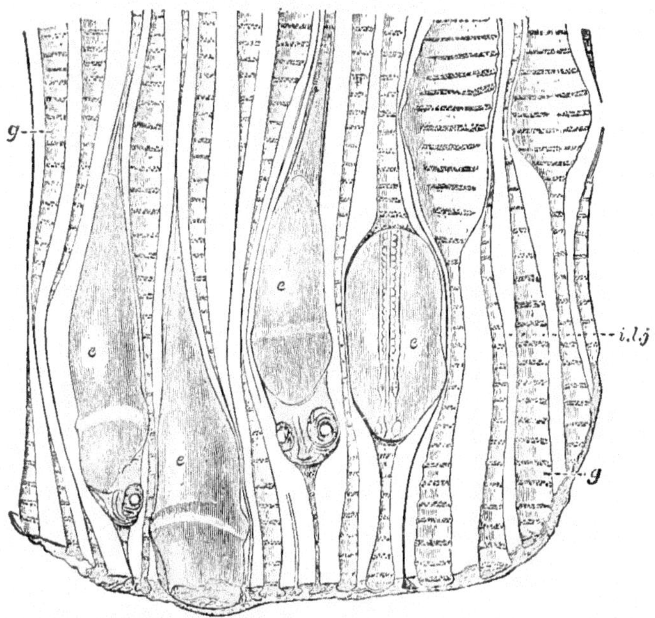

Fig. 73.—Showing the embryos of the fish *Rhodeus amarus* living parasitically in the gill cavities of the clam, *Unio*. *e*, embryos; *g*, interlamellar cavities; *i. l. j*, an inter-lamellar junction. (From Bridge, after Olt.)

none of these characters are found. The teleosts illustrate better than almost any other vertebrate group the principles of adaptive radiation (Fig. 51). They started modestly in the *Jurassic*, increased rapidly through both *Lower* and *Upper Cretaceous*, while in the Tertiary they had radiated adaptively into all of the principle types of body form that characterize the modern condition. When we say that the teleosts are a modern group we do not forget that the group originated many millions of years ago and has been a dominant group for at least three or four millions of years. The teleosts constitute now a *climax group*, at the height of its adaptive radiation; a group in which specialization for strange and limited environments has gone

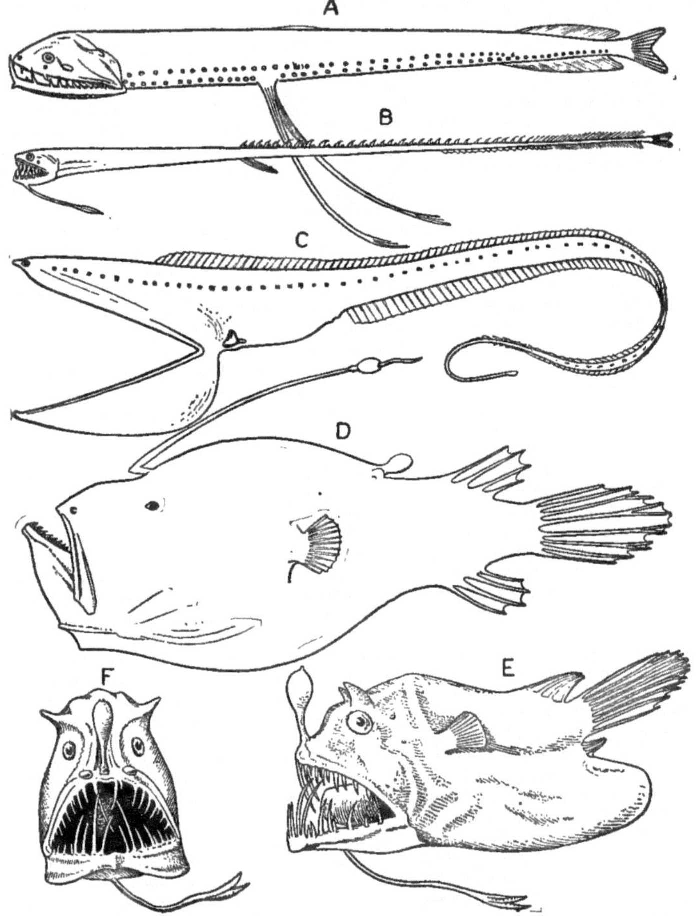

FIG. 74.—Deep-sea fishes. A, *Photostomias guernei*, length 1.5 inches taken at 3500 feet; B, *Idiacanthus ferox*, 8 inches, 16,500 feet; C, *Gastrostomas bairdii*, 18 inches, 2300–8800 feet; D, *Cryptopsarus couesii*, 2.25 inches, 10,000 feet; E, F, *Linophryne lucifer*, 2 inches. (From Lull, after Goode and Bean.)

to extremes; in which structures, especially integumentary features such as spines, scales, and skeletal elements, tend to run to excesses; in which pigmentation and general color characters develop into the most elaborate color schemes. Many teleosts are characterized by exaggeration in the development of the fins, both median and paired; others have the snout and jaws over- or underdeveloped for some peculiar feeding processes. Peculiar breeding habits are accompanied by odd and unique specializations, like the brood-pouches in male pipe-fishes and sea-horses, and the ovipositor by means of which the female *Rhodeus amarus* deposits her eggs in the mantle cavity of the clam *Unio*, where they develop safely and in a well aërated environment (Fig. 73). Perhaps, however, the most extreme special adaptations are those seen in deep-sea fishes of a number of teleost families (Fig. 74). Two main types are developed, the swift moving types that hunt their prey and the sluggish forms that lie in wait. These fishes, through the use of certain physical principles not yet fully understood, are able to resist the pressure of thousands of tons of water and to maintain life at temperatures just above freezing and in the practically total absence of light. In compensation for this life in the darkness they are provided with a great variety of phosphorescent organs that enable them to find their way about.

The Teleost Sub-Orders

The classification of the immense order Teleostei is a matter about which there is no consensus of opinion. The various subdivisions such as the Acanthopterygii or Malacopterygii are given full ordinal value by Jordan, and merely subordinal value by Boulenger. The present writer, on the basis of extensive hybridization experiments with numerous species of teleosts, is inclined to believe that these suborders should be given, at the most, family value; for, the fact that any two species of teleost can be crossed without artificial aids indicates that they are fundamentally not very distantly related. The thirteen sub-orders of Boulenger will be briefly surveyed, especial emphasis being given to those groups that are economically important or phylogenetically significant. Certain types that are either markedly generalized or strikingly specialized will receive attention to the exclusion of a large number of types interesting primarily to the specialist.

Primitive Teleosts (*Malacopterygii*).—This sub-order contains a very large number of species belonging to the Isospondyli and Scyphophori of Cope. About the Isospondyli Jordan says:—"Of the various subordinate groups of fishes, there can be no question as to which is the most primitive in structure, or as to which stands nearest the orders of Ganoids. Earliest of the bony fishes in geological time is the order *Isospondyli*, containing the allies, recent or fossil, of the herring and the trout." There are twenty-one families of Malacopterygii of which the best known are the *Elopidæ* (Tarpons), the *Salmonidae* (Salmon and Trout), and the *Clupeidæ* (Herrings).

The **Tarpon** (Fig. 75) is probably the noblest member of the sub-order and is often called the "Silver King." It is the favorite

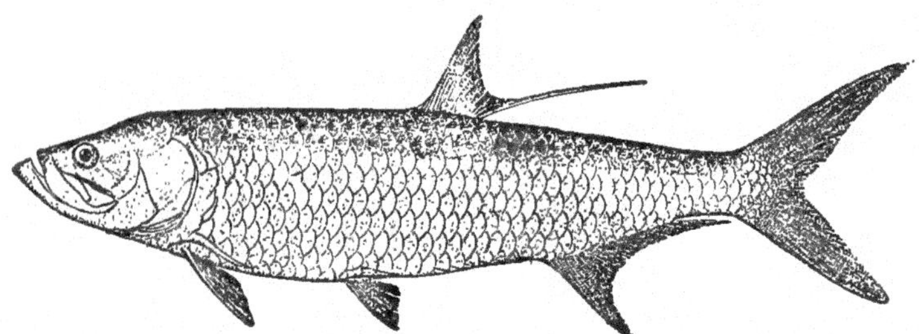

Fig. 75.—Tarpon, *Megalops atlanticus*, much reduced. (From Boulenger, after Goode.)

game fish along the Florida and Carolina coasts; for it is a great fighter and gives the sportsman the fullest scope for the exercise of his skill and experience. It reaches a length of six feet and weighs over one hundred pounds. Its very large silvery scales containing ganoin are used extensively in ornamental work.

The **Salmon** and **Trout** tribes are of all fishes the gamiest and the most sought after by the devotee of the rod and fly. They are characterized by the presence of an adipose dorsal fin. "Of all families of fishes," says Jordan, "the most interesting from every point of view is that of the Salmonidæ, the salmon family. As now restricted, it is not one of the largest families, as it comprises less than a hundred species; but in beauty, activity, gameness, quality as food, and even in size of individuals, different members of the group stand easily first among fishes." The **Salmon** (Fig. 76) is a marine fish, but spawns far up among the small streams near the sources of large rivers. This habit has given rise to the "Parent Stream Theory," according to which the young Salmon go down-stream and out to sea, where they remain for five years until sexually mature, and then return to spawn in the same parent stream. This does not necessarily imply any marvelous homing instinct or geographic sense, for it has been found that when the Salmon goes to sea it does not wander very far from the mouth of the particular river down which it has come. The instinct to spawn in the smaller streams must, nevertheless, be extremely impelling, for they frequently wear themselves out and die owing to the arduous up-stream journey of often more than a thousand miles, through rapids and even over water-falls of considerable height. It

Fig. 76.—Salmon, *Salmo fario*. *a. l*, adipose lobe of pelvic fin; *an*, anus; *c. f*, caudal fin; *d. f. 1*, first dorsal fin; *d. f. 2*, second dorsal or adipose fin; *l. l*, lateral line; *ap*, operculum; *pct. f*, pectoral fin; *pv. f*, pelvic fin; *v. f.* ventral fin. (From Parker and Haswell, after Jardine.)

is said that few, if any, survive to go down-stream and out to sea again; a statement that seems to be out of accord with the fact that some very large specimens, evidently much over five years old, are captured in every salmon river.

The great **Herring family** (*Clupeidæ*) consists of fishes of decidedly generalized proportions and characters. They are diagrammatic teleosts. Fossil Herrings practically like those of the present have been found well preserved in Cretaceous rocks. The family includes also Shad, Anchovies, and White-fishes, which are among the most important of the world's food fishes.

By no means all of the Malacopterygii are generalized types, for there has been a very considerable adaptive radiation within the group. Among the specialized and senescent types are: the mormyrids of the Nile, remarkable electric fishes that were pictured by the early Egyptians; eel-like types such as *Gymnarchus;* proboscis-fishes such as *Gnathonemus;* and several deep-sea forms with degenerate and otherwise aberrant characters.

The Cat-fishes, Carps and Their Kin (*Ostariophysi*).—This is one of the most clearly defined of the sub-orders, and one that possesses certain primitive characters that suggest ganoid affinities. The *Cat-fishes* (*Siluridæ*) are very common fishes of cosmopolitan distribution, distinguished by the presence of spines and barbels about the mouth (Fig. 77). They are as a rule sluggish and mud-loving. Some of the Cat-fishes reach a very large size, growing to be ten feet

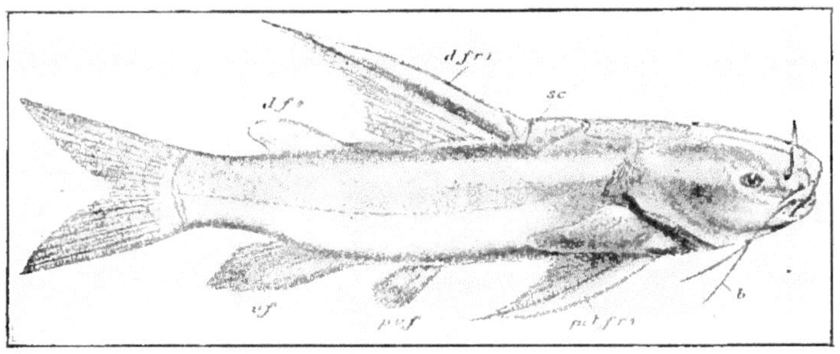

FIG. 77.—A siluroid fish, *Rita buchanani*. *b*, barbel; *d. f. r. 1*, first dorsal fin-ray; *d. f. 2*, adipose fin; *pct. f. r. 1*, first pectoral fin-ray; *pv. f*, pelvic fin; *v. f*, ventral fin. (From Parker and Haswell, after Day.)

long and weighing in the neighborhood of four hundred pounds. The *Gymnotidæ* are the *Electric Eels* of South America, the best known species being *Gymnotus electricus*, a large eel-like fish about eight feet in length, and much feared by the natives on account of the severity of shock it is capable of delivering. The *Characinidæ* comprise about five hundred species of African and South American fishes, that on the whole are the most generalized representatives of the sub-order; their most highly specialized feature is their rather elaborate dentition, which is associated with their carnivorous habits. The *Cyprinidæ* (*Carps*) are also very generalized, so much so that some authorities place them at the bottom of the scale of modern teleosts. A number of senescent armored types of Ostariophysi are known, most of which live in unusual habitats.

The Symbranchii.—This small sub-order of eel-like fishes is doubtless a senescent derivative of either the Ostariophysi or the Apodes. They are without paired fins, have the gill openings united into a single ventral slit, and have no air-bladder. For our purposes no further characterization is necessary.

The Eels and Their Kin (*Apodes*).—This sub-order comprises a very large number of eel-like fishes which are in all probability polyphyletic in origin and have been placed in the same group on account of possessing in common the attributes generally associated with eels. The present writer would hazard the suggestion that they may be degenerate derivatives of several groups of fishes that have undergone the same type of racial degeneration; thus the Apodes present a situation comparable with that of the perennibranchiate Amphibia, a group now adjudged to be of polyphyletic origin and pædogenetic in character. It seems not unlikely that the eels are the victims of racial retardation, due to some kind of developmental defect that inhibits the normal differentiation of the anterior parts of the primary axis and that of the bilateral appendages. Some

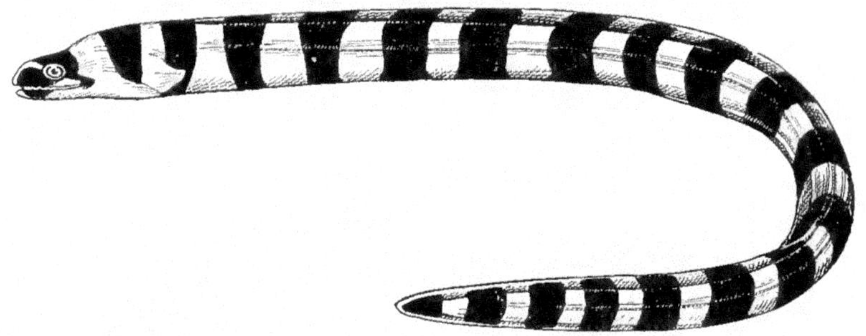

Fig. 78.—Eel or Moray, *Gymnothrax waialuæ*, from Hawaii (after Jordan and Evermann.)

of the eels show still more radical distortions of the generalized fish proportions in having excessively elongated bodies, as is the case in the *Thread-eel* (*Nematichthys*), which one can hardly believe to be a fish at all. Almost as remarkable are the *Gulper-eels* (Fig. 74, C), abysmal forms with enormous head and mouth and a much attenuated body ending in a filamentous tail. Of these Dr. Gill says:—"The entire organization is peculiar to the extent of anomaly, and our old conceptions of a fish require to be modified in the light of our knowledge of such strange beings." The *Morays*, a great family of marine eels, are predaceous fishes of great efficiency, with highly developed teeth and often with color patterns strikingly elaborate and brilliant. These patterns often simulate those of snakes, as in the banded species, *Gymnothorax* (Fig. 78); it is said that they are often quite poisonous.

Pikes and Killifishes *(Haplomi)*.—The **Pikes** (Esocidæ) are well-known fresh-water fishes with long body and large mouth. The finest of the Pikes is the Muskalunge (*Esox masquinongy*), the largest of our inland game fishes. It reaches a length of six feet and a weight of as high as eighty pounds; but the fisherman that gets a good strike from a twenty-five pounder has about as much as he can handle.

Fundulus heteroclitus (Fig. 79), the common **Killifish** or **Mud-minnow,** is perhaps worthy of special mention on account of its

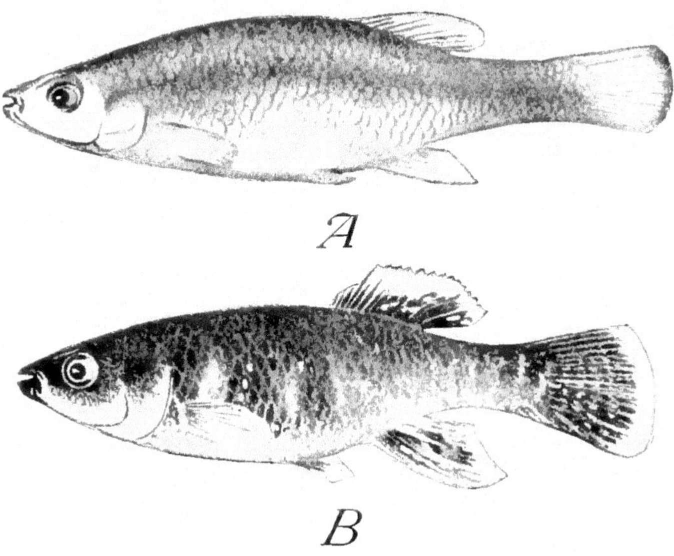

Fig. 79.—*Fundulus heteroclitus,* the Killifish. A, female, B, male; showing sexual dimorphism in fins and in markings. (From Newman.)

important contribution to experimental biology. The eggs of this species have furnished more material for investigation than those of any other fish. If the literature on *F. heteroclitus* were collated it would constitute many large volumes. The family (*Cyprinodontidæ* or *Pœciliidæ*) to which *Fundulus* belongs, contains several viviparous species, in which the anal fin of the male is used as an intromittent organ for introducing sperm into the oviducts of the female.

The Heteromi.—This comparatively small order of degenerate or senescent fishes may be passed over with little comment. Most of the species are deep-sea forms. One genus, *Fierasfer* (Fig. 80), is remarkable on account of its commensal life as a lodger in the various cavities of echinoderms and mollusks. One species, according to Boulenger, enters its host, a holothurian, through the anal opening, and lies with the head only protruding. From time to time it dashes out after its prey and returns to its shelter to eat it. This could

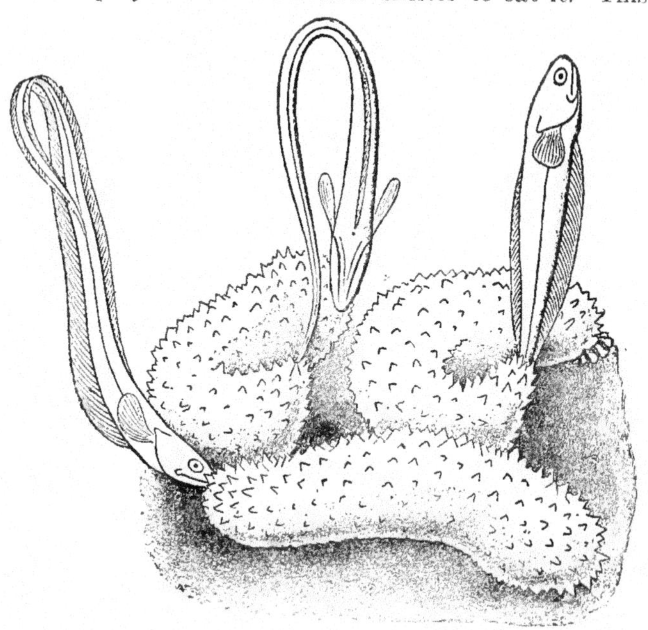

Fig. 80.—*Fierasfer acus*, penetrating the anal openings of holothurians, 2/3 natural size. (From Boulenger, after Emery.)

hardly be interpreted as a case of symbiosis, for there can be no mutuality in the arrangement.

Stickle-backs, Pipe-fishes, and Sea-horses (*Catosteomi*).—This well-defined group of peculiar fishes exhibits a wide range of specialization and senescence. The *Stickle-backs* themselves (Fig. 81), apart from their side-armor and prominent spines, are quite generalized in their proportions. Nothing less fish-like, however, could well be imagined than some of the extreme Sea-horses, which look more like gargoyles than real animals. The typical Sea-horse (Fig. 82) might be compared with a knight of a set of chessmen, with a long, coiled tail instead of a base. The Pipe-fishes may be considered as the "eels" of the Catosteomi, for they are much like greatly attenuated Sea-horses.

The **breeding habits** of all members of the sub-order are peculiar.

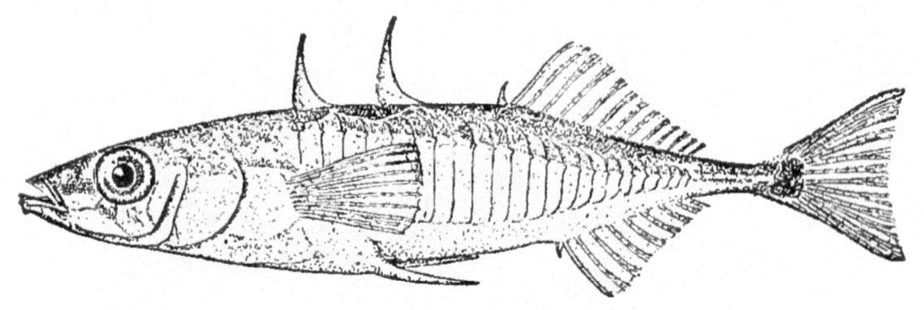

Fig. 81.—*Gasterasteus aculeatus.* x1. (From Boulenger, after Goode.)

In the case of the Stickle-backs the male builds a nest out of green grasses and kindred materials, leaving a front and a rear entrance. When the nest is complete he goes a-wooing and induces a female to enter his nest and lay her eggs there. As soon as she leaves by the back door he enters by the front and fertilizes the eggs. Usually several other females are employed in the same way until the nest is filled with a sticky mass of eggs. He then watches over the nest until the eggs are all hatched. Sea-horses carry to a higher degree of specialization this paternal solicitude for the welfare of offspring, for, instead of building a nest and guarding the eggs, the male uses a part of his body, a brood-pouch on the abdomen, as a nest. According to Jordan, the female lays her eggs on the sea-bottom, and the male, after inseminating them, transfers them to the brood-pouch and carries them about until they are hatched, thus making of himself an animated incubator. Some of the Sea-horses are provided with an elaborate camouflage in the form of leaf-like processes (Fig. 83) colored like sea-weed and are practically invisable in their native haunts. The *Sea-moth*, another member of the Catosteomi, is almost as fantastic as the Sea-horses. It is covered

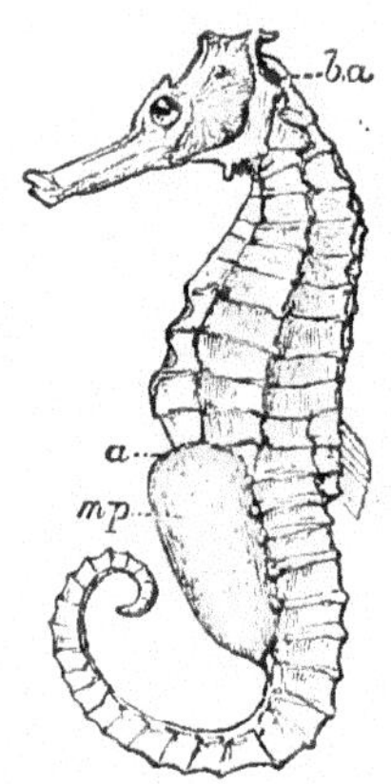

Fig. 82.—*Hippocampus guttulatus.* Male, showing brood-pouch (*mp*). *a*, anus; *b. a*, branchial aperture. (From Boulenger.)

with heavy armor and has enormous pectoral fins that give it a moth-like aspect.

The Flying-fishes and Their Allies (*Percesoces*).—While flying-fish types are found in several other sub-orders of fishes, those of this group are perhaps the most highly specialized. One of the best types is *Exonautes* (Fig. 84), which, when it leaps out of the water, parachutes for some distance by means of its very large pectoral fins.

Fig. 83.—*Phyllopteryx eques*, ½ natural size. (After Boulenger.)

The sub-order is not believed to be homogeneous, for it contains such aberrant forms as *Belone*, a form resembling superficially the Gar-Pike, and sometimes given that name.

The Cod and Their Kin (*Anacanthini*).—Apart from the *Cod-fishes* the members of this group are rather unfamiliar and of no especial interest for us. The Cod (Fig. 85), however, is one of the most important of the world's food fishes. It is one of the most

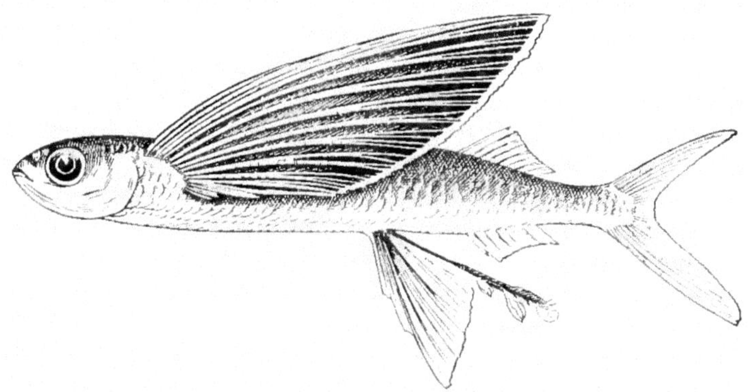

Fig. 84.—Flying fish, *Exonautes gilberti*. (After Jordan and Evermann.)

voracious and omnivorous of fishes. It breeds far out at sea, and its tiny pelagic eggs are almost inconceivably numerous; as many as nine million eggs have been estimated as being laid by a single large

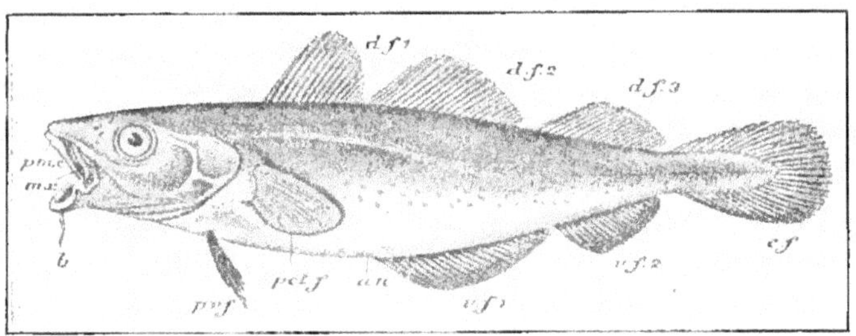

FIG. 85.—*Gadus morrhua* (Cod). *an*, anus; *cf*, caudal fin; *df1-3*, dorsal fins; *mx*, maxilla; *pct. f*, pectoral fin; *pmx*, premaxilla; *pv. f*, pelvic fin; *vf. 1* and *2*, ventral fins. (From Parker and Haswell, after Cuvier.)

female in one season. Most of us who have been children need hardly be reminded of the somewhat unpleasant fact that the liver of the Cod is the source of a highly nutritious and readily digested oil. Relatives of the Cod are the Haddock, the Pollock, the Burbot, the Hake, and some aberrant and degenerate deep-sea forms.

The Spiny-Rayed Fishes (*Acanthopterygii*).—This tremendous assemblage of modernized fishes reminds one of the passerine birds, because they are the most modern of the sub-orders and show a more extensive adaptive radiation than do any of the other groups. The Acanthopterygii comprise no less than thirty-six families including such familiar forms as the Bass, Perch, Flounder, Goby, etc., and a host of less familiar types.

The common Perch is as good a type to illustrate the sub-order as any, though it is perhaps the most generalized member of the group. Many of the others tend to become high, compressed, and short-bodied, such as the little fresh-water sun-fish. Other forms that live in the open seas have carried out this line of development till the dorso-ventral axis appears to overshadow the primary and the bilateral axes, as is the case in some of the *Acanthiuridæ*, a family which is strikingly exemplified by *Zanclus* (Fig. 86). It is from such types as this that the members of the curious sub-order *Plectognathi* are thought to have been derived.

The **Mackerel** family (*Scombridæ*) is one of the most generalized members of the sub-order and one of the most important as food fishes.

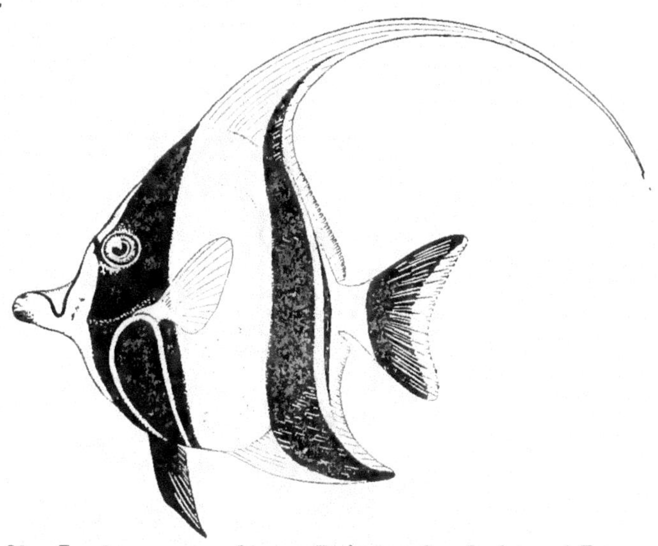

FIG. 86.—*Zanclus canescens*, Linn. (Redrawn after Jordan and Evermann.)

The **Flounders** or **Flat-fishes** (*Pleuronectidæ*) are among the most aberrant of all fishes (Fig. 87). So unique are they in their peculiarities that Jordan sees fit to place them in a separate sub-order,

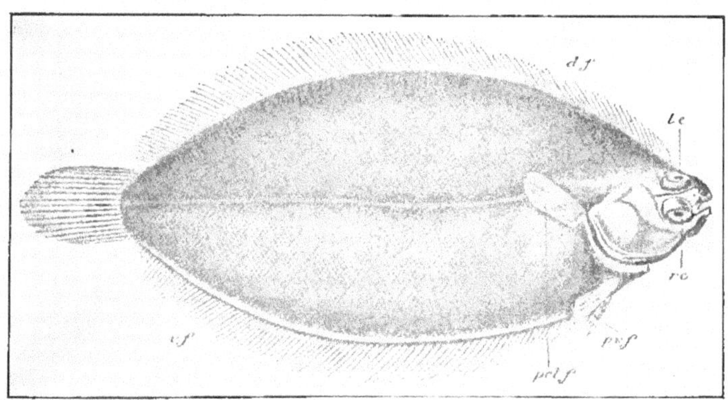

FIG. 87.—*Pleuronectes cynoglossus* (Flounder), from the right side. *d. f*, dorsal fin; *l. e*, left eye; *r. e*, right eye; *pct. f*, pectoral fin; *pv. f*, pelvic fin. (From Parker and Haswell, after Cuvier.)

which he calls *Heterosomata.* The Flounders are bottom-fishes that lie on the side instead of on the belly as do most other bottom-fishes; some species lie on the right, others on the left side. To adapt themselves to this position there is a remarkable twisting of the cranium that results in bringing both eyes on the same side of the head, and makes the whole region of the head decidedly asymmetrical. The upper side of the body becomes variously pigmented in harmony with almost any background; experiments, involving the use of the

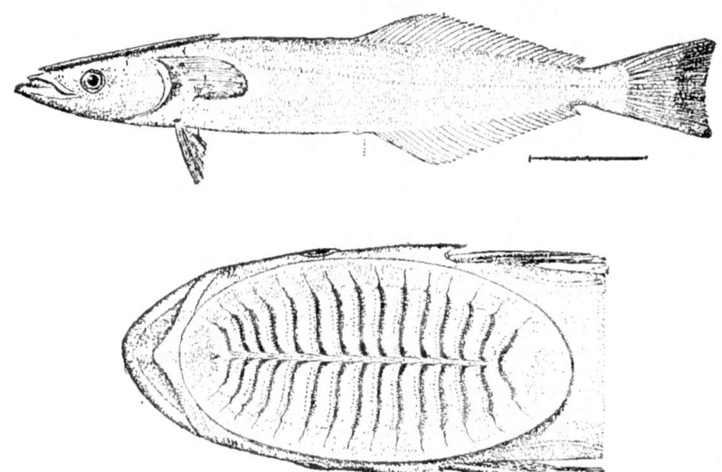

FIG. 88.—The Shark Sucker, *Remora brachyptera*, Lowe (above). (From Jordan and Evermann.)

FIG. 89.—Sucking disc of *Remora brachyptera*, Lowe (below). Dorsal view. (From Jordan and Evermann.)

most elaborate of artificial backgrounds, having proven their extraordinary capacity for imitation. The lower side normally remains unpigmented, but, if it is artificially illuminated by growing the fishes on an elevated glass floor in an aquarium, it acquires an appearance much like the upper side. The young flounder is bilaterally symmetrical and begins the head-twisting process some time before it takes up the bottom-living habit. Flounders are believed to have been derived from some one of the high compressed types, which adopted the bottom-feeding habit and was forced to modify itself in a peculiar way to meet the new conditions.

The **Gobies** (*Gobiidæ*) are surface-dwelling fishes characterized by elaborate fin structures and brilliant colors, reminding one of some of the brilliant birds. They might be designated the "humming-birds" among fishes. The Gobies are excessively numerous in tropical waters. The East Indian Goby has large muscular pectoral fins, which it uses like feet; its habit is to hop about over the mud-flats at low tide, feeding upon stranded crustaceans.

The **Shark-suckers** (*Remoras*) are especially noteworthy on account of the peculiar modifications of the anterior dorsal fin into a lamellated sucker (Figs. 88 and 89), by means of which they adhere to the body of a shark or some other smooth-bodied fish. By means of this sucking disk they obtain free transportation without exertion,

Fig. 90.—Type of teleost with over-specialized fins. *Dendrochirus hudsoni.* (Redrawn after Jordan and Evermann.)

dropping off when they reach a desired destination and awaiting another accommodating conveyance when they wish again to travel. They appear to do no harm to the host fish, except in so far as they somewhat impede its movements.

The family *Scorpœnidæ* (**Mailed-Cheeked Fishes**) are among the most elaborately finned of the Spiny-Rayed fishes. A good type is *Dendrochirus* (Fig. 90), in which the fins, both paired and median, appear to have run riot like the plumage of some of the highly specialized birds. The color pattern is in striking accord with the background.

A number of highly specialized and exaggerated types of Spiny Rayed Fishes might be mentioned. The **Deal-Fish** or **Ribbon-Fish** is an extremely elongated and laterally compressed type, that is said to hold one side obliquely upward when swimming. Some of these fishes attain a length of twenty feet. There are also several types of degenerate abysmal species, both of the eel-type and the head-fish types. Some of the best types of flying-fishes, notably the "**Flying Gurnards**" belong to this sub-order, fishes, which, when they leap from the water, flutter their large pectoral fins like a flying grasshopper. One of the most extreme specializations is seen in the case of *Anabas*, the **Climbing Perch** (Fig. 91), a species in which there is a superbranchial lung-like organ, by the use of which the fish is enabled to live for long periods out of water. It climbs up low trees using the spines on its gill-covers and pectoral fins like claws. It is thus able to capture insect and other food that would be unavailable for a strictly aquatic fish. The *Toad-fishes* are rather sluggish, large-headed, wide-mouthed forms that are especially noteworthy for their extreme ugliness. It is believed that they represent the ancestral group from which the sub-order Pediculati (Anglers) has been derived.

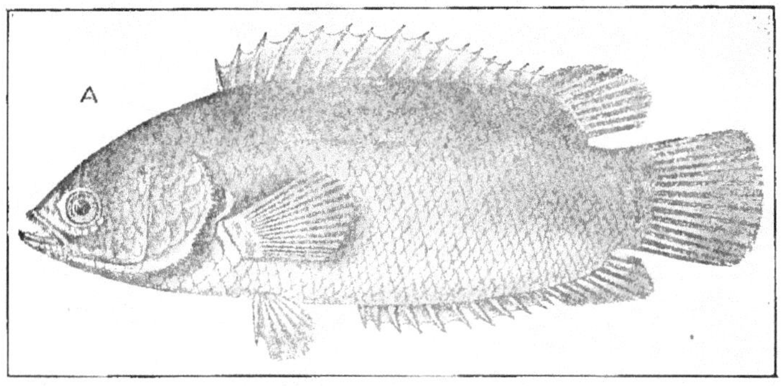

FIG. 91.—*Anabas scandens* (Climbing Perch). (After Parker and Haswell.)

The Opisthomi.—This is a small sub-order of degenerate eel-like fishes, believed to have been derived from the Acanthopterygii. For our purposes they may be passed without further comment.

The Anglers (*Pediculati*).—This sub-order of highly specialized, senescent fishes is believed to have been derived from the Toad-Fishes, an aberrant family of Acanthopterygii. They seem to be little more than a head with an enormous gaping mouth (Fig. 74, E and F). The anterior spine of the dorsal fin is modified so as to form a sort of "fishing-rod" which hangs over the mouth and has pendant from its tip a fleshy pad or bait. The Anglers are almost scaleless and are bottom-feeders, both of which characters are taken to be evidences of senescence. Some of them are abysmal forms that have been called "Bathymal Sea Devils"; these are even more degraded in structure than are the more typical Anglers. One of the strangest of all fishes is the Sargassum Fish, a form that lives a drifting life among the masses of Sargassum weed; its camouflage of color pattern and raggedly weed-like fins make it merely a part of the general floating mass. The Bat-Fish represents perhaps the climax of senescent degeneration in this group; it is a very broad, flat bottom-feeder, with wing-like pectoral fins.

Fool-Fishes, Trunk-Fishes, Porcupine-Fishes, Puffers, and Head-Fishes (*Plectognathi*).—This final sub-order of teleosts is

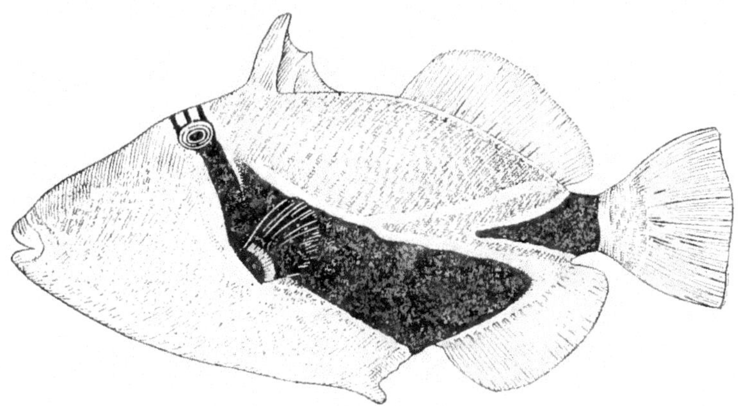

Fig. 92.—Hawaiian Trigger-Fish, *Ballistapus rectangulus*. (Redrawn after Jordan and Evermann.)

perhaps the most highly specialized of all, and is believed to be derived from the family Acanthiuridæ of the Acanthopterygii, which are characteristically short, high, compressed fishes, typified by *Zanclus* (Fig. 86). The Plectognathi comprises a collection of the strangest creatures that the sea affords. Of these the **Trigger-Fishes** are the most moderate in structure, not unlike some of the Acanthopterygii in their high, compressed proportions; a modified spine of the dorsal fin on top of the head looks like a trigger and gives them their name. The **File-Fishes** or **Fool-Fishes** (Fig. 92) are still more flatly compressed and the scales are reduced to vestigial structures; the small mouth with its protruding teeth and the funny staring expression of the eyes have given them an uncomplimentary name. The **Trunk-Fishes** (Fig. 93) are big-headed fishes inclosed in a heavy immovable armor, composed of closely united plates, with a large posterior opening that allows the curious little tail to waggle, and smaller openings for the pectoral, dorsal, and anal fins. **Puffers** or **Globe-Fishes** are unarmored forms, shaped, when de-

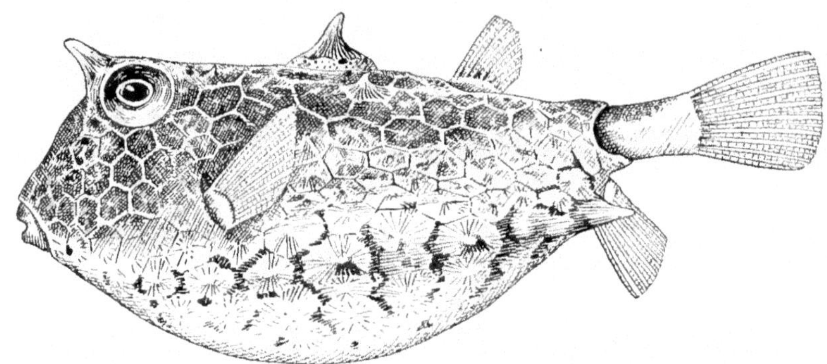

FIG. 93.—Hawaiian Trunk-Fish, *Ostrachion schlemmeri*. (Redrawn after Jordan and Evermann.)

flated, much like **Trunk-Fishes**, but capable of blowing themselves up with water to several times their normal dimensions, thus making themselves difficult to swallow. If taken out of water these strange little fellows suck in air till they are of a drum-like tightness. Some of the Globe-Fishes are said to be extremely poisonous. The **Porcupine-Fishes** are shaped much like the Puffers in a deflated or partly deflated condition, some being much rounder than others; but they are covered with a heavy spiky armor that has suggested their name. They also have the reputation of being decidedly poisonous. The **Head-Fishes** or **Sun-Fishes** (Fig. 94) represent the climax of relative increase of head over body, a character exhibited by the whole group; they are little more than animated fish heads. The body is so abbreviated that the dorsal and anal fins appear to be attached to the upper and lower parts of the head. They inhabit the tropical and sub-tropical seas, living a sluggish, floating life that is almost sedentary. Large specimens reach a giant size, being about eight feet in diameter and weighing as much as twelve hundred pounds. The skeleton is largely cartilaginous and there is a very heavy dermal cartilaginous armor. The skin is smooth and scaleless. All of these characters will readily be recognized as criteria of senescence. Indeed it is a question as to

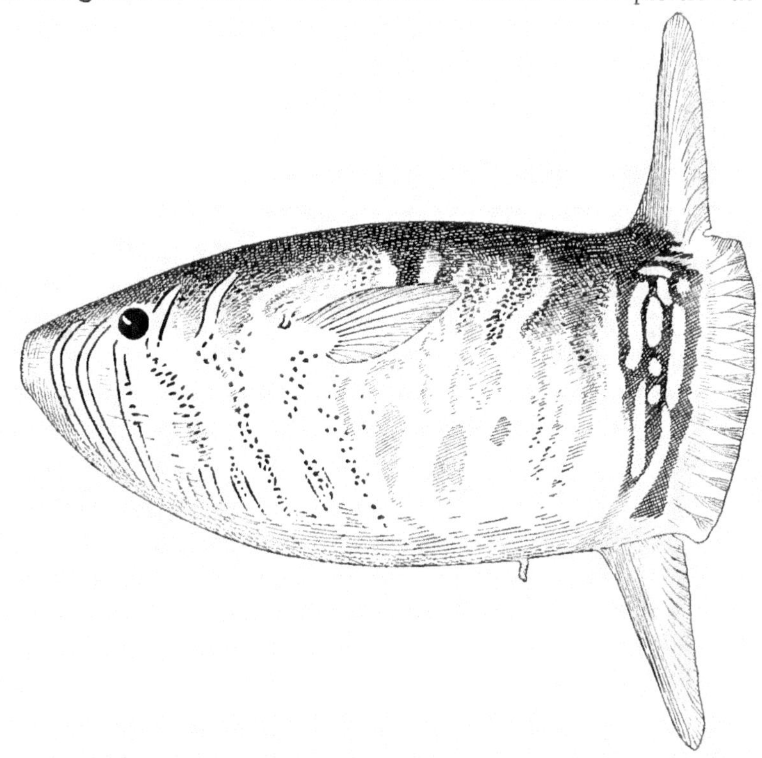

FIG. 94.—Hawaiian Head-Fish, *Ranzania makua*, Jenkins. (Redrawn after Jordan and Evermann.)

whether these creatures or the most degraded of the Pediculati are the most senescent members of the entire class of Fishes. The Pediculati appear to be the logical culmination of the rather broad, flat type of head-fish architecture, while the Sun-Fishes have carried out to the extreme the evolution of the high, compressed head-fish type. Both of these tendencies appear to have been manifest in the groups of Acanthopterygii from which these two sub-orders have respectively been derived.

SUB-CLASS III. DIPNEUSTI (*Dipnoi*). **The Lung-Fishes**

This group of fishes has acquired an unusual interest because of the belief, rather general until recent years, that from it the Amphibia took their rise. A number of writers still appear to hold this view, or at least to maintain that a close affinity exists between the Lung-fishes and the Amphibia. Goodrich, for example, in the volume on Cyclostomes and Fishes in Lankester's Treatise on Zoölogy, says:—"The Dipnoi are among the most interesting of fish. On the one hand, they have a close affinit to the Osteolepidoti (extinct Crossopterygii); on the other, they present many striking points of resemblance to the Amphibia, which cannot all be put down to convergence."

The group is distinguished by the following characters:—the paired fins are rather slender and pointed; the scales are cycloid in form and overlapping; the caudal fin is diphycercal; the upper jaw is firmly fused with the base of the skull, making a holostylic skull; teeth are largely lacking, and their place is taken by large tritoral dental plates, supported by the palato-pterygoid and splenial bones; the premaxillary and maxillary bones are absent, and the dentaries, usually absent, are vestigial when present. It would appear from this summary of characters that the Dipneusti show more evidences of being a degenerate and senescent group than one likely to give rise to a new and successful class like the Amphibia. Most of the characteristics of these fishes lead away from rather than toward amphibian conditions.

The question arises as to whether the characters mentioned were more or less primitive in extinct than in modern lung-fishes. The modern Dipneusti appear to have retained certain primitive characters, such as the diphycercal tail fin and the entirely cartilaginous condition of the primitive cranium. An examination however of the earliest fossil Dipneusti, in which the tail is heterocercal, the median fins broken up into isolated dorsals and ventrals, the chondrocranium extensively ossified, and scales large and close to the surface, show that the modern Dipneusti are degenerate and not truly primitive. Additional degeneration is seen in the loss of maxillary, premaxillary, and dentary bones, which are present in Amphibia. The paired fins are also almost vestigial in some modern lung-fishes and are so constructed as to make the derivation from them of anything like an Amphibian limb impossible. Certain specializations are also present, such as the fusion of teeth into massive grinding or tritoral plates, the development of respiratory filaments on the pelvic

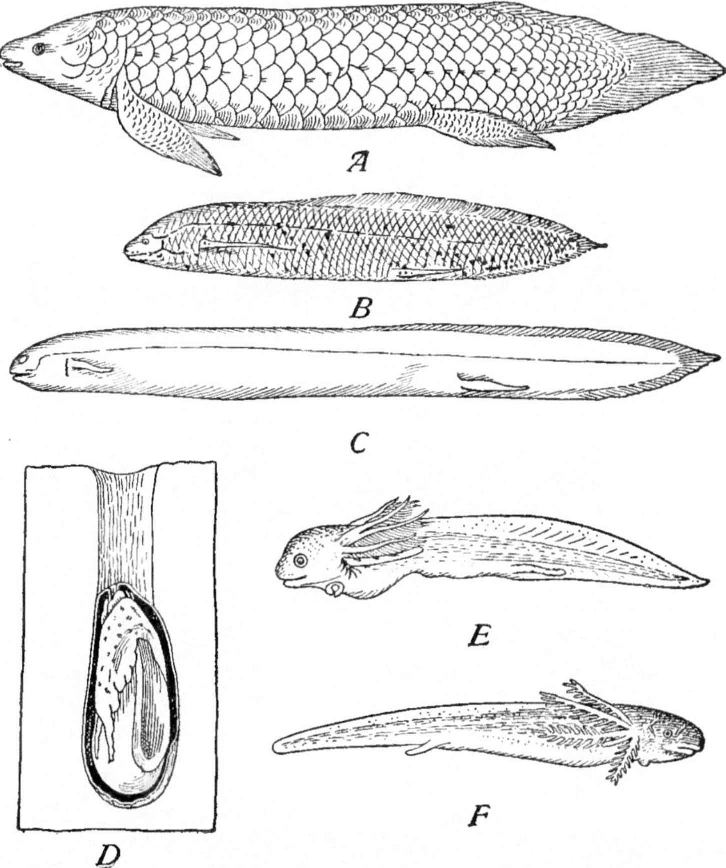

Fig. 95.—Group of Lung-Fishes (Dipneusti). A, *Neoceratodus forsteri*, Queensland. B, *Protopterus annectans*, Gambia. C, *Lepidosiren paradoxa*, Paraguay. (The lozenge-shaped markings in B do not represent scales, but areas of skin outlined by pigment cells. In a fresh specimen the scales are completely invisible, as in C.) D, Diagram of *Protopterus* aestivating in the mud, showing the body coiled up and the mucous sac with tube leading to mouth. E, Larva of *Protopterus* on the seventh day, showing cutaneous gills, cement organ under head, and narrow paired fins. F, Larva of *Lepidosiren* thirty days after hatching, showing some characters as E. (Redrawn from Bridge, A, after Günther; B, and C, after Lankester; D, after Parker; E, after Budgett, and F, after Kerr.)

limb of *Lepidosiren*, the mud-dwelling habit, and the highly specialized air-bladder, used as a lung.

The oldest representative of the Dipneusti is the genus *Dipterus* which occurs along with the earliest known Crossopterygii and the most primitive Actinopterygii in Devonian times. A comparison between *Dipterus* and the early "lobe fins," such as *Osteolepis*, shows obvious resemblances. *Dipterus* has acutely lobate paired fins, and the skull bones are typically dipnoan in their lack of premaxillary, maxillary, and dentary, and the presence of tritoral plate instead of teeth.

The **larval stages of modern dipnoans** are remarkably like those of modern Amphibia. Those of *Protopterus* and *Lepidosiren* are veritable tadpoles (Fig. 95, E and F) with external gills, tails with unsupported median fins, blunt heads, and clinging habits. Just as in amphibian larvæ, they have a sucker or ventral cement organ for adhesion. Evidently this tadpole larva represents an extremely ancient developmental stage present in the common ancestors of these two groups and retained with few modifications by the modern classes. Indeed it seems certain that the whole early development of Amphibia represents a more primitive evolutionary stage than does that of any modern fish. It is customary to deal with amphibian embryology as a step in advance of Amphioxus embryology and to derive the various types of fish development from some condition like that of the amphibian. It is of some interest to know, therefore, that the dipnoan fishes have a more primitive embryology (Fig. 96) than any other living fishes, that the bony ganoids are next in primitiveness, and that elasmobranchs, the more primitive ganoids, and the teleosts show a specialized development, in some respects paralleling that seen in the reptiles and birds.

Habits of Modern Dipneusti

The existing genera of dipnoans are *Neoceratodus* of Australia, *Protopterus* of Africa, and *Lepidosiren* of Paraguay. They are all fishes of stagnant rivers and fresh-water pools. *Neoceratodus* is the most primitive and shows less degeneration; *Lepidosiren* shows the extreme reduction in fins and scales characteristic of the eel-like type; while *Protopterus* is just about intermediate between the other two. Bridge has given good accounts of the lives of these singularly interesting fishes.

Neoceratodus forsteri, a fish that reaches a length of over five feet (Fig. 95 A), "frequents the comparatively stagnant pools or water-holes which alternate with shallow runs and are usually full of water all the year round. In these pools, filled with a rich growth of vegetation, and often the favorite haunt of the Platypus (*Ornithorhynchus*) the Fish is fairly abundant. Inactive and sluggish in its habits, usually lying motionless on the bottom, the Fish is easily captured by the natives with hand nets and baited hooks. *Neoceratodus* lives on fresh-water Crustaceans, worms, and molluscs, and to obtain them it crops the luxuriant vegetation much in the same way that a Polychaet or a Holothurian swallows sand for the sake of the included nutrient particles. Apparently the air-bladder is a functional lung at all times, acting in conjunction with the gills. At irregular intervals the Fish rises to the surface and protrudes its snout in order to empty its lung and take in fresh air. While doing so the animal makes a peculiar grunting noise, 'spouting' as the local fishermen call it, which may be heard at night for some distance, and is probably caused by the forcible expulsion of air through the mouth. Useful as the lung is as a breathing organ under normal conditions, there can be little doubt that its value as such is much greater whenever gill breathing becomes difficult or impossible. This seems to be the case during the hot season, when the water becomes foul from the presence of decomposing animal or vegetable matter. Semon records a striking illustration of this in the case of a partially dried-up water hole, in which the water had become so foul that it was full of dead fishes of various kinds. Fatal as these conditions were for ordinary fishes, *Neoceratodus* not only survived but seemed to be quite healthy and fresh. Such observations are of exceptional interest. Not only do they afford a clue to the conditions of life which, in the course of time, probably led to lung-breathing in *Neoceratodus*, but they also suggest the possibility that a similar environment has been conducive to the evolution of air-breathing vertebrates from gill-breathing and fish-like progenitors. In spite of its pulmonary respiration, *Neoceratodus* more closely resembles the typical Fishes in its habits than any other Dipneusti. It lives all the year round in the water. There is no evidence that it ever becomes dried up in the mud, or passes into a summer sleep in a cocoon, and the well-developed condi-

> tion of its gills suggest that these organs play a more important rôle in breathing than in either *Protopterus* or *Lepidosiren*."

The genus *Protopterus* (Fig. 95, B) has a wide range over the continent of Africa and consists of three species, *P. annectans*, *P. æthiopicus*, and *P. dolloi*. These fishes inhabit the marshes near rivers, living upon frogs, worms, insects, etc., that abound in marshy places. The long slender fins are used probably as tactile organs though they may help in locomotion along the bottom. During the wet season they live and breathe much as does *Neoceratodus*.

> "In the dry seasons," says Bridge, "the marshes in which *Protopterus* lives become dried up, and to meet this adverse change in its surroundings, the Fish hibernates, or passes into a summer sleep, until the next rainy season brings about conditions more favorable to active life. Preparatory to this summer sleep, and before the ground becomes too hard, the Fish makes its way into the mud to a depth of about eighteen inches, and there coils itself up into a flask-like enlargement (Fig. 95, D) at the bottom of the burrow, which is lined by a capsule of hardened mucus secreted by the glands of the skin. The mouth of the flask is closed by a capsular wall or lid, which is perforated by a small aperture. The margins of this aperture are pushed inwards, so as to form a tubular funnel for insertion between the lips of the Fish. While encapsuled in its cocoon the Fish is surrounded by a soft slimy mucus, no doubt for the purpose of keeping the skin moist, and its lungs are the sole breathing organs, the air pouring from the open mouth of the burrow through the hole in the lid directly to the mouth of the animal. The nutrition of the dormant Fish is effected by the absorption of the fat stored about the kidneys and gonads, somewhat after the fashion not unknown in the fat-bodies of Insects and the hibernating glands of Rodents."

Lepidosiren (Fig. 95, C) is just a step more terrestrial in its habits than *Protopterus* and several degrees more degenerate than the latter. It lives in swamps, breathes air more largely, taking several breaths at a time when it comes to the surface. In the dry season it digs a burrow deeper than that of *Protopterus*. It even lays its eggs in deep burrows in the black, peaty soil of the swamps in which it lives, and the male remains in the burrow guarding the eggs till they hatch out into tadpole-like larvæ. While guarding the eggs the pelvic fins of the male act as accessory external gills, the fins being covered with numerous vascular filaments.

GENERALIZED AND SPECIALIZED TYPES OF FISHES AND THE AXIAL GRADIENT THEORY OF STRUCTURAL RELATIONS

In nearly all of the orders or sub-orders of fishes there are species that have retained a well-balanced relation of head, body, and tail; that have the moderately elongated, cylindrical, double-pointed shape; with smooth body, lacking heavy armature; without exaggerations of the fin-system; and with rather dull coloration. The dog-fishes among elasmobranchs, *Polypterus* among the crossopterygians, herring, pike, trout, killifish, cod, perch, mackerel, etc., among the teleosts; all these have, to a more or less complete extent, retained the generalized characters of the ancestral prototype of all the fishes. They all agree quite closely with the very ancient types that have come down to us in fossil form. All of these generalized types are active, predaceous fishes, with an abundant supply of energy; they are youthful in the physiological sense.

Specialization in fishes, as in other groups of vertebrates, follows certain definite lines and results in several types of structural modification. One of the commonest of these is the eel-like type, which appears in all of the dominant sub-orders. In fishes of this type the body is greatly elongated, and the trunk and tail are apt to be proportionately more highly developed than the head. The head remains small (microcephalic) and exhibits a number of degenerate features, such as small eyes and imperfect branchial openings. The paired fins are usually absent and the median fins are of the primitive, undifferentiated diphycercal type. Perhaps the most conspicuous example of reduced head is seen in the Hawaiian eel, *Callechelyx luteus*, a species in which the diameter of the body is about twice that of the head. Modern types of eel-like fishes are usually scaleless, but some of the archaic forms, such as the crossopterygian species, *Calamichthys*, are heavily scaled. In general, the eel-like type may be interpreted as a result of a suppression of the head parts and a consequent relative increase in the development of the body and the tail.

The antithesis of the eel-like type is that in which the head parts are abnormally large (megacephalic) and the body and tail relatively suppressed. Some of the most extreme examples of this condition have just been called to our notice in the description of the last two sub-orders dealt with, the Pediculati and the Plectognathi. These types of structural distortion may be called for convenience *head-fish* (megacephalic) types. Two other types not directly related to the two primary types just mentioned, but more or less closely correlated with them, are: first, the short, high, compressed type; and the second, the low, laterally expanded type. Both of these types agree in having the primary axis foreshortened and the subordinate axes exaggerated. The first involves an exaggeration of the dorso-ventral (secondary) axis at the expense of the primary and tertiary axes. In extreme cases the dorso-ventral diameter exceeds the length; and the dorsal and ventral integumentary elements, such as fins and spines, become greatly lengthened and specialized, as in *Zanclus* (Fig. 86). The second type shows a dominance of the tertiary axis (bilateral) over both primary and secondary axes; and the result is a very wide type, with expanded and specialized pectoral fins, the pelvic fins having been relatively suppressed in the foreshortening process that has affected the primary axis.

All of these specialized and degraded types of fishes may be reduced to two categories:—*a*, those in which there has been a relative suppression of the apical parts of the various axes, especially the primary axis, accompanied by a relative emancipation of the subordinate axes from the dominance or control of the primary axis; *b*, those in which the apical parts of the various axes have become relatively highly specialized or exaggerated, while the basal elements have become relatively suppressed.

In the opinion of the writer, all of these conditions can be readily interpreted as the morphological consequences of growth-inhibiting agents, acting during the ontogeny of the individuals. The morphological equivalents of all of these exaggerated types of natural fishes can be experimentally simulated by the use of inhibiting agents applied to the eggs or young embryos of generalized species of fishes. The writer and other investigators have performed extensive series of experiments with the eggs of *Fundulus* (Fig. 79) and other generalized types of teleosts, using a wide variety of growth-depressing agents, such as anæsthetics, low temperatures, heterogenic hybridization, etc. If, for example, the eggs or early embryos are placed for limited periods in weak sea-water solutions of alcohol or potassium cyanide, the result is a series of monsters, ranging from those in which only the most anterior structures (nostrils and eyes) are abnormal to blind and nearly headless forms. Between these extremes are types in which the eyes are too close together and the mouth narrow and protruding; those in which the eyes are fused into a single median cyclopian eye and the mouth is a sort of extended proboscis; and those in which the merest rudiments of eyes or other anterior structures are developed. All of these types are to be interpreted, according to the nomenclature of Child, as the results of differential inhibition; which implies that the parts of the body that normally have the highest rate of metabolism and are the first to differentiate, are the parts that are most readily inhibited by growth-depressing agents; while the parts that have the lower or lowest metabolic rates are least affected by the same agents.

How then can we explain the type in which the posterior parts are relatively inhibited, and the anterior parts, together with the apical parts of the secondary and tertiary axes, are relatively more highly differentiated? These conditions become instantly intelligible as the result of another type of experiment with *Fundulus* eggs and embryos. If eggs are placed in a weak solution of alcohol or cyanide and are allowed to remain there indefinitely, the rate of metabolism, and consequently of development, is generally retarded for a time, but gradually a process of acclimation or recovery takes place, the result of which, curiously enough, is that the parts that primarily were most seriously inhibited are the first to become acclimated, and recover more completely than the other parts. If the solution be made strong enough to be lethal for most of the embryos, a few of the hardiest of them undergo a very limited recovery, involving only the most apical structures, such as eyes. Several investigators have obtained, in ways similar to that described, embryos that consisted of nothing but isolated eyes; and it is very common to find, as the result of less extreme measures, embryos that consist merely of heads with large rolling eyes and provided with a tiny undifferentiated appendage that stands for the rest of the body. Other embryos that are mainly heads develop in addition large wing-like pectoral fins; still others become broad and flat, like a Skate, or high and compressed like a Head-Fish. In fact a good assortment of experimental monster fish embryos will furnish parallels to most of the stock types of form distortion seen in the specialized and degenerate groups of fishes. Of course none of these embryos differentiate definitive structures or take on hard integumentary coverings, for they live at best for a few weeks and do not acquire the adult characters.

But what *in nature* corresponds to the growth-retarding agents used *in the laboratory?* The inhibitors that are responsible for racial retardation or racial senescence, which are the same, are internal, and are probably associated with an aging of the heredity chromatin or specific germinal protoplasm. The metabolic rate of the germinal elements is believed to lose momentum from generation to generation and from age to age, unless rejuvenated or secondarily speeded up in some way. In the senescent species we must conclude that the progressive slow-down of the germinal metabolic rate is irreversible and that these forms must become extinct when they have gone to the limit of their differentiational excesses. The generalized forms may be looked upon as racially perpetually young or ready for any new processes of differentiation. Whether the inhibiting agents are external or internal the same types of morphological distortion of the generalized condition result.

Some of the specific cases of abnormal structure among present-day fishes may be more directly attributable to the external conditions under which they live and develop. What more reasonable explanation of the blind cave fishes is there than that their embryos have been victims of growth-depressing agencies, such as cold, low oxygen content of the water, or darkness? Similarly many of the abysmal fishes could be explained as the result of the unfavorable developmental conditions of the sea depths (Fig. 74). These creatures are for the most part either eel-like forms or Head-Fish forms, the first of which may be attributed to differential inhibition, and the second, to general inhibition followed by differential recovery of apical structures.

Similarly, racial senescence, as expressed in pædogenetic forms, may mean that certain species have been so slowed down in their developmental momentum that they are unable to push their developmental processes past a larval period, but that the germinal elements, that belong to the part of the axis with the lowest rate of metabolism, go on and become mature, thus enabling these creatures to reproduce while the somatic structures are still larval or juvenile in character.

High, compressed forms appear to be the result of differential inhibition of the primary axis followed by differential recovery of the anterior end of the primary axis and of the secondary or dorso-ventral axis, especially of the dorsal (apical) part of this axis. The broad, flat types are to be thought of as distinctly more senescent types in which the whole system is more profoundly inhibited, but in which the head parts and those nearest to the head recover most completely, while the tertiary axis, especially the apical parts of it, expresses itself at the expense of both primary and secondary axes.

The physiological explanation of the development of inert structures like armor, heavy spines, and even of massive bones and flesh, is that a lowered rate of chemical action allows of the precipitation of inert compounds, which, when the rate is higher, would be used up in dynamic activity, such as rapid locomotion or rapid growth. Giant size then may, on this theory, be just as definitely a product of racial sensecence as is heavy armor.

In this volume it would scarcely be appropriate to pursue this theory of vertebrate morphogenesis further. It is, however, the writer's opinion that the theory is as applicable to all of the vertebrate classes as it is to the fishes. The parallel between the conditions seen in the teleost fishes and in the birds is especially close and would repay detailed examination.

Eggs, Reproduction, and Breeding Habits of Fishes

The eggs of living species of fishes vary within very wide limits both in size and in form, as is well shown in Fig. 96. The largest eggs are those of some Elasmobranchii, which compare favorably in size with those of birds. They contain a large accumulation of yolk and have a hard *chitinous* shell. The smallest eggs are the pelagic eggs of many of the teleosts, which are less than 1 mm. in diameter. Greatest egg-size is found in that group which we have been considering the most primitive; least egg-size, in the most highly specialized group. Are we justified then in believing that the primitive fish egg was of large size and that the course of evolution has been steadily in the direction of a smaller and smaller size of ovum? Certain other considerations demand a negative answer.

Curiously enough both the largest and the smallest fish eggs are decidedly telolecithal and show an advanced type of meroblastic cleavage, while other groups of fishes, such as the Chondrostei, Holostei, and the Dipneusti have eggs of medium size with considerable yolk, and exhibit various degrees of incomplete holoblastic cleavage.

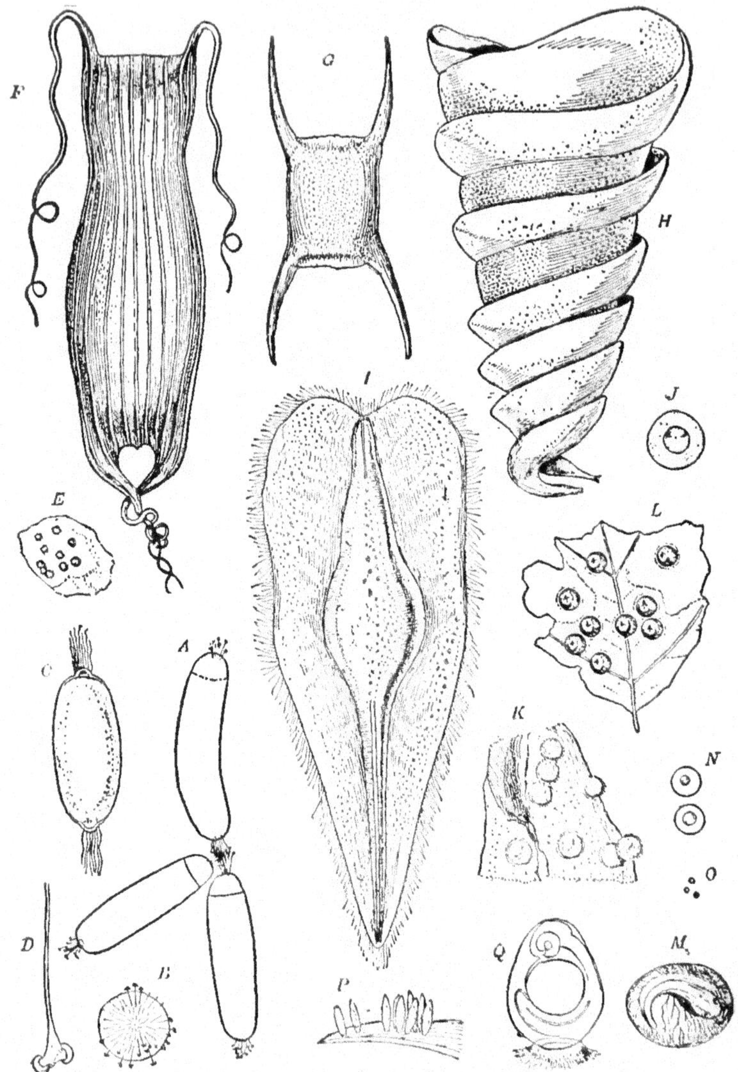

FIG. 96.—Eggs and egg-cases of fishes. A, *Bdellostoma*, egg-case; B, upper pole of same, showing hooks and micropyle (after Ayers); C, *Myxine* (after Steenstrup); D, a process of same; E, *Petromyzon marinus;* F, *Scyllium* (after Günther). G, *Raja;* H, *Heterodontus* (after Günther); J, *Ceratodus* (after Lemon); K, *Lepidosteus;* L, *Acipenser;* M, *Arius*, showing larva (after Günther); N, *Serranus;* O, *Alosa;* P, *Blennius*, egg-capsules attached; Q, the same enlarged (after Guitel.) (From Lankester, after Dean.)

The eggs of *Amia* (a holostean ganoid) show perhaps the nearest approach to complete holoblastic cleavage. In that species the cleavage furrows run meridionally practically from pole to pole, but the equatorial cleavage furrows in the vegetal region are very slow in appearing, giving to the early cleavage stages quite a meroblastic appearance. The egg of *Lepidosteus*, another holostean ganoid, shows an incomplete cleavage of the vegetal pole, furrows running only about to the equator or a little beyond. Subsequently, however, the furrows deepen and the entire egg is broken up by cleavage into true cells. Transitional cases of this sort indicate that the primitive pro-fishes had small isolecithal eggs, like that of Amphioxus, and that, when the first fishes invaded the region of the rapid currents, they had to lay their eggs on the bottom, probably attached to vegetation, or in a prepared nest of some sort. The primitive eggs of the ganoids and Dipnoi are covered with a coating of sticky jelly and are laid on the bottom. Accompanying this habit there was an increase in yolk accumulation and a tendency toward the telolecithal condition and meroblastic cleavage. That the modern teleosts are all meroblastic, even when the eggs are very small and have only a minimum amount of yolk, is doubtless due to their derivation from ancestors that had a very large yolk, larger probably than that of the amphibian eggs of to day. Although pelagic fish eggs are small enough readily to admit holoblastic cleavage, they retain the extreme type of meroblastic cleavage characteristic of their large-yolked ancestors.

The eggs of elasmobranchs, the largest of fish eggs, are laid singly or in pairs at varying intervals over a long breeding season. In some teleosts the number of tiny eggs laid by a single female in the course of a short breeding season of a few days reaches into the millions, the extreme case on record being that of a fifty-four pound Ling that laid over twenty-eight million eggs. Between these two extremes there are all intergrades. When the eggs are large and few are laid, chance methods of fertilization cannot be relied upon. It is therefore significant that in the elasmobranchs a sort of copulation occurs during which the male, by the use of claspers, introduces milt into the oviducts of the female and thus accomplishes internal insemination of the eggs. In some cases gestation is also internal, but this is doubtless a secondary adaptation. When, on the other hand, the eggs are small and extremely numerous, fertilization is external and haphazard.

The males and females simply swim about in schools, emitting eggs and sperm. The eggs are fertilized in large numbers and float about near the surface of the sea. Only a small percentage of them complete their development to hatching and large numbers are eaten as larvæ by enemies. Between these extremes again there are numerous habit intergrades. Some teleosts such as *Gambusia* and *Anableps*, with comparatively few large eggs, practice pairing and intromission of sperm, with resultant viviparity. Members of the same family such

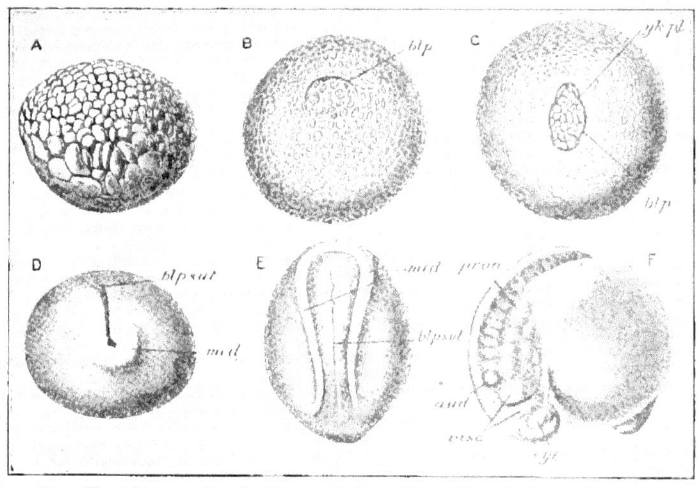

Fig. 97.—Development of *Neoceratodus forsteri*. A, lens-shaped blastula; B, stage with semicircular blastopore (*bl. p.*); *c*, later stage in which the blastopore (*bl. p.*) has taken the form of a ring-like groove enclosing the yolk-plug (*ylk. pl.*); D, stage in which the narrow medullary groove (*blp. sut.*) has appeared with the rudiment of the medullary folds (*med.*); E, stage in which the medullary folds (*med.*) have become well developed; F, later stage with well-formed head and two visceral arches (*visc.*) and rudiments of eye (*eye*) and ear (*aud.*); *pron*, mesonephros. (From Parker and Haswell, after Semon.)

as *Fundulus* and *Cyprinodon*, that have smaller and more numerous eggs, practice pairing, the males clasping the females with their dorsal and anal fins during the process of egg and sperm emission. Thus there is less chance of a failure of insemination. The female of some fishes such as the black bass, sticklebacks, etc., lay eggs in a nest and the male follows her into the nest and inseminates the eggs. The male pickerel follows the female closely and fertilizes the eggs as soon as laid. Many grades and modifications of this habit occur that gradually lead up to the promiscuous fertilization of eggs when numbers of males and females spawn in schools.

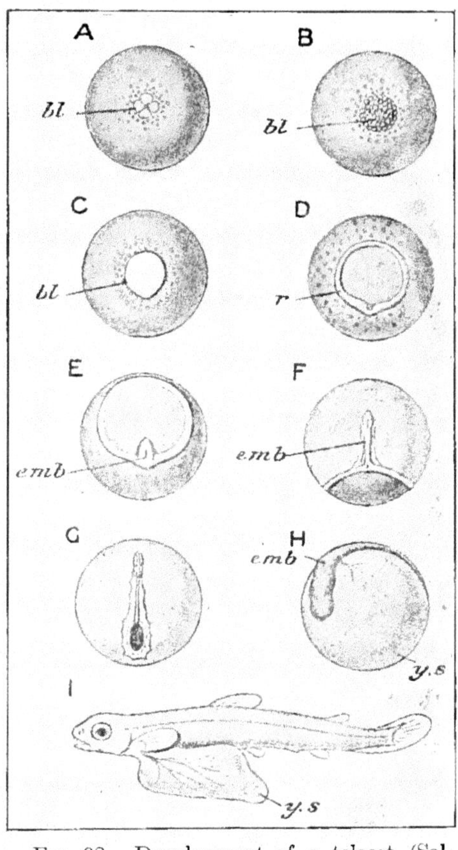

Fig. 98.—Development of a teleost (Salmon.) A, four-cell stage; B, multicellular blastoderm (an early blastula stage); C, blastoderm (*bl.*) beginning to overgrow the yolk; D, gastrulation beginning and germ-ring (*r*) formed; E, and F, embryo formed by concrescence of germ ring and germ ring one-third and two-thirds around yolk; G and H, early and advanced embryos (*emb.*) with blastoderm surrounding yolk-sac (*y. s.*); I, just hatched larva wth remains of yolk-sac (*y. s.*). (From Parker and Haswell, after Henneguy.)

It seems highly probable that the habit of nesting, such as is seen in *Amia* and the Dipneusti is close to the primitive condition and that there has been a specialization, in one direction, of few large eggs and internal fertilization, and in the other direction, as in fishes living in open waters, of an increase in numbers and decrease in size of eggs and consequent haphazard fertilization. For primitive breeding habits therefore, I would be inclined to look to *Amia* and the Dipneusti, where the conditions are not so very different from those in Amphibia.

The fundamental embryological changes following holoblastic and those following meroblastic cleavage are decidedly different, and an example of each, chosen from the fishes, may serve to illustrate some important developmental principles.

The case of *Neoceratodus*, the most primitive of the modern Dipneusti may be taken to illustrate *holoblastic cleavage* and its appropriate type of gastrulation. As described by Semon for the relatively small egg of this fish, the first two cleavages are meridional and divide the egg completely into four equal blastomeres. The third cleavage is also meridional; but the fourth is equatorial, cutting off eight micromeres and eight macromeres. The micromeres multiply more rapidly than the macromeres and produce a blastula with numerous small cells at the animal pole and fewer large cells at the vegetal pole (Fig. 97). The cavity contains many rather loose yolk cells. Gastrulation occurs by invaginating the endoderm to one side of the yolk mass. Embryo formation is very like that of the Amphibia.

The extreme type of meroblastic cleavage is seen in the teleosts (Fig. 98). Here the protoplasmic parts of the egg during the processes of maturation migrate to the animal pole and round up into a yolk-free hemisphere. This hemisphere undergoes cleavage, forming a lens-shaped cap of cells. This so-called embryonic disk then proceeds to surround the yolk by a process of overgrowth or peripheral spreading of the germinal disk, meanwhile differentiating an embryonic head. The germ-ring proceeds past the equator of the egg and then grows together as it becomes narrower to form the embryonic axis. Finally the ring closes completely, its substance having concresced to form the embryonic body. This process of embryonic concrescence is characteristic of all vertebrate embryos in which meroblastic cleavage occurs. There are all connecting stages between the type of development seen in *Neoceratodus* and that seen in the typical teleost.

APPENDIX TO FISHES

The Ostracodermi

Incidental mention has already been made to the Ostracodermi as an early specialized group of pro-fishes. They are Palæozoic forms which have a wide range of characters, so wide in fact that it is doubtful whether it is justifiable to place them in a single group. If we give the group as a whole the value of a class coördinate with Cyclostomata and Pisces, we may be justified in dividing this "class" into a number of orders.

Order I. Heterostraci

Of these forms we have little knowledge except of their external features. They were evidently broad, flat creatures, more like our present Skates and Rays than anything else and probably having similar but more sluggish habits. *Thelodus* and *Lanarkia*, the restored

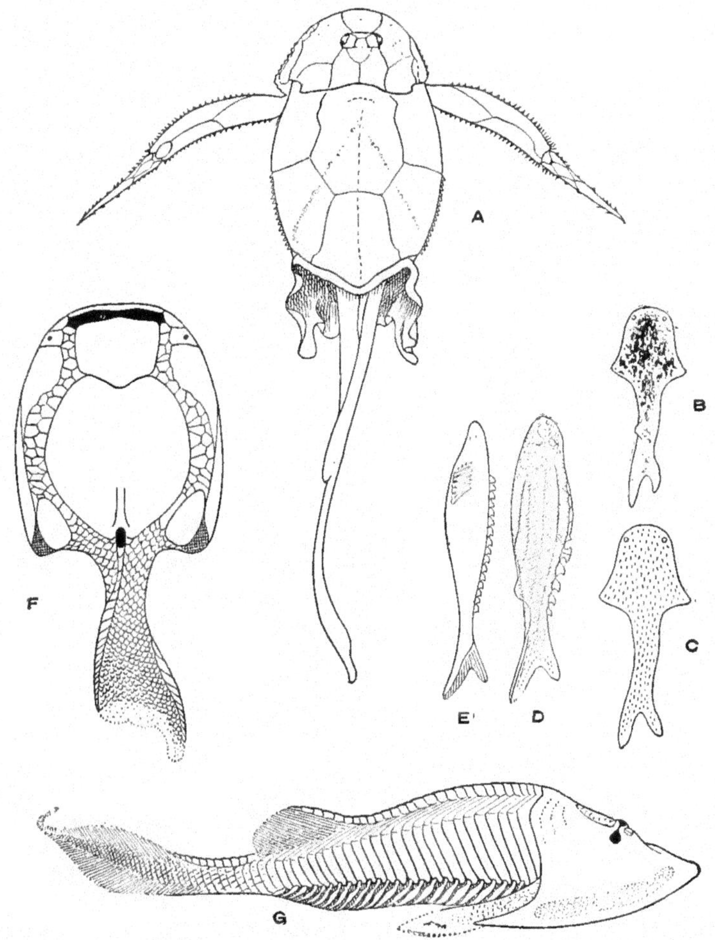

FIG. 99.—A group of ostracoderms and their kin. A and G directly after Patten; the rest from Patten, after Traquair. A, *Bothriolepis canadensis;* B, *Thelodus;* C, *Lanarkia;* D, *Birkenia;* E, *Lasanius;* F, *Drepanaspis;* G, *Cephalaspis.* (Redrawn after Patten.)

outlines of which are shown in Fig. 99 B and C, seem to have had an exoskeleton composed of "a uniform covering of hollow pointed spines, devoid of basal plate, and open below." These spines are composed of dentine covered with ganoin. The forward part of the body has lateral fin-folds of a very primitive type and the tail is provided with a very primitive heterocercal tail. Elasmobranch characters are quite obvious here and it is believed that this group represents a specialized bottom-feeding adaptive radiation from the most primitive shark types of the lower Ordovician or Cambrian times.

Another genus that has been placed in this order is *Drepanaspis* (Fig. 99, F). This creature is much more highly specialized in its exoskeleton than is *Lanarkia* and furnishes a transition between the latter and the more heavily armed condition of the Pteraspidæ, a group that reaches the climax of armature in this order. *Drepanaspis* is also a broad flat form with lateral fin-folds and a heterocercal tail. The whole body is covered with a continuous armature of tile-like plates, some of these plates, especially a median dorsal, a median ventral, and paired laterals, being especially conspicuous. The rest of the body is covered with smaller tessellated plates of various sizes.

In the genus *Pteraspis* the armor over the cephalothorax is much simplified by dropping out all but the largest plates, and by the development of a rostral plate. The tail is decidedly fish-like and covered with rhomboidal plates much like those of the lobe-finned ganoids or the modern gar-pike. A strong dorsal spine is a conspicuous feature of this species.

Order II. Osteostrachi

This order resembles the Heterostraci in having the anterior body covered with a solid armor and the tail free to move. They differ from the Heterostraci in having bony plates instead of mere calcifications, in having a dorsal fin, and in having eyes median instead of lateral in position. The carapace reminds one strongly of that of the King Crab (*Limulus*) and its extinct relatives, but the resemblance is probably merely a superficial one due to similarity of habits. They apparently had "a grovelling bottom-feeding, sluggish habit of life," in contrast with the active predaceous life of their free-living, shark-like ancestors. They played out their string of specialization and became extinct during Devonian times.

Order III. Anaspida

This order is established to contain two genera of fish-like forms, *Birkenea* (Fig. 99, D) and *Lasanius* (Fig. 99, E) about which there is only fragmentary information, and which are placed among the Ostracodermi only provisionally till more knowledge of their characters is forthcoming.

Antiarchi

This group, placed originally among the Ostracodermi, is now given class value and separated from the latter. It is from such forms as *Bothriolepis* (Fig. 99, A) that Patten would derive the vertebrates through the connecting link of the extinct sea-scorpions. Like the Ostracodermi, the Antiarchi have a heavily armored carapace and a free fish-like tail. The carapace is, however, differentiated into a head-shield and a thoracic shield. According to Patten this creature, which is fish-like in most of its characters, has the lateral paired jaws of the Arthropoda and has a brain more arthropodan than chordate. Perhaps the most characteristic feature of the group is the pair of appendages that is jointed to the cephalic carapace and reminds one at the same time of arthropodan appendages and of vertebrate limbs. A lateral-line system of chordate type is quite clearly shown. It has been suggested that the paired appendages may be the forerunners of the vertebrate jaws. This, however, seems rather far-fetched.

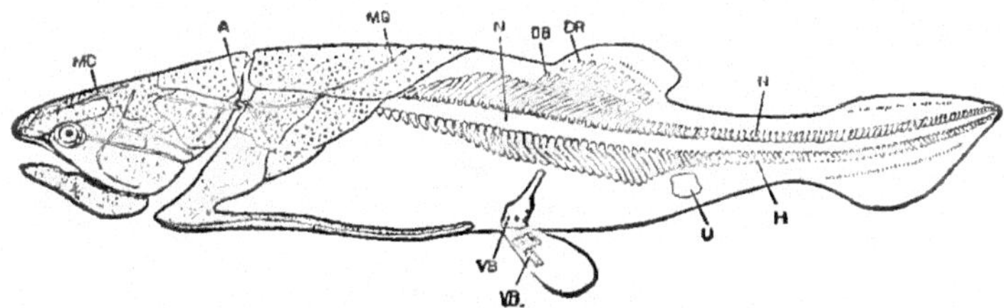

Fig. 100.—*Coccosteus decipiens.* Side view, restored. A, articulation of head with trunk; DB, cartilaginous basals of dorsal fin; DR, cartilaginous radials of dorsal fin; H, hæmal arch and spine; MC, Mucous canals; N, neural arch and spine; V, median unpaired plate (?) of hinder ventral region; VB, basals of pelvic fin; VR, radials of pelvic fin. (From Dean, after Smith-Woodward.)

As has been previously stated this group is not only a highly specialized one, but, in its heavily armed body and evidently sluggish habits, it bears evidences of senescence. Hence it is not at all the kind of group which we would expect to give rise to the generalized vertebrates of the early period. It is rather to be thought of as a precociously specialized, bottom-feeding derivative of some early fish-like group.

Arthrodira

The *Arthrodira* are real armored Fishes, whose exact relations are unknown. It is not unlikely that they represent an adaptive radiation from the primitive lobe-fin ganoids. They are provided with a heavy cephalic and a separate thoracic shield and have a typical fish body of primitive structure. The genus *Coccocteus* (Fig. 100) is a good example of the group.

CHAPTER VI

CLASS III. AMPHIBIA

Present and Past Status

The question of the origin of the Amphibia involves the whole problem of the beginnings of land life among the vertebrates and the radical evolutionary changes that have occurred as adaptations for an entirely new mode of life. While the aquatic habitat may be said to be a comparatively uniform and constant one, only slightly influenced by seasonal changes, the terrestrial life, especially in temperate regions, involves a wide range of changing conditions.

It has already been noted that the fishes had shown marked tendencies to adopt various methods of invading the air-breathing realm, some for the sake of avoiding the respiration of too much CO_2 and other poisonous gases in stagnant waters, some to tide over periods of drought, and still others for the purpose of enabling them to climb out of the water for food (the climbing-perch, etc.). It must have been in association with conditions resembling these that the first true land vertebrates were evolved.

The Amphibia are undoubtedly the most primitive land vertebrates, but it is coming to be believed that the first Reptilia trod closely upon their heels. The Reptilia were much more truly land vertebrates than were the Amphibia, for the Amphibia are tied down to the aquatic medium during at least the developmental period, in most groups, and during the entire life cycle, in others. Fundamentally the Amphibia are aquatic because their developmental processes are aquatic. Only a few of the most highly specialized modern Amphibia lay their eggs out of water, and these have adopted various unique brooding habits, which are at best mere developmental makeshifts as compared with the methods employed by the reptiles with their amnion and allantois.

The Amphibia have never attained the heights of success and of dominance in nature that has been attained by fishes, reptiles, birds, or mammals. Possibly this lack of complete success has been the result of their somewhat anomalous lives, involving the necessity of an amphibious environment. They are forced to occupy a narrow strip of territory between the waters and the dry land, a prey to the dominant denizens of the waters (fishes) on the one hand and to the various more vigorous enemies on the land (reptiles, birds, and mammals) on the other. If hard pressed in one environment the amphibian may seek the other; and this has saved him from complete extinction.

The Amphibia to-day are represented largely by a single highly specialized order, the *Anura* (frogs and toads), that have undergone within comparatively recent times a wonderfully elaborate adaptive radiation into a great variety of habitat complexes. But for the Anura the modern Amphibia would be largely unknown, for the salamanders, newts, and cæcilians are furtive, inconspicuous forms that have sought safety in the hidden nooks and crannies of the world environment and persist through their extremely retiring habits.

At one time, however, the Amphibia occupied a comparatively honorable place in nature. They reached in some cases almost giant size and evidently were active and predaceous creatures. Their wane began with the rapid rise of the Reptilia, which, as a group, became much more completely adjusted to land life than did the Amphibia.

The Origin of the Amphibia

It is now generally admitted that the Amphibia arose as a lateral branch from a very early group of "lobe-fin" ganoids (*Crossopterygii*). The time of emergence of the first Amphibia appears to have been about Middle Devonian, the period when the fishes gained their earliest pronounced ascendancy and when all of the available habitat complexes were occupied. The earliest trace of amphibian life is a single footprint (Fig. 101) of a three-toed species (*Thinopus antiquus*) found in the Upper Devonian shales of Pennsylvania. This foot though primitive was a true foot; not a fin. It is therefore probable that there were many transitional stages from the fin to the foot which are beyond our ken, and that the transition occupied at least thousands of years. The skeletal structure of the lobe-fin ganoid paired fins, especially that of the pectorals, is quite hand-like in arrangement; so that a dropping off of the fringe-like fin portion would leave a structure quite like a hand with three or more fingers (Fig. 102, C, D, E). The dropping of the fin-fringe may have happened quite suddenly in the process of evolution of some group—possibly by a single mutation. The rest of the change would be one of gradual functional adjustment. It was noted that the present-day "lobe-fins" use the pectoral fin like a land limb in that they support the weight upon it while resting on the bottom; so a functional change may readily have preceded the radical structural change. This theory of the origin of the amphibian pentadactyl limb is well shown in Fig. 102, A, B.

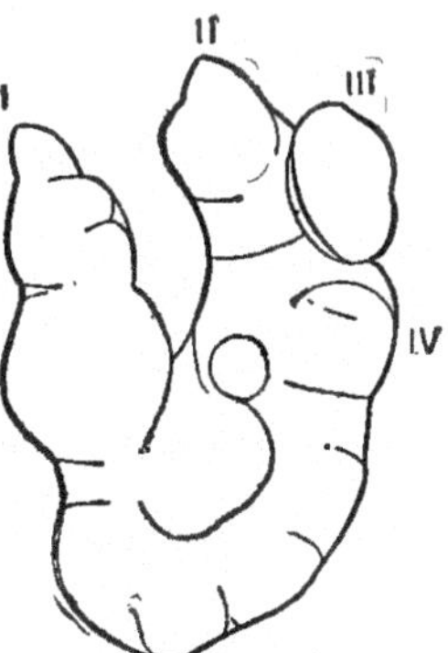

FIG. 101. — Earliest known fossil foot-print, *Thinopus antiquus*, with two fully formed digits, I and II, a budding third, III, and a possible rudiment of a fourth, IV. Upper Devonian of Pennsylvania. ½ natural size. (From Lull.)

Palæographers inform us that the climatic conditions of the Upper Devonian were such as to encourage the development of land life on the part of fishes living in inland waters. There were periods of warmth and heavy rainfall followed by long periods of drought, which became progressively more prolonged. Such conditions would tend to drive a large proportion of the non-air-breathing fishes from the fresh waters and to give their place to the air-breathing crossopterygians and their kin. With increasingly prolonged dry seasons the æstivating habits of the early lung-breathing fishes proved inadequate and it became necessary for the animals to live an active life in the air and to get their food on the land. It is probable that although many early lung-breathing fishes made the beginnings of adaptation to true land life, only one type fully succeeded and became the first true Amphibia, the ancestors of all of those to-day living.

Adaptive Changes Incident to Life on the Land.—The change from aquatic to terrestrial life has been the greatest evolutionary crisis in vertebrate history. No other environmental change possible for animals requires so radical an alteration of developmental and nutritional (in the broadest sense) mechanisms. Changes from salt to fresh water, from shallows to abysses, from surface to subterranean, arboreal or aëreal life, involve much less fundamental alterations than does that from water to land; which is, strictly speaking, rather a change from water to air. Naturally the most important changes had to do with respiration, circulation, and locomotion. Changes of secondary value concern the altered specific gravity, the more pronounced changes in temperature, the tendency toward dessication, the differences in visibility through water and air, and the differences in conduction of sound waves. In the true terrestrial forms (reptiles, birds and mammals) special adaptations for making possible embryonic development in the air have been acquired, but not so in the Amphibia, which develop through to the

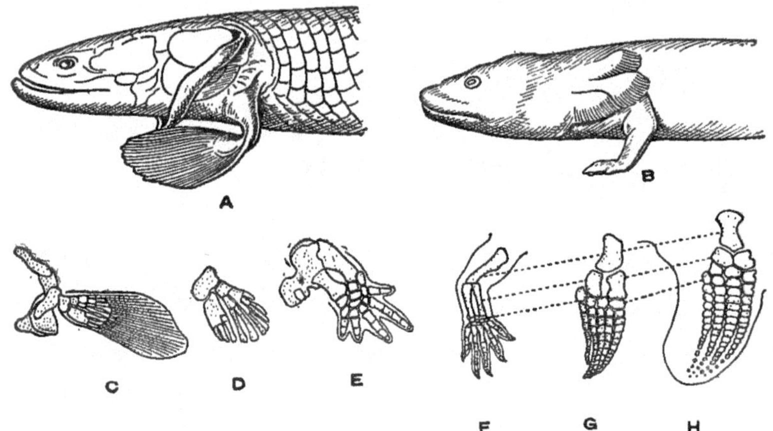

FIG. 102.—To illustrate the change from fin-type of appendage to the foot-type, and the reverse or secondary adaptation of the foot-type into the fin-type. The upper figures (A and B) represent the theoretic mode of metamorphosis of the lobe-fin of the Crossopterygian fish (A) into the foot of the amphibian (B) through the loss of the dermal fringe border and rearrangement of the cartilaginous supports of the lobe. C, D, E, show the skeletal support of the two types of fin; C, the lobe-fin with the fin-fringe; D, the lobe-fin without the fringe; and E, the foot-stage as seen in an early Carboniferous amphibian. F, G, H, show the secondary reversed evolution of the five-rayed limb (F) of a land reptile into the fin or paddle of an ichthyosaur (G, H). (Redrawn after Osborn.)

adult condition in the water, undergoing a rather sudden metamorphosis from the aquatic to the terrestrial physiology. These various primary and secondary changes in structure will, when listed, serve to indicate the **differences between the Fishes and the Amphibia**:—

1. **Respiration.**—If we go back to the lobe-finned ganoids for the ancestors of the Amphibia, we find a double respiratory system, branchial and pulmonary, with the pulmonary playing an accessory or secondary rôle. In times of extreme drought or extreme foulness of the water the branchial respiration was held in abeyance and the pulmonary used almost exclusively. The branchial respiration functions entirely in early life and it is only with assumption of adult

form that the air-bladder comes to be greatly in demand. In the ontogeny of modern Amphibia we find this sequence repeated, for the young amphibian is purely aquatic, and air-breathing comes only when the larva metamorphoses into the young adult.

2. **Circulation.**—The fish type of circulation is built primarily along lines laid down by branchial respiration. The heart is purely venous in its blood content and pumps blood forward and through the branchial arches. This involves as many pairs of branchial arches as there are paired functional *afferent* vessels carrying blood to the gills, and *efferent* vessels carrying the aërated blood from the gills to the dorsal aorta. In the evolution from the fish to the amphibian type the principal changes in the circulation have to do with the branchial arches, which cease to have a value as such, and their profound remodelling into blood vessels that fit into an air-breathing physiology. The branchial vessels of lobe-finned ganoids and of larval amphibians consist of four pairs; the first pair becomes the *carotid arteries* that supply the head, the second becomes the *systemic arches* that supply most of the body, the third disappears, and the fourth becomes mainly the *pulmonary arches.* It is of interest to note that in all lung-breathing fishes the lungs are supplied from a branch of the fourth branchial arch. In most Amphibia a branch of the fourth arch becomes *cutaneous,* for the skin respiration is almost as important as the pulmonary. The *heart* becomes *three chambered,* the auricle dividing into a *systemic* half and a *pulmonary* half. The single ventricle receives both arterial and venous blood, but there is very little admixture of the two.

3. **Locomotion.**—In fishes the chief locomotor organs are the tail and the median fins. The paired fins are used largely for balancing; in lobe-finned ganoids they are used to support the head when resting on the bottom. Naturally then we should expect the most radical change to concern the loss of tail fins and other median fins, on the one hand, and the development of feet, on the other. The median and caudal fins appear in the amphibian larvæ and persist in a reduced form in certain persistently aquatic amphibia, which are probably no more than permanent larvæ (pædogenetic). When the fins do appear they are mere soft folds of the skin without any true skeletal supports, and they never become regionally specialized, but remain in the ancient diphycercal form. In the land salamanders the tail-fin is lost but the tail persists. In the Anura, as well as in the

cæcilians, the tail, formed fully in the larvæ, is secondarily resorbed during metamorphosis.

The change from paired fins to paired limbs is not so radical as it was once supposed. Thanks to the discovery of the limb-skeleton

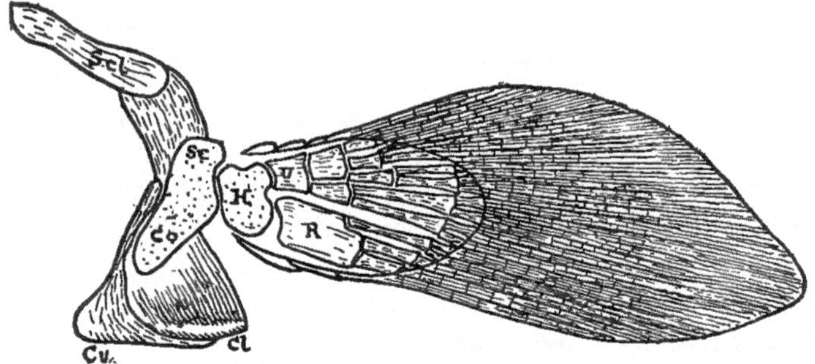

FIG. 103.—Pectoral fin of extinct crossopterygian, *Sauripterus taylori*. *cl*, clavicle; *co*, coracoid; *H*, humerus; *R*, radius; *Sc*, scapula; *Scl*, supra clavicle; *u*, ulna. (From Lull, after Gregory.)

of some of the early crossopterygians, *e. g.* the pectoral fin of *Sauripterus taylori* from the Upper Devonian (Fig. 103), it is not difficult to see how a fin could become a foot. Note that this fin, minus the fringe, is a hand-like structure, with humerus, radius, and ulna, wrist, and several fingers. The shoulder girdle of *Sauripterus* is also part for part homologous with that of an amphibian. Some changes in relative sizes of the elements and a reduction in the number of repeated parts would give

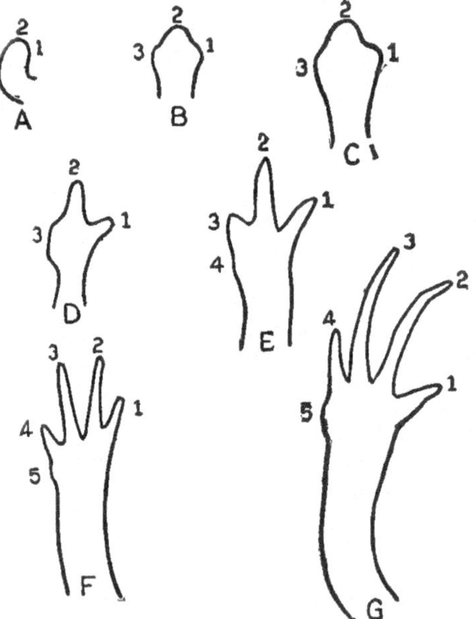

FIG. 104.—Development of the hind foot of a salamander, *Triton tæniatus*. 1–5, first to fifth digits; A–F, seven stages from the simple limb-bud to the definitive foot. (From Lull, after Rabl.)

an amphibian hand. The footprint of *Thinopus*, the earliest amphibian trace, does not reveal any of the skeletal parts, but it is likely that only two fingers were fully developed and two others partially separated. In the development of the amphibian foot as shown in Fig. 104 the three-fingered condition persists till rather late in development; then a fourth finger appears well down on the ulnar side of the hand and a rudiment of the fifth (the little finger) appears as a mere bump. The thumb, index, and second finger seem to be phylogenetically the oldest digits, and this is important in connection with the loss of fingers in other vertebrates; for it is always the latest to develop that is first lost. With the substitution of a foot for a fin the kind of movement becomes profoundly altered. Instead of a mere paddling back and forth, a variety of movements are necessary and thus the old myotomic musculature becomes decidedly modified until nearly all traces of the segmental arrangement of muscles are lost.

4. **Changes Due to Increased Specific Gravity.**—When the animal comes from the water to the land it is relatively heavier; hence there is need of a more rigid skeleton, stronger limb girdles and limb skeleton. This is accomplished by more complete ossification of the parts of the skeleton that bear the most weight. There is likely also to be a reduction of dead weight, such as exoskeletal parts. In modern Amphibia the exoskeleton has disappeared completely (except in cæcilians, where it is rudimentary), but in the stegocephalians the head armor persisted, while that of the rest of the body largely disappeared.

5. **Responses to Seasonal Changes of Temperature** are much more necessary on land than on water. Nearly all Amphibia hibernate during winter either by burying themselves in the earth or in water.

6. **Changes for Avoiding Dessication.**—If a fish is taken out of water it soon dries up on the surface and becomes stiff. Amphibia, however, have abundant skin glands, secreting moist mucus which keeps the skin in proper condition to perform a respiratory function and helps it to retain its flexibility. Some of the Amphibia have rudimentary lungs and respire almost exclusively through the skin.

7. The **eyes** change, especially in the shape of the lens, which becomes flattened instead of spherical.

8. An **external sound receptor** appears in the form of a *tympanic membrane*, which is in communication with the inner ear through a *columella*, a bony apparatus that vibrates with the ear drum.

THE EXTINCT AMPHIBIA

All of the early Amphibia of the coal-measures (Upper Carboniferous) are classed as *Stegocephali*, and it is customary to divide the Class Amphibia into two sub-classes: (1) *Stegocephali* (with dermal armor, especially on the head); (2) *Lissamphibia* (modern and extinct Amphibia with smooth bodies, devoid of heavy dermal armor).

Sub-Class I. Stegocephali

The earliest actual skeletal remains of Amphibia were found in the Edinburgh Coal-Measures, which belong to Lower Carboniferous

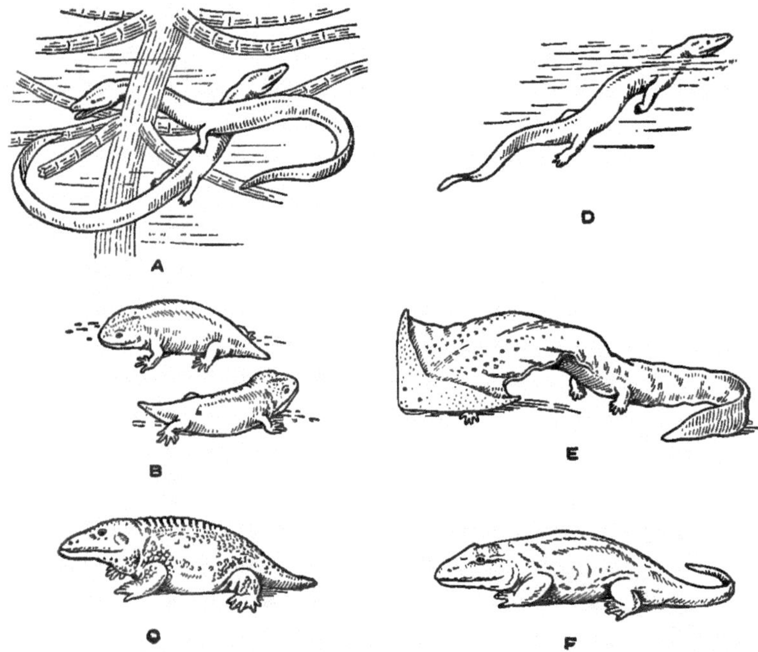

Fig. 105.—Group of Extinct Amphibia, A and B from the Carboniferous; C and F, Permo-Carboniferous. A, *Pytonius;* B, *Amphibamus;* C, *Cacops;* D, *Cricotus;* E, *Diplocaulus;* F, *Eryops*. (Redrawn after Osborn, following restorations of Gregory and Deckert.)

times. These forms, *Loxomma* and *Pholidogaster*, are not transitional but fully modified Amphibia. There is every reason to believe therefore that the earliest Amphibia arose back in the Devonian.

The Amphibia apparently did not find the climatic conditions of the Lower Carboniferous especially adapted to them and did not really become a successful and dominant race till during the Upper Carboniferous and Permian times, when their great deployment and first adaptive radiation ccurred.

The first Amphibia (Fig. 105, A) were probably small-headed, long-bodied forms with fish-like appearance, resembling, doubtless, our modern newts and salamanders. During the Upper Carboniferous, however, there was an adaptive radiation resulting in the development of large-headed, short-bodied types, more or less resembling our frogs and toads (Fig. 105, B), but without jumping legs. There also appeared some broad, flat types (Fig. 105, E) with reduced limbs that must have been bottom-feeders (*e. g.*, *Diplocaulus*).

Three orders of Stegocephali are distinguished:

Order I. Stegocephali Leptospondyli.—This group is characterized by pseudocentrous vertebræ, by which is meant that a thin shell of bone surrounds the notochord. Two types of these animals are distinguished, one in which the form was evidently much like our modern newts. They were broad-headed, had several pairs of gills, at least in the young (Fig. 106). As an example of the skull structure of the group, that of *Branchiosaurus* (Fig. 107), one of the most generalized of vertebrate skulls, is shown. The other type of this order was a snake-like form, without limbs, evidently a precociously senescent type in which the ribs reach about halfway round the body.

Order II. Stegocephali Temnospondyli.—The vertebræ are composed of three separate pieces, two dorsal and one ventral. These animals had rather long ribs and their armor was chiefly ventral. Some of the types were: *Chelydosaurus*, a turtle-like form; *Dissorophus*, a sort of "Batrachian Armadillo;" *Archegosaurus*, a thoroughly terrestrial form about five feet in length.

Order III. Stegocephali Stereospondyli.—The three components of the vertebra unite into one solid amphicœlous vertebra. This group has been given the name of *Labyrinthodonta* on occount of their much folded teeth. The vertebræ are sometimes one-headed, sometimes two-headed. The limb-girdles are very primitive and strikingly resemble those of crossopterygian fishes. The foot skeleton is extremely generalized. It now seems likely that this is the group of early Amphibia that is most nearly related to the crossopterygian fishes. Watson has studied the skulls of the Carboniferous labyrinthodonts

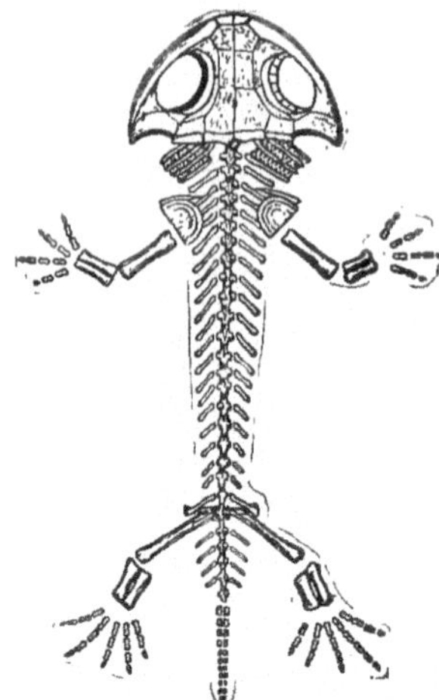

FIG. 106.—Stegocephalian, *Branchiosaurus amblystomus*. (From Eastman-Zittel.)

Loxomma and *Pteroplax* and has pointed out their many striking resemblances to those of the Carboniferous crossopterygian *Megalichthys*. These resemblances are carried out in so many finer details that one cannot escape the conviction that the two groups are closely related.

It may be said in concluding this very much abbreviated account of the extinct Amphibia, that recent discoveries of early land vertebrates of the Texas and New Mexico Permian by Williston and his colleagues, has revealed a number of genera that show a combination of amphibian and reptilian characters. Sometimes it is difficult to decide readily whether the creature belongs to one or the other group. These forms are evi-

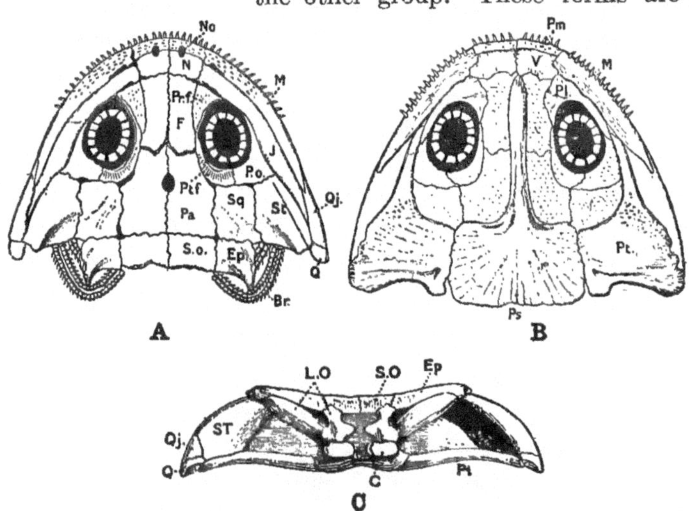

FIG. 107.—A, dorsal and B, ventral views of the cranium of *Branchiosaurus salamandroides* (after Fritsch). C, posterior view of cranium of *Trematosaurus* (after Fraas). *Br*, branchial arches; *C*, condyle; *Ep*, epiotic; *F*, frontal; *J*, jugal, *L. O*, lateral occipital (exoccipital); *M*, maxillary; *N*, nasal; *No*, nostril; *Pa*, parietal; *Pl*, palatine; *Pm*, premaxillary; *P. o*, postorbital; *Pr. f*, prefrontal; *Ps*, parasphenoid; *Pt*, pterygoid; *Ptf*, postfrontal; *O*, quadrate; *Oj*, quadratojugal; *So*, supraoccipital; *Sq*, squamosal; *St*, supratemporal; *V*, vomer. (From Gadow.)

dently Amphibia that have developed certain dry-land adaptations. Good examples of this transitional group are *Cacops* (Fig. 106, C), *Eryops* (Fig. 106, F), and *Cricotus* (Fig. 106, D).

The Permian was the period in which the amphibians passed their climax. From that time on the Amphibia have lived a hard life in competition with the Reptilia which are better adapted for land life, and with the fishes which are better adapted for the waters. Only a few rather small groups have survived and these largely through their retiring habits and inconspicuous appearance. The group of Anura has recently gained a secondary dominance through a remarkable adaptive radiation into various land habitats.

PRESENT-DAY AMPHIBIA

Sub-Class II. Lissamphibia

The recent Amphibia are the Cæcilia or Gymnophiona, newts and salamanders, frogs and toads. The Amphibia are the least numerous of the vertebrate classes, except the Cyclostomata. In all there are only about 1,000 species (nearly 900 of which are frogs and toads). This is to be compared with the nearly 10,000 species of birds, nearly 8,000 species of fishes, about 3,500 species of reptiles, and about 2,700 mammals. As a class the Amphibia have always been relatively unimportant numerically, possibly because it is essentially an "in-between" group, as has been shown.

The Characters of the Amphibia (after Gadow)

1. The vertebræ are (a) acentrous, (b) pseudocentrous, or (c) notocentrous.
2. The skull articulates with the atlas by two condyles which are formed by the lateral occipitals (exoccipitals).
3. There is an auditory columellar apparatus fitting into the fenestra ovalis.
4. The limbs are of the tetrapodous, pentadactyle type.
5. The red-blood corpuscles are nucleated, biconvex, and oval.
6. The heart is (a) divided into two atria (auricles) and one ventricle, and (b) it has a conus provided with valves.
7. The aortic arches are strictly symmetrical.
8. Gills are present at least during some early stage of development.
9. The kidneys are provided with persistent nephrostomes.

10. Lateral line sense organs are present at least during the larval stage.
11. The vagus is the last cranial nerve.
12. The median fins, where present, are not supported by spinal skeletal rays.
13. Sternal ribs and a costal or true sternum are absent.
14. There is no paired or unpaired medio-ventral copulatory apparatus.
15. Development takes place without amnion and allantois.

None of these characters is absolutely diagnostic, except 1 (c), and this applies only to Anura and most of the Stegocephali.

Numbers 1 (b), 1 (c), 2, 3, 4, and 12 separate the Amphibia from the Fishes.

Numbers 1, 6 (b), 7, 8, 9, 11, 13, 15, separate them from the Reptiles, Birds, and Mammals.

Number 2 separates them from Fishes, Reptiles, and Birds.

Number 6 (a) separates them from the Fishes (excl. Dipnoi), Birds, and Mammals.

Gadow says: "*Amphicondylous Anamnia* would be an absolutely correct and all-sufficient diagnosis" of Amphiba, but concludes that "Amphicondylous animals without an intra-cranial hypoglossal nerve," is a more practical diagnosis.

ORDER I. APODA (GYMNOPHIONA)—LIMBLESS AMPHIBIA

The Apoda or cæcilians, sometimes called "blind worms," constitute a small group of about forty species, living in the warmer parts of the world, but widely distributed. They are worm-shaped, burrowing creatures (Fig. 108, A) with habits somewhat like those of earthworms and not unlike them in appearance. They have no limbs nor limb-girdles and there is also the merest rudiment of a tail; hence the anal opening appears to be terminal. The skin is folded into numerous ring-like folds and is smooth and slimy. Small deep-set dermal scales occur, which are believed to be an inheritance from stegocephalian ancestors. The cranium is very solid and compact in appearance, more like that of a reptile than that of other modern Amphibia, but the same bones as in other Amphibia are present in a broadened-out form. The vertebræ are pseudocentrous and extremely numerous, being as many as 200 to 300 in some species. The eyes are rudimentary and practically functionless. They feel their way about by means of a protrusible tentacular organ that lies in a groove between the eye and nose. The eggs are meroblastic and are fertilized internally by means of an eversion of the cloaca of the male which becomes a tube-like copulatory organ. Some species are oviparous, others viviparous.

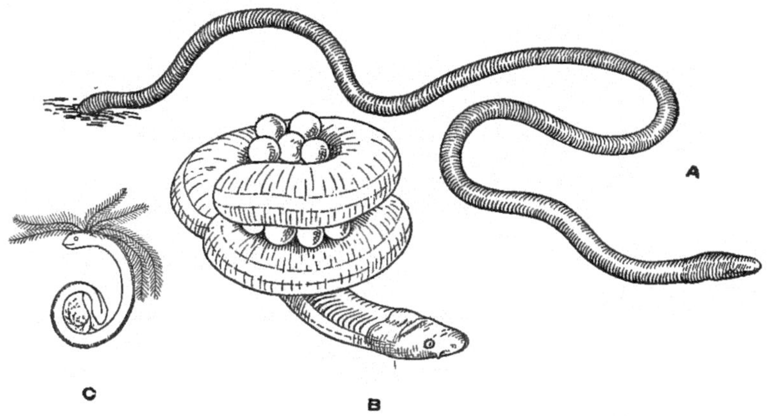

Fig. 108.—Group of Apoda. A, *Caecilia*, emerging from burrow; B, *Ichthyophis glutinosus* (nat. size), female guarding her eggs, coiled up in hole in the ground; C, a nearly ripe embryo, with cutaneous gills, tail-fin, and still a considerable amount of yolk. (Redrawn after P. and F. Sarasin.)

Natural History of Ichthyophis glutinosa.—This species is chosen as an example of Apoda because it has been adequately studied and described by the Sarasins. The species extends from the foot hills of the Himalayas to Ceylon, the Malay Archipelago, and Siam. It reaches a length of about a foot. In color it is dark brown or bluish black with a yellow band along the side. The ovarian egg is oval, about 4 x 6 mm. There is a heavy coat of albumen with chalazæ, much as in the birds, these chalazæ uniting the eggs in bunches. The egg bunch is laid in a shallow hole near the water. The female coils herself about the glutinous mass (Fig. 108, B) to protect it from ground-burrowing animals. The gilled larval period is passed through in the egg before hatching. The three pairs of larval gills (Fig. 108, C) are of the external type and are very large and finely branched. The gills are lost when the larva hatches. The larva swims about in the water for a time like an eel, but comes frequently to the surface to breath air. The larval period is a long one, but at length the two gill-clefts close, the skin changes its character, the tail-fin disappears, and it emerges upon the land and lives a burrowing life. So exclusively terrestrial does it become that it drowns if after metamorphosis it is put in water for any length of time. Several other genera of Apoda are viviparous, the embryos becoming several inches in length before birth.

The Apoda constitutes a very degenerate group. In some respects they are more primitive than other living Amphibia, but life in burrows has caused a profound degeneration of structure. They are to be included among the eel-like type of senescent, degenerate forms.

Order II. Urodela—(Tailed Amphibia)

This order is represented by about 100 species of mud-puppies, salamanders, newts, and efts. They range in habitat from forms living permanently in the water and breathing with external gills in addition to lungs, to forms that live after metamorphosis entirely on land, favoring moist woods or other sheltered places. Some authors group all forms with permanent external gills in a separate family, *Perennibranchiata;* but this arrangement is believed by the best authorities to be artificial in that the retention of the aquatic habit and larval gills is probably due to arrested development and may have taken place in more than one family. The classification given here is taken from Gadow and is based on fundamental anatomical characters.

Family I. Amphiumidæ.—Without gills in the definitive stage; gill-clefts vestigial, consisting of one pair of small openings, or entirely absent; maxillary bones present; teeth on both jaws; vertebræ amphicœlous; both fore and hind limbs present, but small; small eyes without lids.

The family is represented by two genera, *Cryptobranchus* and *Amphiuma. Cryptobranchus alleghenien*sis (Fig. 109, A) occurs in the mountain streams of our Eastern States. Another species, *C. japonicus,* is the giant salamander of Japan. There is only one species of *Amphiuma* (Fig. 109, B), which is also an American species confined to the southeastern States, from Carolina to Mississippi.

*Cryptobranchus alleghenien*sis, the "*hellbender,*" is a comparatively large salamander reaching a length of nearly two feet. An account of the development of this species and certain excellent descriptions of the larvæ and the process of metamorphosis have been furnished by B. G. Smith. The eggs are fertilized externally, males emitting sperm masses near the egg-laying female. Batches of eggs were found in the shallow parts of a rather large stream lying on the gravelly bottom. They were arranged in festoon-like strings. A single female lays from 300 to 400 eggs. The cleavage of the egg is especially clean-cut and illustrates the transition between holoblastic and meroblastic cleavage. The animal is a voracious feeder, capturing fish in considerable numbers, and is therefore unpopular with fishermen. The larvæ are much like the adults of *Necturus*. *C. japonicus*, is very much like

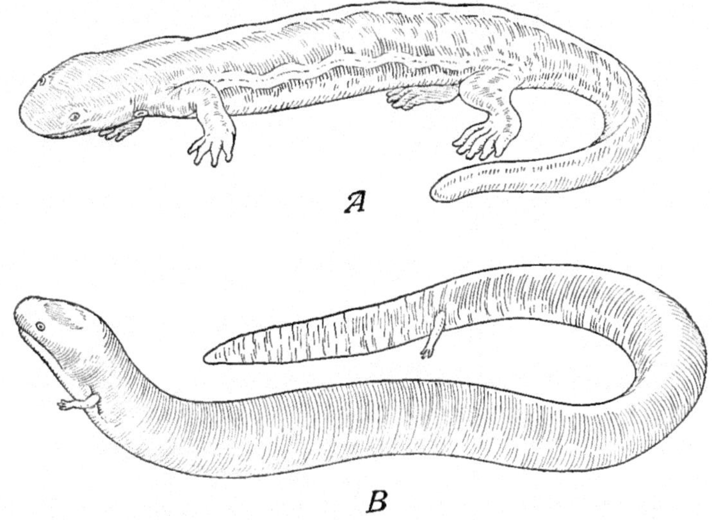

FIG. 109.—A, *Cryptobranchus allegheniensis;* B, *Amphiuma means.* (After Lydekker.)

C. allegheniensis in appearance and in habitat, but reaches a large size, the largest specimens being about five feet three inches in length. This is the extreme size reached by modern Amphibia, a size which almost rivals that of some of the giant land Amphibia that became extinct during the Permian. In Japan these animals are used for food and are caught with a baited hook, the hook being thrust into the retreat of the animal by means of a pole, which is not attached to the line, and may be removed when the animal seizes the bait.

Amphiuma means (Fig. 109, B) is an eel-shaped salamander with limbs very much reduced, both in size and in numbers of digits (2 or 3 being the characteristic number). One pair of small inconspicuous gill-clefts is present, guarded by skin flaps. They reach a length of about three feet, live in swamps and muddy water, often invading the rice fields of the Mississippi lowlands. The rather hard-shelled eggs are laid in festoons and are protected by the female, which lies about them in a coil. The larvæ have well-developed external gills and legs relatively larger than those of the adult.

Family 2. Salamandridæ. (Salamanders and Newts.) These urodeles are without gills in the adult stage; maxillaries are present; teeth occur in both jaws; eyes have movable eyelids; fore and hind limbs present, but sometimes much reduced. Nearly three-fourths of the tailed Amphibia belong to this family. Only a few typical species can be mentioned here.

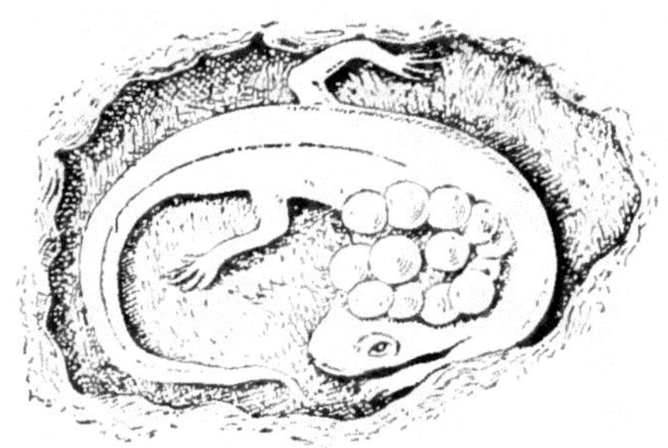

Fig. 110.—*Desmognathus fuscus;* female with eggs in hole underground. (From Gadow, after Wilder.)

Desmognathus fuscus (Fig. 110) is one of our commonest American newts. It is a small type, about four inches in length, living a nocturnal life, hiding in the daytime under stones or concealed along the edges of mountain streams. The color is brown suffused with pink and gray. They are, strange to say, lungless, the process of respiration being carried on in the skin and possibly also by the mucous lining of the intestine. The eggs are laid in a bunch, each egg attached by a string, the whole group looking like a bunch of toy balloons. The female lays the eggs in a hole in the mud and coils her body partly about them. After a period of incubation the eggs hatch and give forth larvæ

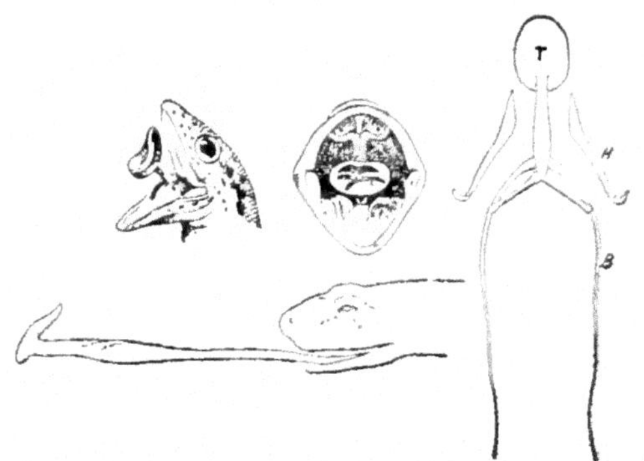

Fig. 111.—*Spelerpes fuscus*, showing the position and shape of the partly protruded tongue and the tongue skeleton on the right. *T*, tongue; B, branchial arch; *H*, hyoid. (From Gadow, after Berg and Wiedersheim.)

which are nearly definitive in form. *Spelerpes bilineatus*, another newt of the Atlantic States, has been described by H. H. Wilder. It is about four inches in length, brownish yellow in color above, with a black lateral line, and brightest yellow beneath. It lives a nocturnal life and hides under stones or logs. The eggs are laid in bunches of thirty to fifty on the under sides of submerged stones. *S. fuscus* (Fig. 111) has an extremely extensible tongue, capable of being shot out nearly two inches. With this it captures insects by means of a sticky disk on its end.

Amblystoma tigrinum (Fig. 112, A) the "tiger salamander" is the commonest and most widely distributed of North American salamanders. It occurs from the Atlantic Ocean to Minnesota and well into the Southern States. The ground color is nearly black, with large yellow spots and blotches, which sometimes merge into broad stripes or bands. It lives in damp situations on land, under stones or logs, and is not infrequently found in cellars. The large prominent golden eyes, the very broad head and large mouth, are characteristic features. The length varies from five to nine inches. This species is best known because of the fact that it exhibits a classic case of *pædogenesis* or *neoteny*, the capacity to become sexually mature while still retaining the larval body. The larva of *Amblystoma* is the classic *Axolotl* (Fig. 112, B), which has for a long time been abundant in the lakes near Mexico City. It was supposed to be a perennibranchiate species and was called "*Siredon axolotl*." The true situation, however, was revealed when some of these larvæ were kept in aquaria in Paris. Some of them lost their gills and other aquatic adaptations and metamorphosed into the well-known *Amblystoma tigrinum*, a purely terrestrial type. It was found possible by experimental means to control the metamorphosis so as to keep the animals permanently as Axolotls, or to cause them to metamorphose promptly into adult Amblystomas. Even after metamorphosis had begun it was possible to check it and cause the animals to revert to the Axolotl condition. This situation throws light on the significance of some of the perennibranchiate species that never metamorphose; for it has often been suggested that these forms are permanent larvæ or that they exhibit a type of racial pædogenesis that has become so fixed that metamorphosis is no longer possible.

Salamandra maculosa, the "spotted" or "fire salamander" is one of the commonest salamanders of Europe, having a wide range over the whole of central, southern, and western Europe. It lives under moss or rotten leaves, in cracks in the ground or in the roots of old trees; in fact, in almost any moist place. It is about the same size

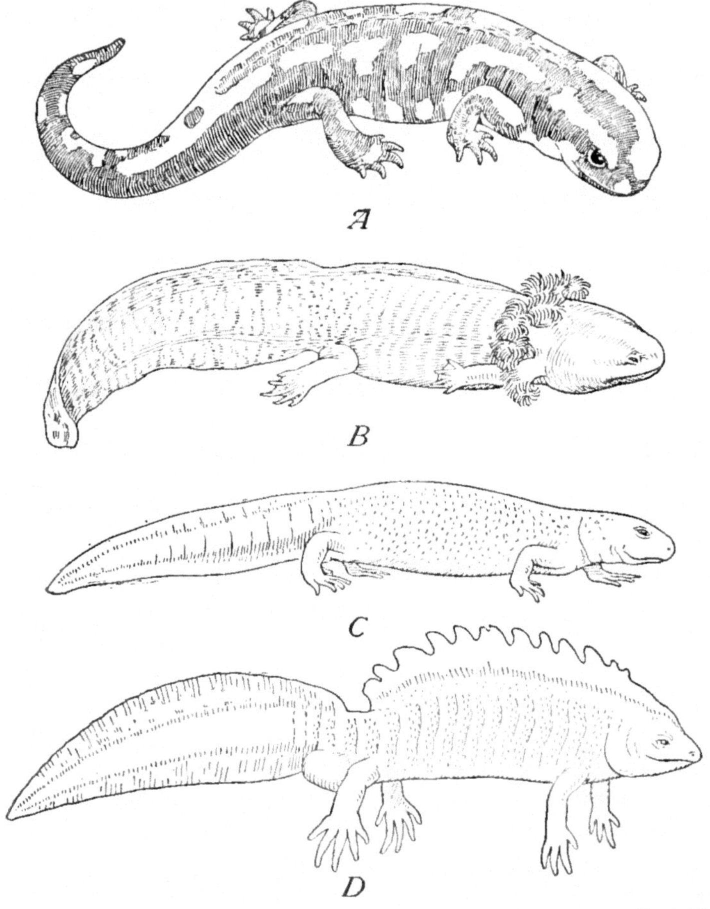

FIG. 112.—Group of Urodela. A, *Salamandra maculosa* (after Lydekker.) B, *Axolotl* larva of *Amblystoma tigrinum;* C, female and D, male of *Triton cristatus* (male in nuptial dress) x $^2/_3$. (After Gadow.)

as *Amblystoma* and resembles it in form and habit. The color pattern is also similar, with its black groundwork and yellow spots. Fire salamanders are poisonous, as is shown by the quick death of bull-frogs, snakes, or warm-blooded animals that have eaten them; the poison is due to a cutaneous secretion. The breeding habits are rather odd. In July after a preliminary exciting performance between the sexes on land, both males and females go into the water, but leave the heads out. The male deposits a spermatophore, or package of sperm, which the female partly takes into the cloaca. Fertilization is internal and slow development occurs in the uterus, taking about ten months to complete itself. The well-developed young, to the number of about fifteen or so, are born in the water. The species is therefore truly viviparous. The larvæ have external gills and live in the water for about four months and then very slowly metamorphose into the terrestrial adult form.

Salamandra atra is an alpine form like *S. maculosa* but much darker. It occurs in mountain lakes at an altitude of 2,000 to 9,000 feet above sea level. It produces only two young at a birth. These, while still in the uterus, feed upon the other eggs found there and metamorphose completely before birth. Kammerer claims that *S. atra* can be changed into *S. maculosa* by bringing them into the lowland waters and that after they have been kept there for a few generations they tend to retain the breeding habits of the lowland form though transferred back to the Alpine environment. This has often been cited as evidence in favor of the inheritance of acquired characters, a doctrine which is quite generally unacceptable to biologists, but is strongly advocated by a small but growing minority.

Diemictylus viridescens is a good example of the "efts," sometimes also called "newts." It is commonly called the "vermilion spotted eft." It has a prolonged life history, taking several years to reach full maturity. For the first three years it lives in the water, being green in color and having external gills. It then leaves the water and becomes yellow with vermilion spots. After some time it again returns to the water, becomes green, and lives an aquatic life during the breeding season, after which it once more takes on the terrestrial features and migrates to land. The life cycle of this species illustrates as well as any other the extreme plasticity of the group and the delicate equilibrium that exists between the aquatic and terrestrial phases.

Triton cristatus (Fig. 112, C and D), the "crested newt," received its name from the fact that in the male (Fig. 112, D) there is a pronounced dorsal crest during the breeding or nuptial season. The color at this time is also very striking, the top of head being marbled black and white, the under side yellow with black spots, and the side of the tail has a broad bluish-white band. The female has quieter colors, a general brownish-black ground with a yellow line down the middle of the back. This is the most pronounced instance of sexual dimorphism among the Urodela and possibly within the class Amphibia. The crested newt has a wide distribution, occurring in England and Scotland and through Central Europe. Many other species of the genus *Triton* occur, two of which, *T. torosus* and *T. virescens*, occur in North America, the latter being common through the Eastern United States.

Family 3. Proteidæ. The Mud-Puppies.—These animals have three pairs of fringed external gills throughout life (perennibranchiate); both fore and hind limbs are present; eyes are without lids; maxillaries are absent; teeth occur on premaxillaries, vomers, and mandible; vertebræ are amphicœlous. Only three genera, each represented by a single species, occur, two in America and one in Europe.

Necturus maculatus (113, A) is the common American "mud-puppy," so called because the fringed gills look something like the pendant hairy ears of a water spaniel. They are found all over the eastern part of the United States and in eastern Canada. They are about a foot in length, of a muddy-brown color, mottled with blackish spots. Behind the dark red external gills are paired gill-clefts. These amphibians impress one as rather dull, stupid animals, living a sluggish life on the muddy bottoms of lakes and rivers. At times, however, they move about quite smartly with graceful eel-like motion. They are active chiefly at night, when they swim about in search of frogs, Crustacea, worms, fishes, and insects. They are often found in very cold water and seem to be well adapted to low temperatures. In general it may be said that these animals resemble closely the larvæ of the Salamandridæ, especially that of *Amblystoma* (Axolotl); a fact that has given rise to the idea that they are not truly primitive aquatic Amphibia at all, but simply a species that, after possibly thousands of years of pædogenetic habit, has lost its plasticity and is no longer able to metamorphose from the larval to the adult condition. It would be interesting to try thyroid feeding experiments upon the larvæ with the idea of inducing metamorphosis into a true adult type.

Proteus anguineus (Fig. 113, B), "olms" as the Germans call them, are blind cave mud-puppies. The whole body is white or nearly so. If kept in places where light is not absolutely excluded they become at first grayish and later jet-black. The eyes are rudimentary and completely hidden beneath the opaque skin. It has been suggested that the conditions for development are such as to inhibit certain

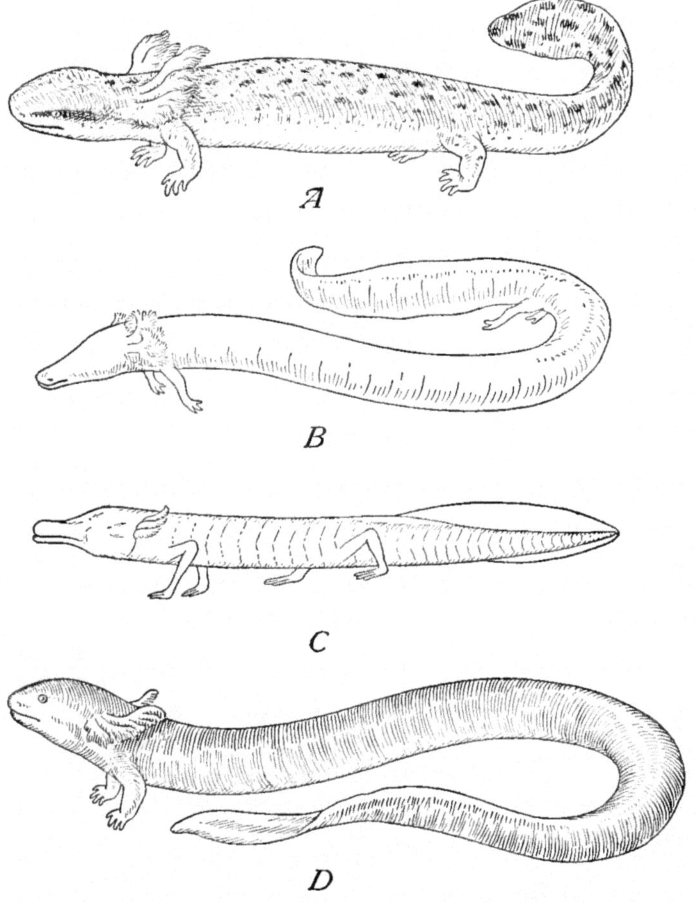

FIG. 113.—Group of perennibranchiate urodeles. A, *Necturus maculatus;* B, *Proteus anguineus;* C, *Typhlomolge rathbuni;* D, *Siren lacertina.* (Redrawn after Lydekker and others.)

structures such as the eyes and pigment from developing. Possibly this might also be tied up with the permanently larval condition. They require very little nourishment, probably being accustomed to only a minimum amount in their natural habitat.

Typhlomolge rathbuni (Fig. 113, C) a form very much like *Proteus*, is a native of Texas. It inhabits subterranean caves and is sometimes brought to the surface in water from artesian wells. It is likely that both of these cave mud-puppies arose independently, in response to similar conditions, from some form like *Necturus*.

Family 4. Sirenidæ. (The Sirens.)—They have three pairs of permanent fringed external gills; the body is eel-like, and there are no hind limbs; maxillaries are absent; no teeth are present, except some small ones on the vomer; jaws are furnished with a horny sheath like that of frog larvæ; there are no eye-lids. There are two genera, each represented by a single species.

Siren lacertina (Fig. 113, D), the "mud-eel," has three pairs of gill-clefts and external gills in the adult. It reaches a length of two and a half feet. The tail is strongly compressed and with well-developed fins. The color is blackish above and lighter below. The animal lives in the mud at the bottom of ponds. It is found in the southeastern parts of the United States.

Pseudobranchus striatus is very much like Siren but is smaller, seldom exceeding seven inches in length. It has only one pair of gill-clefts and only three fingers. A broad yellow band along the side relieves the somber coloration.

The Sirenidæ are considered the most degraded of urodeles. That they are not truly primitive is borne out by the observation of Cope that the young lose the external gills which then redevelop in the adult; so the adult condition is a renewed or secondary larval condition. This and other observations tend to confirm the theory of pædogenesis as applied to both Serenidæ and Proteidæ.

Which of the Anura are to be considered the most primitive, the nearest approach to the first ancestral Amphibia? It seems certain that the perennibranchiate forms are not ancestral but merely retain, or return to, a larval condition. Probably *Cryptobranchus* represents a condition more nearly primitive or ancestral than any other living amphibian.

Pædogenesis or Neoteny

There is perhaps no better group of vertebrates for illustrating the phenomenon of pædogenesis or the retention of larval structures during sexual maturity. It is a common phenomenon in a number of groups of invertebrates and especially so among the most highly specialized orders. In the Diptera, for example, there are several species in which young larvæ or pupæ produce young. The genus *Miastor* is a classic case. Here the young while still within the mother become sexually mature.

The prolongation of larval life was noted for the Ammocœtes larva of the lamprey, which requires from three to four years to reach the time of metamorphosis. The larva is much like Amphioxus, and if pædogenesis should occur in this species we would have a case quite parallel, I believe, to that seen in the perennibranchiate urodeles. A permanent Ammocœtes would be classed as an extremely primitive chordate not very distantly related to Amphioxus; in fact, before Ammocœtes was discovered to be the larva of *Petromyzon*, it was so considered. May it not be possible that Amphioxus itself is a permanent larva of some cyclostome more primitive than the myxinoids or the petromyzonts? Possibly it is, but such a view would seem to discredit the Amphioxus theory of the ancestry of the vertebrates, a view dear to our hearts and one that we should be reluctant to abandon.

The causes of pædogenesis are obscure, but we are at least justified in attributing the foreshortening of development of the soma to certain growth-inhibiting agents or to the absence of certain stimuli to metamorphosis. In general it may be said that low temperatures retard development and produce defective organisms; and the perennibranchiate Amphibia live in water that is low in temperature. Possibly these animals are merely senescent and have lost so much growth momentum that they are unable to push through to complete adult differentiation.

Certain experiments on anuran larvæ seem to throw some light on this matter. It has been shown by several writers, in the case of certain species of frog which have a prolonged larval period, that prompt metamorphosis may be induced by thyroid feeding. It is also possible to induce precocious metamorphosis in other species by similar methods. This suggests that the underlying cause of pædogenesis may have something to do with the failure of the thyroid to function or to a deficiency in its secretion. The recent experiments of B. W. Allen and his pupils are of interest in this connection. It was shown that the early extirpation of the hypophysis in tadpoles prevents the development of the thyroid gland and that the operated individuals remain permanent larvæ.

Order III. Anura (Tailless Amphibia)

The frogs and toads are the characteristic Amphibia of the present age. They are represented by about 900 species and exhibit a very pronounced adaptive radiation. They are the most highly specialized of modern Amphibia and so much specialization exists within the order that there is difficulty in listing characters that apply to all of its members. Not only has there been adaptive radiation in the order, but all of the large families exhibit a radiation into terrestrial, arboreal, aquatic, and burrowing types. Since the frog is a favorite type vertebrate and is used in nearly all elementary courses in zoölogy, it will save time and space to omit any detailed anatomical description of a type form. We shall therefore proceed to give an abbreviated systematic survey of the various groups of Anura. Two sub-orders are distinguished: the *Aglossa,* which are without a tongue and have eustachian tubes united in one pharyngeal opening; and the *Phaneroglossa,* in which a tongue is present. The first is a coherent natural group, but the second may be an artificial assemblage.

Sub-Order I. Aglossa

This is a small group of frogs not well known to the layman. Three genera, *Pipa, Xenopus,* and *Hymenochirus* occur. Of these *Pipa* is a South American tropical form and the other two belong to Africa.

Pipa americana (the Surinam toad) is a classic object to the zoölogist on account of its unique breeding habits (Fig. 114, A). The creature is an odd, ugly aquatic toad, with exceedingly large hind feet and a very short, broad head. The following description of its spawning is described by Bartlett:—"About the 28th of April the males became very active and were constantly heard uttering their most remarkable metallic call-notes. On examination we then observed two of the males clasping tightly around the lower part of the bodies of the females, the hind parts of the males extending beyond those of the females. On the following morning the keeper arrived in time to witness the mode in which the eggs were deposited. The oviduct of the female protruded from the body more than an inch in length, and the bladder-like protrusion being retroverted, passed under the belly of the male on to her own back. The male appeared to press tightly upon the protruded bag and to squeeze it from side to side, apparently pressing the eggs forward one by one on to the back of the female. By this movement the eggs were spread with nearly uniform smoothness over the whole surface of the back of the female to which they became

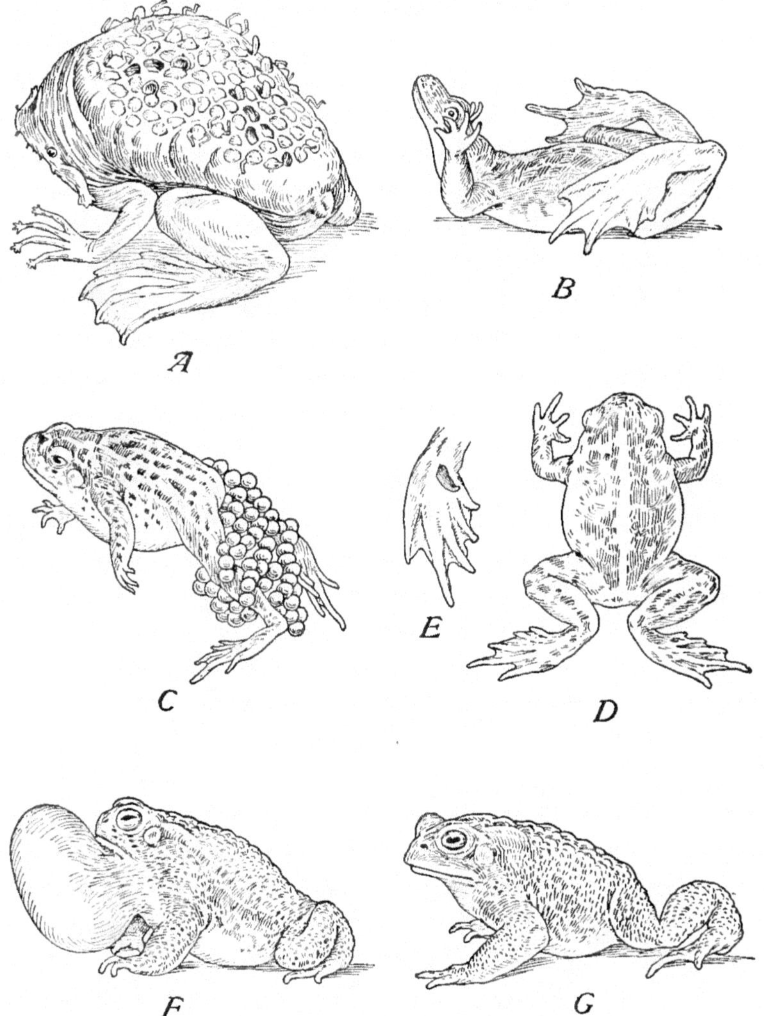

FIG. 114.—Frogs and Toads (Anura) I. A, Surinam Toad, *Pipa americana;* B, Fire-bellied toad, *Bombinator igneus;* C, Midwife Toad, *Alytes obstetricans;* D, Spade-foot Toad, *Pelobates cultripes;* E, foot of *Pelobates* showing tarsal spur; F, Common Toad, *Bufo lentiginosus s. americanus,* with vocal sac inflated. G, same stalking its prey. (A and C, redrawn after Lydekker; B, D, and E, redrawn after Gadow; F and G, redrawn after Dickerson.)

firmly adherent." The eggs then sink into pockets in the skin. Each pocket develops a sort of hinged lid, which the young toad pushes open from time to time, as in Fig. 114, A. The habits of the African Aglossa are less specialized and have no features of especial interest for the general student.

Sub-Order 2. Phaneroglossa (Tongued Anura)

There are seven families of these frogs and toads and only the most general distinguishing characters of these groups can be given. As in any other group that has undergone pronounced adaptive radiation, the chief points that interest the student are the special structural adaptations and peculiar habits. Under each family only the best known and most interesting species will be described.

Family 1. Discoglossidæ.—The tongue is disk-shaped and non-protrusible; the vertebræ opisthocœlous; the upper jaw and vomers have teeth; and the male has no vocal sac.

Bombinator igneus, the "*fire-bellied toad*," is a poisonous form with pronounced warning coloration and a special method of displaying it. The under surface is colored a purplish black with conspicuous orange-red patches. They are decidedly aquatic, floating at the surface with legs extended so that the conspicuous color is well displayed to all aquatic enemies. They also rest on land and when surprised there they make a strong effort to bring the under surface to view by turning the legs over the back and throwing back the head as in Fig. 114, B. Besides their coloration they are interesting because of the weird noises they make. The voice is described as like "hoonk, hoonk" or "ooh, ooh," and the males join in a mournful concert of sound during the breeding season.

Alytes obstetricans (Fig. 114, C), the "midwife-toad," is in general appearance quite ordinary. It occurs in France and Switzerland. The interesting feature of the species is the method of caring for the eggs by the males and the latter's odd habit of relieving the female of her eggs. The male assiduously massages the cloaca of the female with the paws. After a considerable time the female suddenly and with great apparent effort expels the eggs all in a bunch. The male then clings to the female's head and fecundates the eggs, after which he carries the bunch of eggs off with him, attached to the hind legs, to a hole in the ground. He moistens the eggs with dew and occasionally takes them into the water with him. When the eggs are nearly ready to hatch he betakes himself to the water for the period of hatching.

Family 2. Pelobatidæ.—The tongue is oval, hitched in front but free behind, so that it can be thrown out; upper jaw and vomers with teeth; vertebræ procœlous. There are seven genera consisting of about 20 species. *Pelobates cultripes* (Fig. 114, D), the "spade-foot toad," is the best known of the family. They are typically burrowing toads, digging rather deep holes in sand and resting there during the day. The "spade" is a modified hind foot, which has a strong spur (Fig. 114, E) on its under surface which aids in digging. The species are nocturnal in feeding habits.

Family 3. Bufonidæ. The Common Toads.—They have no teeth in upper or lower jaws; vertebræ procœlous and without ribs. They are for the most part decidedly terrestrial, some of them occupying arid territory.

Bufo vulgaris (Fig. 114, F and G), the common toad of North America and other palearctic regions, is typical. The color is variable and changeable, highly protective in its resemblance to the background. They are nocturnal in habits, feeding on worms, insects, and snails. One sees them frequently under electric lights waiting for dazed insects to drop to the ground. They hop quickly to the fallen insect and snap it up suddenly. Earthworms are crushed and squeezed till comparatively quiet before swallowing occurs. In the daytime toads hide under stones or in dark corners. They breed in temporary pools in the early spring. The newly hatched toads are surprisingly small and require nearly five years to reach maturity.

Toads are almost without enemies on account of their noxious skin secretions. About the only agency in keeping down their numbers appears to be parasites and epidemics of disease.

Family 4. Hylidæ. (Tree-Frogs or Tree-Toads.)—Upper jaw and vomers with teeth, lower jaw also toothed in one species *Amphignathodon;* vertebræ procœlous without ribs; fingers armed with adhesive pads; tongue protrusible to varying degrees. They are all climbing arboreal frogs, many, but not all, being green in color. They are very widely distributed and have in all about 150 species; hence this is one of the largest families. The genus *Hyla* is the most generalized and wide-spread genus. *H. versicola* (Fig. 115, A) is the common tree-toad of the Northern United States. It emits a "clear, loud, thrilled rattle" quite familiar to most naturalists. These tree-toads

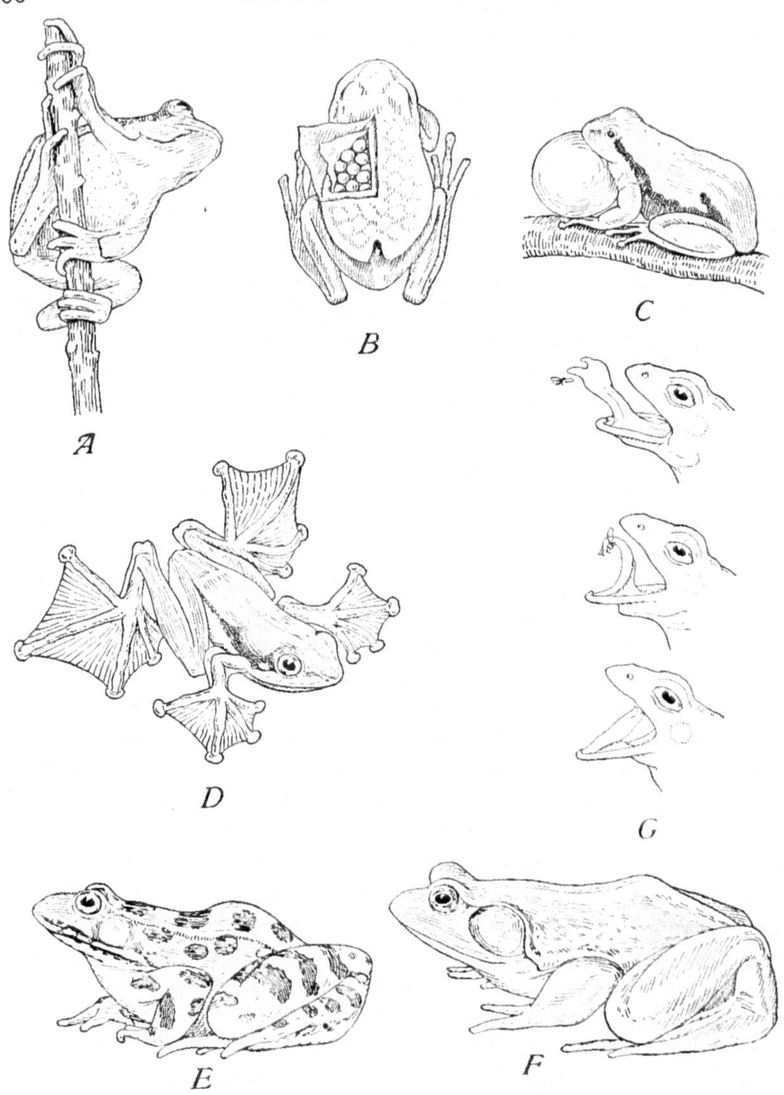

Fig. 115.—Frogs and Toads, (Anura) II. A, Tree-Toad, *Hyla versicola;* B, *Nototrema marsupium* with brood-pouch laid back to show inclosed eggs; C, *Hyla arborea*, with vocal sac expanded; D, Javan Flying Frog, *Rhacophorus pardalis;* E, Leopard Frog, *Rana pipiens;* F, Bull-Frog, *Rana catesbiana*. G, *Rana esculenta*, showing the movement of the tongue in capturing a fly. (A, E, and F, redrawn after Dickerson; B and G, after Gadow; C and D, redrawn after Leydecker.)

change color quickly from dark brown to a delicate gray. In the daytime they hide quietly in sheltered crevices of bark or the crotches of limbs, but at night they become lively and noisy, jumping about and busily catching insects. They breed in shallow pools in May, passing through a regular tadpole stage, and metamorphose into small perfect frogs in about seven weeks.

Hyla faber, a native of Brazil, is one of the most interesting of the tree-frogs, on account of its remarkable voice and extraordinary nest-building habits. Its voice is said to resemble that of a mallet beaten against a copper plate. When caught it utters a cry like that of a wounded cat. It makes an aquarium-like nest for its eggs and young, by digging a basin of some depth in the bottom of shallow pools and building a mud wall about it. The whole inside of the basin is most carefully smoothed off by rubbing the belly over it. The male takes no part in this building operation, but raises an unearthly racket all the while.

FIG. 116.—*Hyla gœldii* x1. Female with eggs in incipient dorsal brood-pouch. (From Gadow.)

Hyla gœldii (Fig. 116) is remarkable on account of the fact that the female carries the eggs on the back in a shallow depression till they are almost ready for metamorphosis.

The genus *Nototrema* (Fig. 115, B) differs from *Hyla* in that the female has an egg pouch or marsupium on the back, which is merely a fold of the skin. It has been suggested that the marsupium may be a specialization of the simple pocket seen in *Hyla gœldii*.

The tiny *Acris gryllus* (Fig. 115, C), or "Cricket Frog" of eastern and central United States, is one of our smallest frogs. It is described as a merry little frog, chirping constantly even in captivity. It frequents the borders of pools, jumping into the water if disturbed and quickly burying itself in the mud at the bottom.

Family 5. Cystignathidæ is another of the large anuran families, having like the Hylidæ about 150 species. They play about the same rôle in the southern continents (Notogæa) as the Ranidæ do in the northern continents (Arctogæa). Some of them are difficult to distinguish from the Ranidæ. The whole family is ill-defined, being specialized in a great many directions. They sometimes have adhesive pads on the toes. Only one genus need be mentioned.

The genus *Hylodes*, of which there are nearly 50 species, is one of the best known. They are tree-toads much like those of the genus *Hyla*. *Hylodes martinicensis* is a tiny frog in which the pairing and egg laying take place on land, the eggs in a foamy mass being glued to a leaf. The large eggs, about 4–5 mm. in diameter, develop practically to metamorphosis before hatching, the aquatic larval period being omitted (Fig. 117).

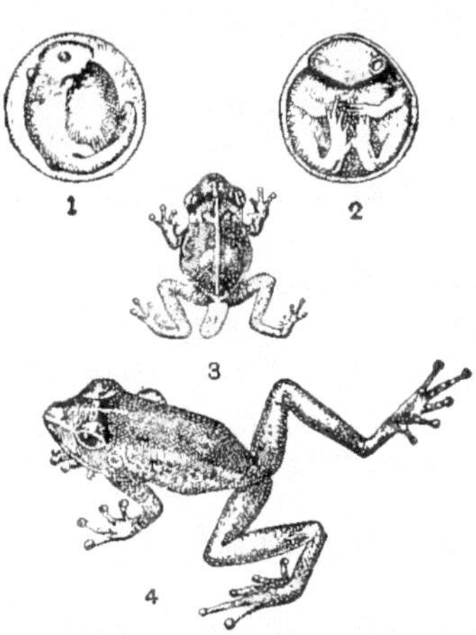

Fig. 117.—*Hylodes martinicensis*. *1*, an egg with embryo about seven days old; *2*, another, twelve days old; *3*, the young frog just hatched; all by ¾; *4*, adult male x1. (From Gadow, after Peters.)

Family 6. Engystomatidæ (*Narrow-Mouthed Toads*).—They are sometimes called "toothless toads." A representative of this family in the United States is *Engystoma carolinense*, a family common in the old South. They are sharply distinguished by the dilation of the sacral diapophyses. Perhaps the most significant feature of the family has to do with their *ant-eating habits* and adaptations for it. It is well known that the ant-eating habit in various groups of vertebrates is associated with rather definite changes in structure. These frogs have a narrow mouth, protruding snout, toothless condition, hidden tympanum, modified feet and shoulder girdle for digging; these are also characters of ant-eaters in other vertebrate classes such as Reptilia and Mammalia.

Family Ranidæ.—These are the "true frogs" and, to dwellers of the northern hemisphere, the most familiar of amphibians. Two small families, one consisting of one species, the other of about a dozen species, differ from the Raninæ (the typical frogs) chiefly in the teeth.

Sub-Family 1. Ceratobrachinæ, with teeth in upper and lower jaws. *Ceratobrachus guentheri* is a native of the Solomon Islands.

Sub-Family 2. Dendrobatinæ.—Arboreal frogs of small size without any teeth. The toes have adhesive disks. Members of the genus

Dendrobatinus carry their tadpoles on the back while going from a pool that is drying up to one with plenty of water.

Sub-Family 3. Raninæ, comprises the typical frogs, to which our common bull-frog, leopard frog (Fig. 115, E), grass-frog, etc., belong. These common frogs are very well known to every student of zoölogy and will not be dealt with here. The Raninæ are remarkable for the range of their adaptive radiation. They range from purely aquatic frogs like *Rana catesbiana* (Fig. 115, F), the "bull-frog," to terrestrial frogs like *R. temporaria*, the European brown frog; from those living on the ground in woods to those living in trees, and even to fairly good flying frogs. These so-called "flying frogs" have fully webbed large feet (Fig. 115, D) with which they parachute to the ground or from tree to tree. There is, however, considerable doubt as to their "flying" ability. Their air leaps cannot be very great, possibly from 20 to 30 feet being the maximum. Wallace's exaggerated account of these activities has been too widely accepted.

The frog's usual method of capturing insect prey is shown in Fig. 115, G. Any statement as to special breeding habits would be largely a repetition of what has been said about other groups.

The Development of the Frog

The early embryology of the Amphibia is in general the most generalized found among the vertebrate classes, and that of our commonest frogs is as primitive as can be found. Why the development of the Amphibia is more primitive than that of the fishes is not an easy question to answer. It appears probable, however, that the earliest fishes, such as the lobe-fin ganoids, had a type of egg and a process of development even more like that of the Amphibia than have the modern fishes, and that the amphibian descendants of these ancestral fishes have retained more nearly than the fish descendants the primitive features of development. A study of comparative embryology of chordates usually begins with the development of Amphioxus and then proceeds directly to that of the frog. Then follows the development of the chick, as an example of conditions in the Sauropsida, and finally that of a eutherian mammal.

The life history of the frog may conveniently be divided into four periods:—

1. The period of germ-cell formation, which terminates with spawning.

2. The period of embryonic development, which begins with fertilization and ends with hatching.
3. The larval period, which extends from hatching to the completion of the process of metamorphosis.
4. The period of adolescence, extending from the end of metamorphosis to sexual maturity.

1. The **period of germ cell formation** involves both ovogenesis and spermatogenesis, together with the processes of maturation. These stages are quite regular and require no special comment. The eggs are laid in a string, attached to one another by means of a continuous gelatinous envelope, which is at first dense and viscous, but soon absorbs sufficient water to cause it to swell to several times its original thickness. This jelly, which is laid down in two layers, has the double function of conserving heat for incubation purposes and of preventing the attacks of bacteria.

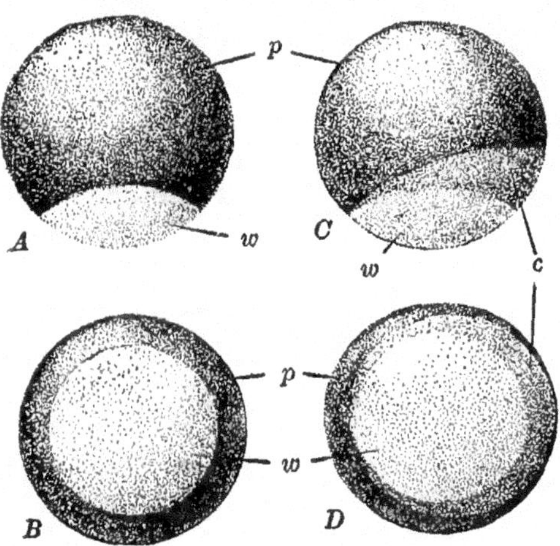

Fig. 118.—Frog's egg before and after fertilization, showing symmetry relations. A, unfertilized egg, from side; *B*, Unfertilized egg, from vegetal pole; *C*, Fertilized egg just before cleavage, from side; D, same from vegetal pole. *C*, gray crescent; *p*, pigmented animal pole; *w*, unpigmented vegetal pole. (From Kellicott.)

The fertilized egg (Fig. 118) is rather highly organized before cleavage begins, for the various axes of the future embryo (antero-posterior axis, dorso-ventral axis and the axis of bilaterality) are already clearly defined. These relations can be made out readily from the pigment pattern of the peripheral parts of the egg. The upper hemisphere of the egg is covered with black pigment, which is like an obliquely placed cap. A gray crescent, thick at one side and fading out on the other, separates the pigmented area from the pale yellow area at the vegetal pole. Only one line can be drawn around the egg so as to divide it into bilaterally equal halves; this represents the primary axis of the embryo. The yolk is abundant, and only a small region at the apical pole is free from yolk granules. Maturation of the egg occurs partly before laying, one polar body being given off during the descent of the egg in the oviduct. The second maturation division occurs after insemination.

2. **The Embryonic Period** (Fig. 119).—Fertilization occurs while the eggs are being laid, the spear-shaped spermatozoön penetrating the gelatinous envelope and forcing a path through the yolk to the egg nucleus. Cleavage is total or holoblastic in spite of the large accumulation of yolk. The first and second cleavage furrows being meridional and the third unequally equatorial, cutting off four micromeres from the apical and four macromeres from the vegetal pole of the egg. The micromeres cleave much more rapidly than the yolk-laden macromeres, resulting in the formation of a rather thick-walled but fairly typical *blastula*, with numerous small pigmented cells above and comparatively few large unpigmented cells below. The hollow of the blastula, or segmentation cavity, is much reduced in volume because of the thickness of the cells at the vegetal pole. Gastrulation, while not so simple as in Amphioxus, is clearly homologous with the latter. The departure from the diagrammatic condition is due to the accumulation of yolk, which prevents the typical embolic invagination of the very thick layer of vegetal pole cells. The difficulty is evaded by having the invagination take place at the edge of the thickened area, where a flat infolding of surface cells takes place just below the edge of the pigmented area, leaving a crescentic *blastopore* on the surface. The main part of the gastrulation process is accomplished by the overgrowth of the endoderm cells by the ectodermal cap, a process known as *epibolic gastrulation.* The *archenteron,* at first flat and without a lumen, soon expands and largely displaces the segmentation cavity. The *gastrula* is morphologically a two-layered embryo, with a layer of ectoderm on the outside and a layer of endoderm within, though in the frog each of these layers is more than a single cell layer in thickness. Mesoderm formation is accomplished by the ingrowth of a sheet of cells around the blastopore. This zone-like layer soon splits into two layers, an outer *somatic* and an inner *splanchnic* layer, with the primitive cœlomic cavity between. Cœlomic pouches not unlike those in Amphioxus arise from the dorsal lateral regions of the archenteron, and a median dorsal strip of the latter is left over to form the notochord. The development of the central nervous system is decidedly precocious, for even in a late gastrula stage the *medul-*

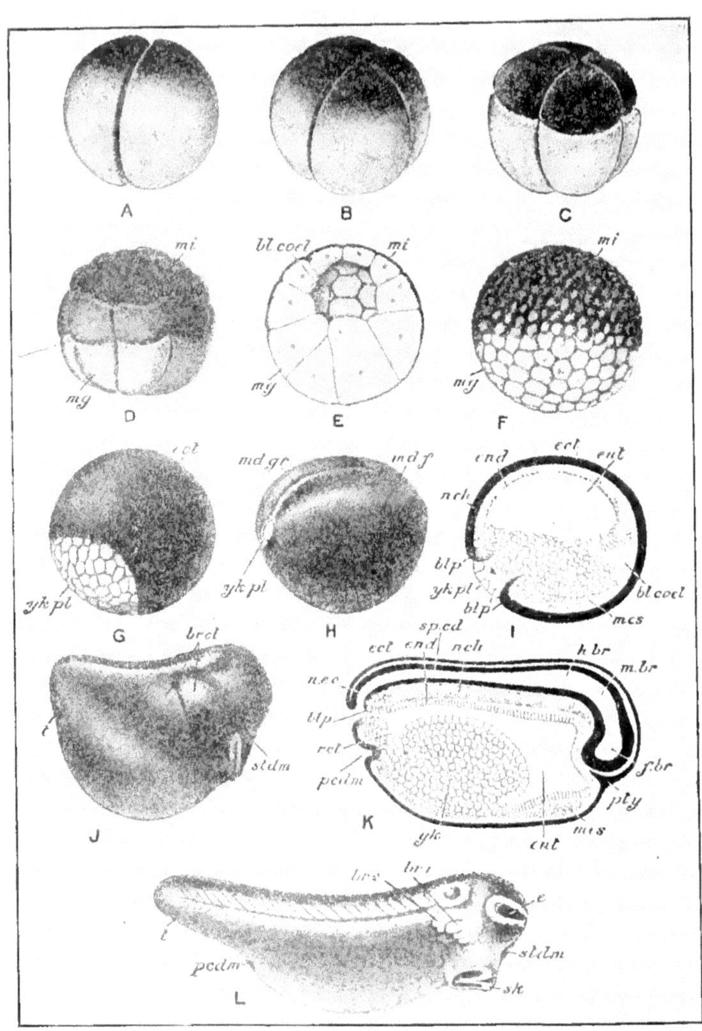

Fig. 119.—Development of the Frog. A–F, cleavage; G, overgrowth of ectoderm; H, I, establishment of germ-layers; J, K, assumption of tadpole-form and establishment of nervous system, notochord and enteric canal (archenteron); L, newly-hatched tadpole. *bl. coel*, blastocoele; *blp. blp'*, blastopore; *br'. br''*, cutaneous gills; *br. cl*, branchial arches; *e*, eye; *ect*, ectoderm; *end*, endoderm; *ent*, enteron; *f. br*, fore-brain; *h. br*, hind-brain; *m. br*, mid-brain; *md. f*, medullary fold; *md. gr*, medullary groove; *mes*, mesoderm; *mg*, megameres; *mi*, micromeres; *nch*, notochord; *n. e. c*, neurenteric canal; *pcdm*, proctodæum; *pty*, pituitary invagination; *rct*, commencement of rectum; *sk*, sucker; *sp. cd*, spinal cord; *stdm*, stomodæum; *t*, tail; *yk*, yolk cells; *yk. pl*, yolk plug. (From Parker and Haswell, after Ziegler and Marshall.)

lary plate is clearly defined. At a time when the blastopore is nearly closed the dorsal parts of the embryo show the broad primitive groove, flanked on both sides by two pairs of medullary folds, inner and outer. The outer folds fade away, but the inner ones arch over the groove and meet in the region of the future neck, the closure proceeding thence backwards. Thus the groove is converted into the *neural tube.* The anterior part of the tube soon becomes differentiated into the primitive brain, with the three primary brain lobes representing the primordia of the fore, mid, and hind brain. During these changes the embryo has been elongating and before hatching has reached a length nearly three times its breadth.

3. **Larval Period** (Fig. 120, 1–10).—At the time of hatching the larva is a somewhat fish-like creature with a fairly long vertically flattened tail. The mouth is ventral in position and is soon surrounded by a chitinous rim or scraper, which is used as a larval organ in scraping off nutritive scum from lily pads, etc. Two pairs of branching *external* (larval) *gills* are the first functional respiratory organs. In addition to the *external gills, internal gills,* homologous with those of adult fishes, are formed and take over the respiratory function for a considerable period. Soon the external gills disappear, and a fold of skin grows backward from in front of their original location, forming an *operculum* under which lie the internal gills. The operculum has but one outlet, a small unpaired *spiracle* on the left side. Some writers have interpreted this operculum as the equivalent of the atrium of Amphioxus, but the homology has not been fully established. The hind limbs are the first to appear, closely followed by the fore limbs, which for some time are concealed beneath the operculum. Only in the later stages of larval life are the lungs developed, and as long as the larva uses the gills the lungs remain very small.

Metamorphosis (Fig. 120, 11–15).—The period of metamorphosis is really a part of the larval period and cannot be sharply marked off from the latter, since the change is a gradual one. Toward the close of the larval period the tail begins to be resorbed and its materials are stored up in the liver. The long, spirally coiled intestine shortens. The mouth loses its chitinous rim and grows much wider. The gills disappear and the lungs grow rapidly in size and the larva comes frequently to the surface to breathe air. When these changes are complete the animal is no longer a larva, but an adolescent frog.

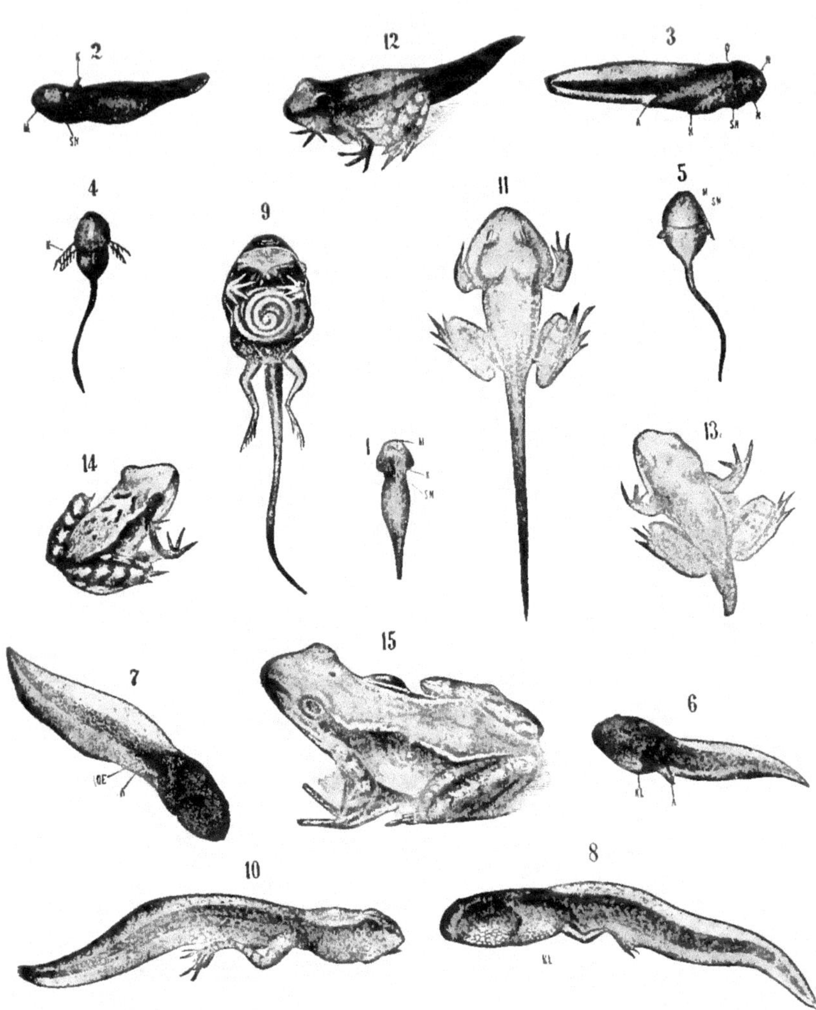

FIG. 120.—Metamorphosis of the Frog. *1*, tadpole just hatched, dorsal aspect; *2–3*, older tadpoles, side view; *4*, later tadpole, dorsal view showing cutaneous gills at maximum development; *5*, later stage, ventral view, showing degeneration of cutaneous gills and development of operculum; *6*, older tadpole, left side, showing single opening of operculum; *7*, older stage, right side, showing hind leg and anus; *8* and *10*, lateral views of two later stages showing development of hind legs; *9*, dissection of tadpole to show internal gills, spiral intestine, and anterior legs developed within operculum; *11*, advanced tadpole just before metamorphosis; *12*, *13*, *14*, stages in metamorphosis, showing gradual resorption of tail; *15*, juvenile frog after metamorphosis. (From Weysse, after Leuckart-Nitsche wall chart.)

Some species of frog go through to metamorphosis in a few weeks, others require months, and some require two or three years.

4. **The Period of Adolescence** has not been very fully studied. It is a long, slow process involving changes in the relative proportions of the parts, elaborations of the histological structure, and ossification of the cartilaginous skeleton. The most significant changes are those that are last to take place, namely, those that have to do with the onset of sexual maturity. Shortly before the beginning of the first breeding season, the cells of the ovaries and testes begin the processes known as *ovogenesis* and *spermatogenesis*, that constitute the chief features of the first period considered in this brief life history. Hence we have completed the cycle for one generation.

CHAPTER VII

CLASS IV. REPTILIA

It is generally conceded that of all the vertebrate classes the Reptilia, past and present, stand foremost in numbers, in size, in range of specialization and in dominance in the organic world. Although the reptiles of the present (crocodiles, turtles, lizards and snakes) play a comparatively unimportant rôle in the realm of nature, those of the past were frequently of giant proportions and were the dreaded tyrants of the earth, of the waters, and to some extent of the air. The golden age of the reptiles was the Mesozoic, which has therefore been called the "Age of Reptiles." After a modest career during the early period of the Mesozoic, several specialized groups arose and underwent a remarkable adaptive radiation into all of the principal life zones. Perhaps the most remarkable of these specialized assemblages was that of the dinosaurs, a group that for millions of years held sway over the earth to an extent equaled only by that of Man to-day. During this period other orders of reptiles played only a secondary rôle.

Dramatic as was this rise to dominance of the greater reptilian orders during the Mesozoic, their sudden extinction at or near the close of this age was even more remarkable. After an unprecedented reign as autocrats of earth and sky and sea for a period of not less than ten millions of years, they abruptly ceased to be. The causes of their extinction are unknown and we can only vaguely conjecture that they died off for no better reason than that they had run their course, had reached the limits of their various lines of specialization, had become stereotyped, senescent, and could evolve no further. To use an idea of Osborn's, they had proceeded to the end of an evolutionary cul-de-sac from which there was no egress.

Only the crocodiles, turtles, lizards and snakes, among reptiles, were sufficiently generalized to weather the crisis and live on into the Cenozoic age to be the contemporaries of the birds and the mammals, which are the dominant orders of that period. These modern groups have evidently carried on a war of destruction against the reptiles, and still the unequal struggle goes on, with the reptiles on the losing side. The reptiles are doomed. Between the birds of the air and the beasts of the field and forest, and especially at the hand of that super-mammal, Man, who seems to have centered his aversion upon the serpents, the reptiles are destined to oblivion, except in so far as Man sees fit to preserve certain types for his own uses. No mere verbal

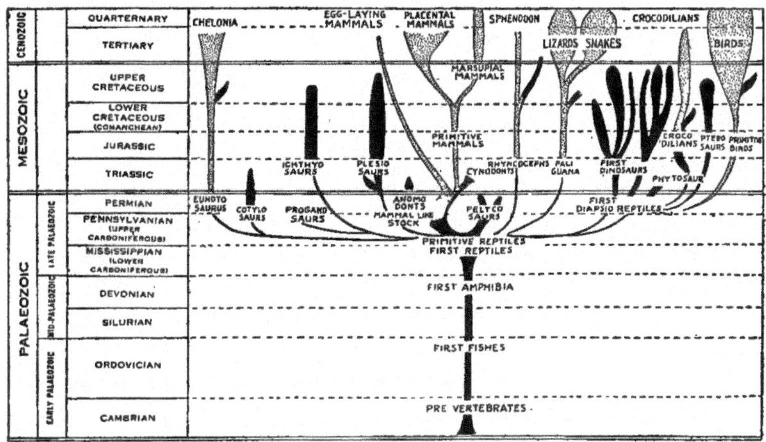

Fig. 121.—Chart, showing origin and adaptive radiation of the reptiles. Dotted areas represent existing groups, black areas, extinct groups. This chart also shows the origin of the birds and mammals from reptilian stock. In the cases of several modern groups (Chelonia, egg-laying mammals, placental mammals, Sphenodon and Crocodiles) the dotted areas should reach the top. (After Osborn's "Origin and Evolution of Life.")

description shows so vividly the origin and adaptive radiation of reptiles as does the accompanying chart (Fig. 121).

From the dawn of the reptiles during the Palæozoic up to the present there have passed between fifteen and twenty millions of years, an immense period as compared with the brief span of Man's life upon the globe. The history of the rise and fall of the reptilian empire is one of giant proportions and of intense dramatic interest. Only the vertebrate palæontologist is in a position adequately to picture this drama for us.

Evolutionary Advances Made by the Reptiles

"The environment of the ancestor of all the reptiles," says Osborn, "was a warm, terrestrial, and semi-arid region, favorable to a sensitive nervous system, alert motions, scaly armature, slender limbs, a vibratile tail, and the capture of food both by sharply pointed, recurved teeth and by the claws of a five-fingered hand and foot."

The essential evolutionary advance which the reptiles made upon the Amphibia had to do with the total abandonment, from the very beginning of development, of the aquatic habit of life. The Amphibia were, and still are, for the most part, dependent upon water during at least a considerable portion of their life cycle; the Reptilia were from the first quite independent of an aquatic environment. This emancipation from the need of an aquatic habitat was accomplished in three ways: by the acquisition of lungs of a more adequate sort; by a scaly covering to prevent dessication; by the development of important embryonic membranes, the *amnion* and the *allantois*, which are merely adaptations for embryonic development in the air.

Undoubtedly the Amphibia had made considerable progress in the direction of adaptations for adult life on land, for they often have as good lungs as do some of the reptiles. Moreover, many of the extinct Amphibia had an adequate scaly covering. There is no evidence, however, that any of the Amphibia have ever acquired the capacity of reproducing on the dry land, except in connection with some peculiar brooding adaptation, such as that seen in some of the tree-toads. The fact that all modern Amphibia have functional gills at some period adds force to this last statement.

The reptiles have entirely given up gill respiration, as is evidenced by the total absence, except for minute transitory traces, of gill tissue at any stage of development. In place of gills the embryo uses the *allantois*, an extensive embryonic lung. To avoid dessication and the possibility of contact injuries, the embryo is surrounded by a fluid-filled sac, made from folds of its own tissues. This veritable private aquarium is called the *amnion*, a structure adopted by all land vertebrates for development in the air.

The **amnion** and the **allantois** then are structures of first rate importance in connection with the origin and evolution of land vertebrates, and especially of the Reptilia. The egg of the reptile is a large object, much like that of the bird. It is provided with a tough shell for protection and a thick layer of albumen for nutriment. The egg-yolk is abundant and serves as the main food supply of the growing embryo. There is no real larval stage, for the newly hatched young is essentially like the adult except in the relative proportions of head and body and in being sexually immature.

The **amnion** (Fig. 122), which is formed very early, results from the fusion together of two lateral folds of the extra-embryonic blastoderm and is a sort of bladder-like membrane containing a watery fluid in which the growing embryo lies. The fluid within the amnion increases greatly in amount and provides an ample space for the further growth of the embryo, and preserves the latter from contacts, mechanical abrasions, or other mechanical injuries.

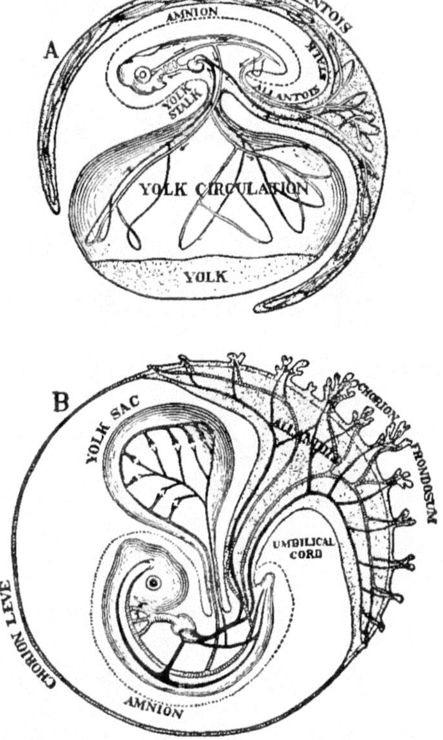

FIG. 122.—Diagrams of the embryonic membranes, amnion, allantois, yolk-sac, of Amniotes. A, Sauropsida (reptiles and birds). B, mammal with primitive allantoic placenta. (From Lull, after Wilder.)

The **allantois** (Fig. 122) starts as a finger-like diverticulum of the embryonic hind-gut and grows out extensively between the amnion and the chorion as a large umbrella-shaped sac, which is highly vascular; and it functions as an embryonic lung. The amnion and the allantois together make development possible even in arid regions and give to the reptiles a decided advantage in available range, since they may live far from the water.

The Earliest Reptiles and Their Origin

"Just when the animals we call reptiles arose in geological history we do not know; certainly it was in early Pennsylvanian (Upper Carboniferous) times, probably Mississippian (Lower Carboniferous). That they arose from what we call the Amphibia, forms with temnospondylous vertebræ, is certain, though there is not much more reason for calling the ancestral stock Amphibia than Reptilia. I prefer to call it provisionally, Protopoda. It was ancestral to both and both classes have advanced since their divergence, the Amphibia some, the Reptilia much. Could we find, as some time we hope that we may, in mid-Mississippian or late Devonian times, a skeleton of those ancestral creatures, we should perhaps not call it by the name of any known order; it would be the old question over again of the difference between animals and plants. At present we know the Protopoda only by their footprints."

This statement of Williston is somewhat radical in that it places the Amphibia and the Reptilia on the same level, neither being ancestral to the other, but both derived from a common ancestor about which we know nothing except the characters revealed by its footprints. Some authors have assumed that this creature was an ancestral amphibian, a view that has gained wide acceptance. For our purposes it seems advisable to think of the creature that made the footprints (Fig. 101) as the earliest known land vertebrate, and to call it the ancestral amphibian.

Palæontologists generally agree that the reptiles go back nearly as far as do Amphibia, and that their evolution has been to a large extent parallel. Both experimented with adaptations for land life and the reptiles were much more successful in these ventures than were their rivals. During the Permian, however, they were neck-and-neck in the race; for every reptilian type of that period there was a parallel amphibian type. The reptiles finally outstripped the Amphibia, largely through their adoption of the amnion-allantois complex and their consequent emancipation from the water.

Permian Reptiles

The reptiles of the Permian were partly archetypal forms and partly precociously specialized and already senescent types. Williston finds four assemblages of Palæozoic reptiles (belonging to the American Permo-Carboniferous). Each of these assemblages has evidently the systematic value of a sub-class and has one or more modern representatives.

a. **Diapsida;** represented to-day by Sphenodon, crocodiles, birds and the great extinct dinosaurs, pterosaurs, parasuchians, etc.
b. **Synapsida;** represented to-day by the mammals, and by the extinct plesiosaurs, anomodonts, therapsidans and theromorphs, etc.

c. **Parapsida**; represented to-day by the lizards and snakes, and by the extinct mososaurs, ichthyosaurs, etc.

d. **Anapsida**; represented to-day by the turtles, and by the extinct cotylosaurs and the extinct semi-chelonian, Eunotosaurus.

Thus we see that the reptiles had undergone an extensive adaptive radiation even in the Palæozoic. Of the several adaptive types of

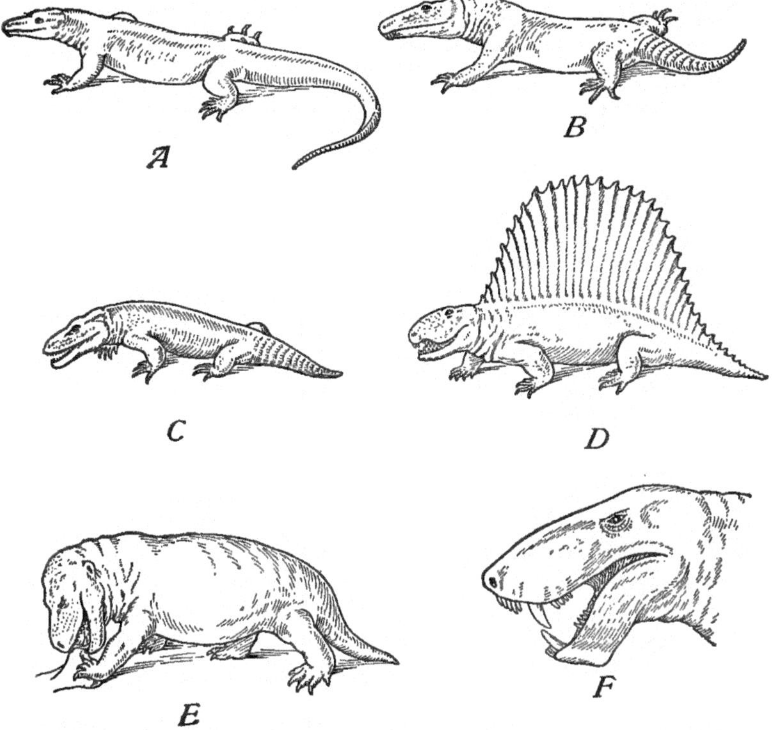

Fig. 123.—Group of Palæozoic Reptilia. A, *Varanops;* B, *Labidosaurus;* C, *Seymouria;* D, *Dimetrodon;* E, *Cynognathus* (a mammal-like reptile); F, head of *Scymnognathus* (a South-African "dog-toothed" reptile). (Redrawn from Osborn, after Williston and after Gregory.)

that period perhaps the most significant is that exemplified by the lizard-like reptile *Varanops* (Fig. 123, A), a creature so generalized that it might well be selected as an archetypal reptile. The fact that it has a skull and skeleton much like that of modern lizards has led Williston to the conclusion that some of our modern lizards are more primitive than the classic Sphenodon, long thought of as the most primitive of living reptiles. *Varanops* and its more slender relatives represent the quick-running or cursorial adaptation as it appeared within the sub-class Parapsida.

As examples of the secondarily semi-aquatic adaptive types we may cite several members of the sub-class Anaspida, such as *Labidosaurus* (Fig. 123, B), *Seymouria*, (Fig. 123, C), and *Diadectes*, three types that probably lived much as do our modern frogs and salamanders. As an example of the heavy-bodied, and heavily armored type we may cite two members of the sub-class Synapsida: *Edaphosaurus* and *Dimetrodon* (Fig. 123, D), reptiles strikingly characterized by a riotous growth of dorsal spines. These so-called pelycosaurs evidently represent an end product of a very early line of specialization and have left no descendants.

The Mammal-Like Reptiles (*Cynodonts*).—Another remarkable group of Permian reptiles which appears to have been purely African in distribution was a group of mammal-like reptiles, called Cynodontia, believed by the authorities to have given rise to the line from which the mammals arose. These cynodonts (dog-toothed reptiles) showed many tendencies toward mammalian conditions, chief among which were: heterodont dentition (a specialization of the teeth into incisors, canines and molars), more effective types of limbs for rapid land locomotion, a tendency for the angulare and articulare bones of the lower jaw to disappear, and a tendency for the skull to become completely roofed over and for the so-called vacuities to disappear. These cynodonts were evidently carnivorous types of which *Cynognathus* (Fig. 123, E) is a good example. The head of another cynodont, *Scymnognathus* (Fig. 123, F) shows clearly the dog-like dentition. These reptiles are once more to claim our attention when we come to discuss the question of the origin of mammals.

There is reason to believe that at least five or six other reptilian orders had representatives in the Permian or Permo-Carboniferous: Chelonia (turtles), plesiosaurs (aquatic reptiles), ichthyosaurs (fish-like reptiles), Squamata (primitive lizards), Rhynchocephalia (beaked reptiles), and Parasuchia (primitive crocodiles). Possibly also the great order of dinosaurs had its beginnings in the Permo-Carboniferous, though as yet there is no direct evidence of their presence during this period.

A number of orders of reptiles not only had their origin during the

Palæozoic, but actually ran out their entire course of specialization and became entirely extinct before the Mesozoic age began. Thus the cotylosaurs, proganosaurs, anomodonts, pelycosaurs, and phytosaurs died out either in Permian or at least not later than early Triassic times. The remaining orders that arose in the Palæozoic were able to weather the climatic crisis at the end of this age and were the ancestors of the great Mesozoic orders of reptiles.

The Golden Age of Reptiles

The reptiles, as we have seen, made a modest start in the Carboniferous, underwent a considerable degree of adaptive specialization during the Permian, and in some lines became senescent and died out. On the whole, however, the Palæozoic reptiles were of generalized or primitive types and gained no great ascendency. It was not until the Mesozoic that the reptiles really came into their own. It was during this "Age of Reptiles," an immense period involving several millions of years, that they gained their world supremacy and came to exercise undisputed sway over the land habitats, and disputed with the fishes the right to rule the waters. The dominance of the reptiles of this period was due largely to five great groups: ichthyosaurs, plesiosaurs, carnivorous and herbivorous dinosaurs, and pterosaurs. Each of these assemblages deserves individual attention.

ICHTHYOSAURIA

No more extreme case of adaptation of a member of an essentially terrestrial class for an aquatic habitat could be given. The first reptiles are believed to have acquired their main characters in adaptation to land life, so we have no alternative than to believe that the ichthyosaurs have been derived from land forms that found the sea a rich hunting ground and developed the habiliments of a fish to facilitate their aquatic activities. A change involving, first, adaptations for land life and, second, a return of aquatic adaptations is referred to as an example of reversed aquatic adaptation, and is by no means uncommon among the higher vertebrates. The external form of an ichthyosaur (Fig. 124, C) is strikingly like that of a sword-fish. The pectoral and pelvic limbs are flipper-like fins, the tail has a remarkable caudal fin externally precisely like that of a fish, a very fish-like dorsal fin plays the same rôle as that in a fish. Unlike other reptiles and like the fishes, these creatures have no real neck, but the head seems to be joined broadly with the trunk. All of these adaptations are utterly fish-like, but the fundamental internal anatomy of the creature is essentially reptilian. It seems probable that the ichthyosaurs were viviparous, as are some purely aquatic reptiles of

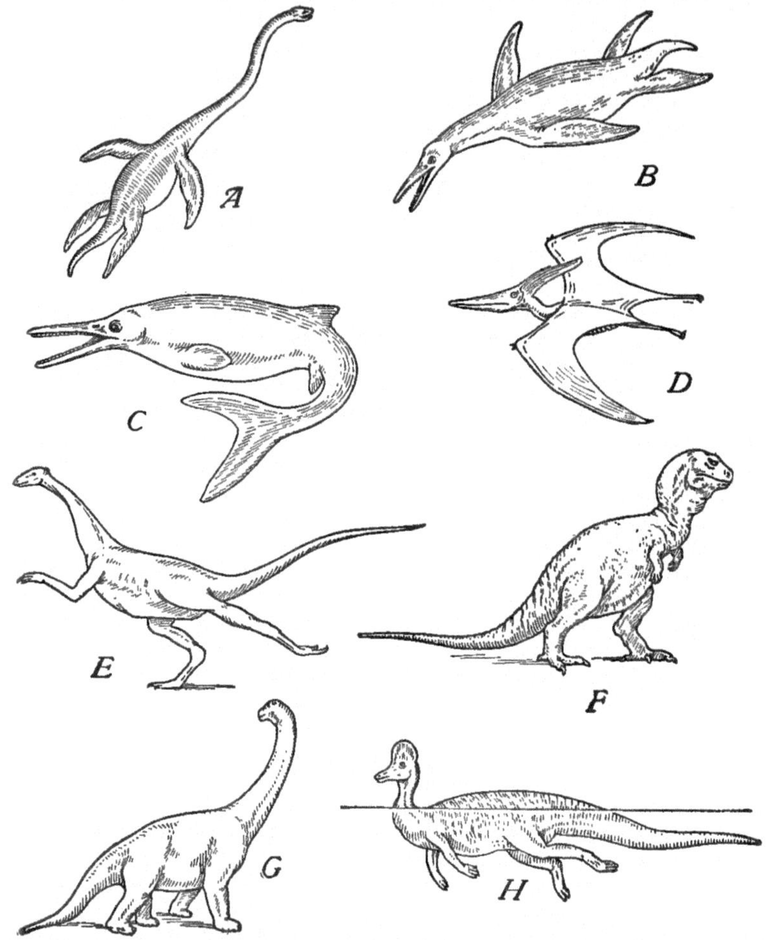

FIG. 124.—Group of Mesozoic Reptilia. A, Long-necked plesiosaur, *Elasmosaurus;* B, short-necked plesiosaur, *Trinacromerion;* C, ichthyosaur, *Baptanodon;* D, pterodactyl; E, "Ostrich" dinosaur, *Struthiomimus;* F, carnivorous dinosaur, *Tyrannosaurus;* G, giant herbivorous dinosaur, *Brachiosaurus;* H, hooded "duck-bill" dinosaur, *Corythosaurus.* (Redrawn after Osborn.)

to-day; for it is inconceivable that creatures so purely aquatic in habits should come ashore to lay their eggs.

PLESIOSAURIA

These marine reptiles furnish less extreme examples of aquatic adaptations than do the ichthyosaurs. Some of the plesiosaurs reached a giant size, being upwards of fifty feet in length. The early members of this group were of a rather generalized type and might properly have been thought of as marine lizards. Later came a type such as *Elasmosaurus* (Fig. 124, A), a slow-moving, long-necked, short-bodied, small-headed type, with long, narrow paddles. The climax of plesiosaurian specialization was reached by such forms as *Trinacromerion* (Fig. 124, B), which is characterized by short body, rather short neck, and fiercely predaceous jaws; a creature with all the earmarks of an aquatic speed demon, and doubtless as much of a terror to the fishes as were the dinosaurs to the smaller denizens of the dry land.

CARNIVOROUS DINOSAURS

The dinosaurs are the last word in terrestrial specialization among the reptiles. Doubtless the ancestors of this great group were rather generalized lizard-like forms that lived in the late Permian, but as yet the palæontologists have not been able to place their hands upon an unequivocal ancestral dinosaur. As has already been said, the group had a dramatic rise to dominance and an equally dramatic extinction at the close of the Cretaceous. While they lasted, their course was an impressive one and far out-shadowed that of all contemporaneous land creatures. On this account the middle and late Mesozoic period has with some justification been called the "age of dinosaurs."

The carnivorous dinosaurs were for the most part *Saurischia* (with lizard-like pelvis), as opposed to the *Ornithiscia* (with bird-like pelvis), a group to which most of the herbivorous dinosaurs belong. The principal evolutionary changes that took place within the group of carnivorous dinosaurs are associated with an absolute increase in body size, relative decrease in the size of the fore limbs and increase in that of the hind limbs, accompanied by a progressive tendency toward bipedal locomotion and speed of running. The culminating types were swift, cursorial creatures, with long tails for balancing, short grasping fore limbs, long neck, and head armed with heavy recurved teeth. One can readily imagine them as able to use their powerful hind legs as effectively as does the ostrich. As a climax type we may cite the great *Tyrannosaurus rex* (Fig. 124, F) of which Matthew says: "It reached a length of 47 feet and in bulk must have equalled the mastodon or the largest living elephants. The massive hind limbs, supporting the whole weight of the body, exceeded the limbs of the great proboscidians in bulk." · It stood about 20 feet high, had a head over four feet long, teeth three to six inches long and an inch wide. The claws on the hind feet were about eight inches in length and of massive proportions. One can readily imagine a scene of carnage, the like of which the modern animal world cannot afford, when such a reptilian dreadnought went into action against one of those huge, heavily armed, monitor-like reptiles such as the herbivorous dinosaur, *Triceratops* (Fig. 128). Such a struggle would decide the question of supremacy between offensive and defensive armaments.

THE HERBIVOROUS DINOSAURS (*Sauropoda*).—The contrast between the carnivorous dinosaurs and the herbivorous dinosaurs involves largely the matter of relative speed, of offensive and defensive equipment. While haste is the essence of success in raptorial life, no time element is involved in securing plant food. It appears to be certain that the early herbivores and early carnivores were quite similar and that both were more or less bipedal. While the carnivores carried this cursorial tendency to such an extreme that the fore limbs were reduced to weak grasping appendages, useless for locomotion, the herbivores eventually underwent a reversed evolution and became secondarily quadrupedal. Evident traces of the bipedal habit, however, are to be noted in the general body form and the relative proportions of the fore and hind limbs, the latter in most forms being much larger. An interesting series of massive forms appeared of which *Brontosaurus* (Fig. 125) is typical. This great creature, with a total length of nearly a hundred feet, has comparatively small fore legs, but they more nearly equal the hind limbs than in some of the earlier members of this group. Evidently the fore limbs later

FIG. 125.—*Brontosaurus*. (From Lull.)

underwent a secondary increase in proportions, for the culminating type of the Sauropoda, *Brachiosaurus* (Fig. 124, G), had the fore legs even heavier and longer than the hind legs. This immense creature rivaled the modern whales for sheer bulk, and was possibly the most ponderous creature of all time; unquestionably it was the largest by all odds of the known terrestrial giants, dwarfing the largest elephants by contrast.

"In the final extinction of the herbivorous sauropod type," says Osborn, "we find an example of the law of elimination, attributed to the fact that these types had reached a cul-de-sac of mechanical evolution from which they could not adaptively emerge when they encountered in all parts of the world the new environmental conditions of advancing Cretaceous time."

The Ornithischia.—While both of the groups of dinosaurs just described are alike in having the typical reptilian pelvis and are therefore grouped together as *Saurischia*, another great group of contemporary dinosaurs had the avian type of pelvis and are called *Ornithischia*. These dinosaurs appear to have been an offshoot of the early herbivorous types and had retained that habit, using probably the harder vegetable foods, as is attested by the development of a heavy, chitinous beak much like that of the modern bird. These Beaked Dinosaurs radiated adaptively into three distinct structural types: Ornithopoda (bird-footed), Stegosauria (armored dinosaurs), and Ceratopsia (horned dinosaurs).

The **Ornithopoda** were unarmed, bipedal forms doubtless capable of great speed. As examples of these forms we may cite the familiar *Iguanodon* (Fig. 126) *Trachodon*, the duck-billed dinosaur, and *Corythosaurus* (Fig. 124, H) the "hooded duck-bill." All of them show well the bird-like feet and pelvis. It is significant in this connection to note that it is from this group that some authors would derive the birds, a theory that is presented in discussing the ancestry of the birds. Of all of the Ornithischia perhaps the most bird-like type is the "Ostrich Dinosaur" *Struthiomimus* (Fig. 124, E).

The **Stegosauria** seem to have reacquired the quadrupedal habit in correlation with the massive weight of their armature. *Stegosaurus* (Fig. 127) represents the culmination of the evolution of the armored types, and has been fittingly chosen as the exemplar of one type of preparedness, distinguished by possessing "all armor and no brain," an equipment as little apt to meet with success in those antique days as in the strenuous present. Though one of the most grotesque of nature's excesses in all of its curious make up, its central nervous

FIG. 126.—Restoration of *Iguanodon*. (From Lull, after Heilmann.)

system presents one of the prize oddities within the field of biology. It is literally practically brainless, for the nervous content of its cra-

FIG. 127.—Restoration of the armored dinosaur, *Stegosaurus*. (From Lull, after Schuchert.)

nium could not have weighed more than about two ounces, a brain that even a two weeks' old kitten might be ashamed of. An elephant of comparable size has a brain-weight of at least eight pounds. Curiously enough the animal's real "brain," if we may call it such, is situated near the base of its enormous tail, where there is a great sacral enlargement of the spinal chord many times as large as the dwarfed brain. Such a creature, with his brain in his rump, must have been nothing but a bulky, ponderous automaton, driven by stimuli arising in the lower nervous centers.

The **Ceratopsia** were creatures whose proportions suggest those of the rhinoceros. *Triceratops* (Fig. 128), a classic example of the group, was about twenty-five feet in length and about ten feet in

FIG. 128.—Restoration of the horned dinosaur, *Triceratops*. (From Lull, after Schuchert.)

height. The head was exceptionally massive, nearly eight feet in length, with a wide frill-like expansion of the skull, which extended like a shield over the neck and shoulders. On the front there were three great horns, one on the snout and two above the eyes. Doubtless such creatures as these had as enemies the great carnivorous dinosaurs, for no other contemporaneous animals would have necessitated such a defensive armament on the part of *Triceratops* and its kin. The ceratopsians were strictly North American and lived for only a brief span, geologically speaking; for they are confined exclusively to the Upper Cretaceous.

The Extinction of the Dinosaurs.—"One of the most inexplicable of events" says Lull, "is the dramatic extinction of this mighty race, for in the rocks of undoubted Tertiary age not a single trace of them remains. One student has argued internecine warfare among the dinosaurs themselves; another, the destructive slaughter, not of adults but of young, possibly while yet in the egg, by small blood-thirsty mammals; yet another, change of climate, either by the diminution of the necessary heat without which no reptilian race may thrive, or of the moisture with an accompanying change of vegetation. These are all conjectural causes of extinction; but this we know, that with the extensive changes in the elevation of land areas which marked the close of the Mesozoic, came the draining of the great inland Cretaceous seas along the low-lying shores of which the dinosaurs had their home, and with the consequent restriction of old haunts came the blotting out of a heroic race. Their career was not a brief one, for the duration of their recorded evolution was thrice that of the entire mammalian age. They do not represent a futile attempt on the part of nature to people the world with creatures of insignificant moment, but are comparable in majestic rise, slow culmination, and dramatic fall to the greatest nations of antiquity."

PTEROSAURIA.—The pterosaurs represent a series of experiments in aviation on the part of the reptiles. They varied greatly in size from tiny forms comparable with our sparrows to flying dragons with a wing-spread of twelve feet. That they were good flyers, able to venture far out over the sea, is indicated by the fact that their remains are found mingled with those of the marine mososaurs miles away from the Mesozoic shore lines. They were scarcely flyers in the strict sense, but must have been effective gliders or soarers; for they did not possess the powerful musculature necessary for active flight. In *Pterodon* (Fig. 124, D), one of the largest of the flying reptiles, the head is prolonged into a great keel-like structure, used as a balancing mechanism. Other types had a long tail with a terminal rudder-like expansion much like that of an aëroplane.

The pterosaurs or pterodactyls might appear superficially to be well suited to be the ancestors of the birds, but anatomically they are quite unsuited for this rôle. They represent simply a highly specialized adaptive radiation, that was short-lived and met with utter extinction during the Upper Cretaceous. They furnish a final chapter in the remarkable adaptive radiation of the Mesozoic reptiles.

MODERN REPTILES

The following list of diagnostic characters, which should be compared with a similar list already given for Amphibia, is taken from Gadow:

Characters of Reptilia

1. The vertebræ are gastrocentrous.
2. The skull articulates with the atlas by one condyle, which is formed mainly by the basioccipital.
3. The mandible consists of many pieces and articulates with the cranium through the quadrate bones.
4. There is an auditory columellar apparatus fitting into the fenestra ovalis.
5. The limbs are of the tetrapodous pentadactyl type.
6. There is an intracranial hypoglossal nerve.
7. The ribs form a true sternum.
8. The ilio-sacral connection is post-acetabular.
9. The skin is covered (a) with scales, but (b) neither with feathers nor with hairs; and there is a great paucity of glands.
10. Reptiles are poikilothermous.
11. The red blood-corpuscles are nucleated, biconvex and oval.
12. The heart is divided into two atria and an incompletely divided ventricle. It has no conus, but semilunar valves exist at the base of the tripartite aortic trunk.
13. The right and left aortic arches are complete and remain functional.
14. Respiration is effected by lungs; and gills are entirely absent even during embryonic life.
15. Lateral sense organs are absent.
16. The kidneys have no nephrostomes. Each kidney has one separate ureter.
17. There is always a typical cloaca.
18. The eggs are meroblastic.
19. Fertilization is internal, and is effected, with the single exception of Sphenodon, by means of copulatory organs.
20. An amnion and an allantois are formed during development.

Numbers 1, 2, 3, 7, 8, 14, 16, 18, 20 separate the reptiles from the Amphibia.

Numbers 9 (b), 10, 12, and 13 separate them from the birds and mammals.

Numbers 3, 8, and 11 separate them from the mammals.

ANATOMY OF A MODERN REPTILE (TURTLE)

For several reasons the turtle or tortoise is chosen as a reptilian type for detailed description rather than a lizard, although the latter is in most respects a more representative form, and considerably less specialized than are any of the chelonians. The first reason for our choice is one of expediency, for the turtles are the most plentiful reptile in by far the greater part of the United States. A second reason is that they are of convenient size for laboratory work and are of compact structure. There is, however, a third and more fundamental advantage gained by the use of the turtle; it is structurally more nearly related to the group of reptiles from which the mammals are believed to have arisen than is any other living reptile.

External Characters

The turtle is a reptile in a box. This box, whether it forms a complete or only a partial housing for the body, head, limbs and tail, has a dome-shaped roof, called the *carapace* and a flat floor called a *plastron.* Paired lateral pillars join the floor to the roof. The box is open broadly in front and behind in order to allow the head, legs, and tail to emerge; but these appendages can all be withdrawn within the shelter of the eaves, and in some cases the front and rear sections of the floor (plastron) are hinged in such a way that they can bend upward and completely close the house after the appendages have been drawn in. The head is of moderate size and somewhat flat; the neck is characteristically long and flexible and capable of being folded up when the head is withdrawn; the mouth is large and toothless, but is provided with a sharp-edged, *horny beak.* The *external nares* (nostrils) are close together near the end of the snout, sometimes protruding into a regular proboscis. The eyes are situated laterally and have three eyelids: a short opaque upper lid, a longer lower lid which makes the turtle shut its eye upwards instead of downwards as a man does, and a third eyelid or transparent *nictitating membrane* which may be drawn across the eye from the inner corner. The *tympanic membrane* is quite similar to that of the frog and is just back of the gape of the jaws. The feet are pentadactyl and each finger is usually armed with a claw. As a rule the feet are webbed as in aquatic birds. The skin of the head is usually smooth and scaleless, as is also the neck in most species; but the rest of the body is usually covered with scales, except the base of the thighs. The tail is as a rule poorly developed, but in the more primitive types, as for example the snapping turtles, it may retain its primitive reptilian proportions.

The Armature

The **carapace** and **plastron** (Fig. 129) are, in most of our modern chelonians, somewhat stereotyped structures; they have settled down

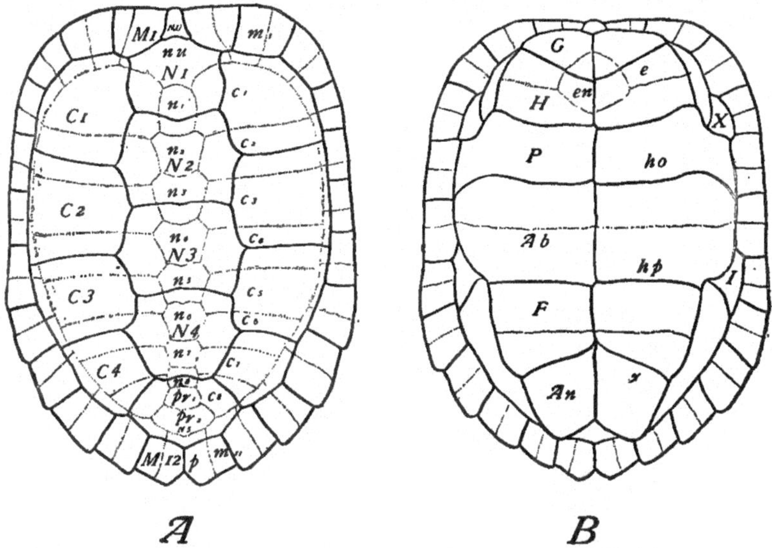

Fig. 129. A, Carapace; B, Plastron of tortoise, *Graptemys*. Capital letters refer to chitinous scales or scutes, small letters to bony plates whether cartilaginous or dermal. *Ab*, Abdominal scute; *An*, Anal scute; *C1–4*, costal scutes, *c1–8*, costal plates; *e*, epiplastral plate, *en*, endoplastral plate; *F*, femoral scute; *G*, gular scute; *H*, humeral scute; *ho*, hyoplastral plate; *hp*. hypoplastral plate; *I*, inguinal scute; *M*, marginal scutes; *m*, marginal plates; *N1–5*, neural scutes; *n1–8*, neural plates; *NU*, nuchal scute; *pr*, *1*, *2*, procaudal plates; *X*, axillary scute; *x*, xiphiplastral plate. (From Newman.)

upon a very definite arrangement of the principal units of structure. The *carapace* (Fig. 129, A) is composed of two kinds of bony elements (dermal and cartilaginous) and corneous scutes or shields. The main part of the bony carapace is composed largely of the much broadened tips of the spinal processes of the vertebræ and of the much flattened ribs; there are usually eight neural plates and eight pairs of costal plates. In front of the first neural is a dermal plate, the *nuchal;* back of the eighth neural are usually three dermal plates, the first and second *procaudals* and the *pygal.* Around the margin of the carapace are usually eleven pairs of dermal plates, the *marginals.* Overlying the bony carapace there is a horny carapace composed of five neural scutes, four pairs of costals, a small anteriorly placed nuchal, and twelve pairs of marginals. This elaborate composition prevails in nearly all of our modern turtles as well as in many species long extinct.

The **plastron** (Fig. 129, B), like the carapace, is composed of two kinds of bony elements covered with horny elements. The bony elements consist of four pairs of plates: the epi-, hyo-, hypo- and xiphi-plastrals. The *epiplastrals* are the modified clavicles, the *hypoplastrals* and *xiphiplastrals* are broadened abdominal ribs, the *hyoplastrals* appear to be dermal elements without homologies. A small median dermal element between the epiplastrals and hyoplastrals is called the *endoplastrals.* There are usually six pairs of horny scutes that break the joints of the bony plastron. The pillars between the carapace and plastron are derived from the hyoplastrals and hypoplastrals.

The conventionalized pattern of bones and scutes in the armature has evidently been arrived at after a long period of evolution. Many evidences indicate that the ancestral condition was much more plastic and variable and that there were originally many more plates and scutes than at present. By dropping out both longitudinal and transverse rows of elements the whole system has been greatly simplified. Most species of turtles to-day show a certain percentage of individuals with supernumerary scutes and plates, that are evidently vestiges of ancestral conditions.

The *vertebræ* in the trunk region are rigidly united to the narrowed bases of the paddle-like ribs. They are not very numerous: 8 cervical, 10 thoracic, 2 sacral, and a variable number of caudal vertebræ, which are procœlous in form.

One of the most puzzling features of the skeleton (Fig. 130) is the peculiar position of the limb girdles. Both *pectoral* and *pelvic girdles* are inside instead of outside of the ribs. How they got inside is a mystery that not even a study of their embryogenesis is able to clear up, for they arise from primordia internal to the ribs. The pectoral girdle consists of a triradiate group of flattened bones: the scapula, the procoracoid and the coracoid, the last being the largest. Together they unite to form the socket which receives the head of the humerus. The pelvic arch is more compact and is composed of the pubis, ischium and ilium, uniting to form the acetabulum for the head of the femur.

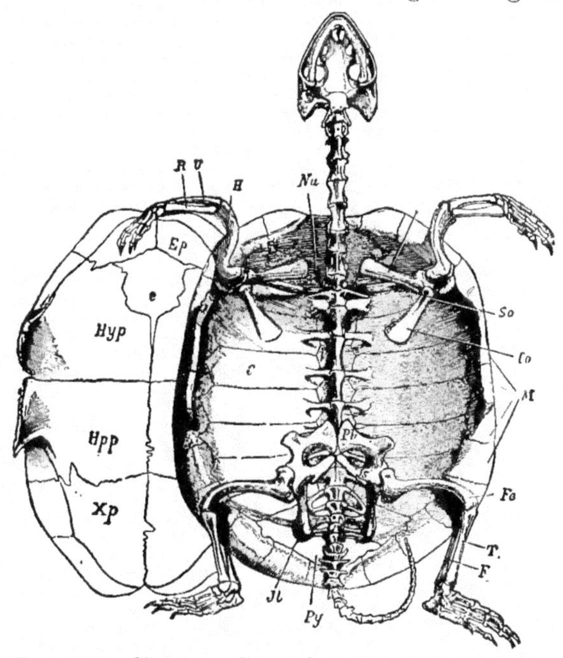

Fig. 130.—Skeleton of tortoise, *Cistudo lutaria*; seen from the ventral side with plastron removed and placed to one side. *C*, costal plate; *Co*, coracoid; *e*, endoplastron, *ep*, epiplastron (clavicle); *F*, fibula; *Fe*, femur; *H*, humerus; *Hyp*, hyoplastron; *Hpp*, hypoplastron; *Il*, ilium; *Js*, ischium; *M*, marginal plates; *Nu*, nuchal plates; *Pb*, pubis; *Pro*, procoracoid process of scapula; *Py*, pygal plates; *R*, radius; *sc*, scapula; *T*, tibia; *U*, ulna; *Xp*, xiphiplastron. (From Parker and Haswell, after Zittel.)

The **skull** is fairly generalized in structure, but has some special features. The jaws are devoid of teeth, and maxillary, premaxillary and dentary bones are covered with hard chitinous sheaths, that form the upper and lower members of the cutting beak; the vomer is a single unpaired median bone; there are no lachrymals nor ectopterygoids; the pterygoids send inwards wings of bone, that, with the aid of the palatines, form a continuous roof to the mouth; the supraoccipital is prolonged backwards into a large narrow process upon which are inserted the heavy neck muscles. All of these bones, even the quadrate, are firmly united into a solid cranium. Further details of the skull are shown in the figure (Fig. 131).

The **digestive system** varies somewhat in carnivorous and herbivorous forms, but in all turtles is comparatively simple. The tongue

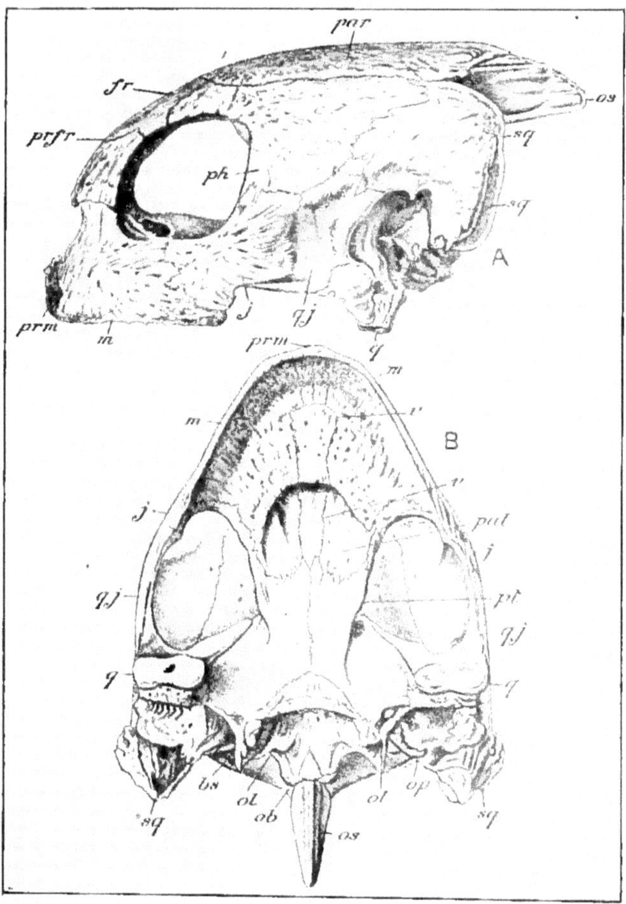

FIG. 131.—Skull of turtle. *A*, lateral; *B*, ventral view. *bs*, basisphenoid; *fr*, frontal; *j*, jugal; *m*, maxilla; *ob*, basioccipital; *ol*, exoccipital; *op*, opisthotic; *os*, supraoccipital; *pal*, palatine; *par*, parietal; *ph*, postfrontal; *prfr*, prefrontal; *pt*, pterygoid; *prm*, premaxilla; *q*, quadrate; *qj*, quadratojugal; *sq*, squamosal; *v*, vomer. (From Parker and Haswell, after Hoffmann.)

is broad and soft and cannot be protruded. The stomach is a simple U-shaped enlargement of the alimentary tract. The intestine is without a cæcum; it is clearly divided into large and small intestine. The cloaca is proportionately large.

The **respiratory organs** (lungs) are large and complicated. Inhalation and exhalation are effected partly by drawing in the neck and thrusting it out again, thus decreasing and increasing the volume of the thoracic cavity. The air is also swallowed into the lungs by filling and then emptying the throat.

The **circulatory system.** The heart is very broad laterally, having two entirely separate atria or auricles, and a ventricle partially divided into two parts by a perforated partition. The right auricle receives the venous blood from two precaval and one postcaval veins; the blood then goes to the right half of the ventricle, and thence through the pulmonary artery to the lungs. From the lungs it returns through the pulmonary veins to the left auricle, thence to the left ventricle, which pumps it out through the paired aortic arches to all parts of the body. There is no renal portal system, but the hepatic portal system is better developed than in the Amphibia.

The **urogenital systems.** The kidneys are metanephric bodies, which pass their excretion through paired ureters directly to the cloaca, thence into a urinary bladder, which in turn empties into the cloaca. The male reproductive organs consist of a pair of testes, a pair of much coiled vasa deferentia, through which the sperm passes to the grooved penis, which is attached to the front of the cloaca. The female organs consist of paired ovaries and large oviducts provided with albuminous and shell glands. The eggs when laid are covered with a tough shell and are usually buried in the ground.

The **nervous system** shows a considerable advance over that of the Amphibia. The cerebral hemispheres are larger and the cerebellum more complete. Many other changes in details will be noted on comparative study. The eyes are small, but the vision is keen; the pupil is round and the iris unusually dark in color. The sense of hearing is not very acute; the tympanic membrane is thin and exposed, and is connected with the auditory organ by a slender columellar bone. The sense of smell is the keenest of the senses in most turtles, both in the water and in the air. In correlation with the keen olfactory sense the olfactory lobes of the brain are highly developed.

THE FOUR ORDERS OF LIVING REPTILES

The order Prosauria is represented to-day by one genus, *Sphenodon,* placed in the family Rhynchocephalia. The order Chelonia is represented by two sub-orders and several large families. The order Crocodilia is to-day a minor order, represented by only a few species of large reptiles. The order Sauria contains a large proportion of living reptiles, since to it belong the lizards and the snakes.

ORDER PROSAURIA (RHYNCHOCEPHALIA)

The only representative of this order now living is *Sphenodon (Hatteria) punctatum* (Fig 132, A), the "tuatara" of the Maories of New Zealand. Gadow refers to this species as "the last living witness of by-gone ages, this primitive, almost ideally generalized type of reptile, this 'living fossil.'" This almost reverential attitude toward the antiquity of this reptile has, however, broken down through the discovery that *Paleohatteria,* the extinct type which was supposed to link *Sphenodon* with the remote past, is really more nearly an ancestral lizard than an ancestral *Sphenodon;* a fact that led such an authority as Williston to assert that some of our modern lizards are more primitive than is *Sphenodon.*

The tuatara is decidedly primitive in some features. There is no penis; the centra of the vertebræ are amphicœlous; the first three basiventralia are quite large; and the skeletal structure of both limbs and girdles is primitive. The skull (Fig. 132, B, C, D) is an almost ideally generalized reptilian skull and well repays study.

Sphenodon is facing extinction at the present time. Already it has disappeared from the mainland and is confined to a few small islands off the coast of New Zealand. For years it has been assiduously hunted by the Maories who consider its flesh an unusual delicacy. Unless protected it will soon go the way of its extinct ancestors and will be represented only by a few specimens in our museums.

The tuatara is nocturnal in habits, living in burrows during the day. It shares its capacious burrow with various kinds of petrels, with which it lives on apparently amicable terms. It will not tolerate any other species of guest, not even other individuals of the same species, and viciously attacks any invader no matter how formidable. Its lizard-like aspect is only skin deep; for it requires only a very casual study of the skeleton and viscera to see the difference between this form and any of the lizards. The illustration (Fig. 132, A) shows better than a verbal description its salient external features.

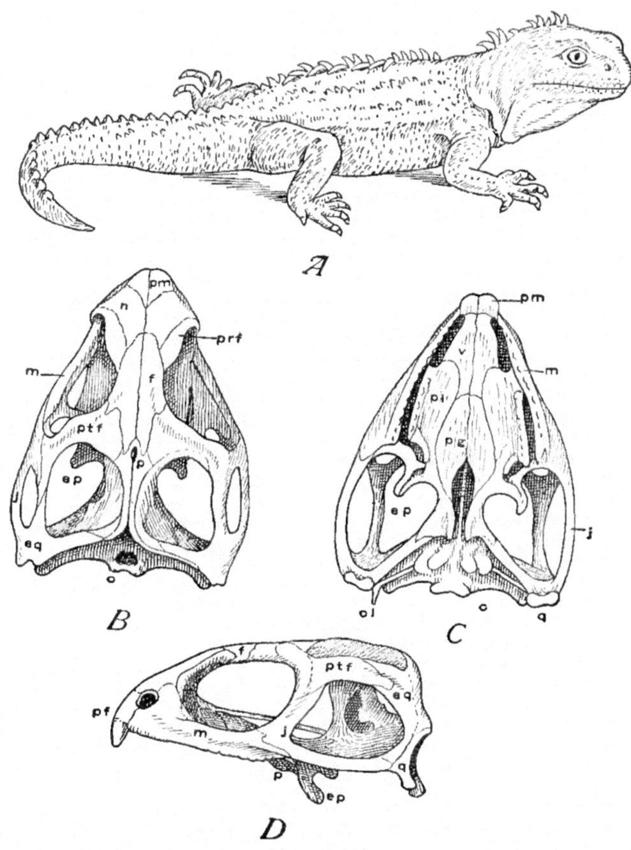

Fig. 132.—*Sphenodon punctatum.* A, Lateral view; B, dorsal view of skull; C, ventral view of skull; D, lateral view of skull. *c*, condyle; *cl*, columella; *ep*, ectopterygoid; *f*, frontal; *j*, jugal; *m*, maxillary; *pm*, premaxillary; *n*, nasal; *p*, parietal; *pl*, palatine; *prf*, prefrontal; *ptf*, postfrontal and post-orbital; *pg*, pterygoid; *Q*, quatrate or quadratojugal; *sq*, squamosal; *V*, vomer. (After Gadow.)

ORDER CHELONIA (TURTLES AND TORTOISES)

The members of this order are so uniquely modified that there is no difficulty in recognizing them and in distinguishing them from all other living creatures. Their short, broad body, covered by the characteristic carapace and plastron, and their horny, toothless jaws, constitute their outstanding characteristics.

They occupy a very wide range of habitat zones without displaying any very radical departures from the typical chelonian form and proportions. They range from the pure marine types which come on land only for the purpose of laying their eggs in the sand; through a whole series of amphibious forms, living in ponds and spending a considerable part of the time on land; culminating in the giant purely terrestrial forms that are found on several groups of oceanic islands. Their adaptive radiation does not include arboreal, cursorial or volant types, for the probable reason that the shape and weight of the armature does not readily lend itself to these modes of life.

Until recently the ancestry of the Chelonia was entirely a mystery, but it is now believed by palæontologists that the *Eunotosauria* of Permian times furnish a connecting link between the Chelonia and still more primitive cotylosaurs of the Permo-Carboniferous. The Eunotosauria have been ably discussed by Watson and there is now very general agreement with his contention that these forms represent a group which was ancestral to the Chelonia.

The order Chelonia is divided into two sub-orders, the *Athecæ* (without a true carapace), and the *Thecophora* (with a carapace).

Sub-Order I. Athecæ

The sole living representative of this sub-order is *Dermochelys coriacea*, the *Leather-back Turtle* (Fig. 133, A). Instead of the usual closely-knit carapace and plastron it has twelve longitudinal rows of dermal plates (5 dorsal, 5 ventral and 2 lateral). The homologues of these can be recognized in the scute rows of some of the Thecophora. The limbs are large, flipper-like paddles of a highly specialized aquatic type. The tail is rudimentary. *Dermochelys* has a wide distribution, ranging over all of the inter-tropical seas, but is nowhere abundant. It is carnivorous, feeding chiefly on mollusks, fishes and crustaceans. One of the most peculiar facts about this species is that only large specimens and "babies" have ever been found. Where they pass the many years of their youth and early maturity is a mystery. Possibly there is in some obscure corner of the world an undiscovered Dermochelys rookery.

It is believed by Dollo that *Dermochelys* is derived from an early terrestrial thecophoran, which lost the primitive armature when it assumed the completely marine habit; a belief that involves the idea of reversed aquatic adaptation. Structurally the "leather-backs" are so different from the other turtles that some authors advocate

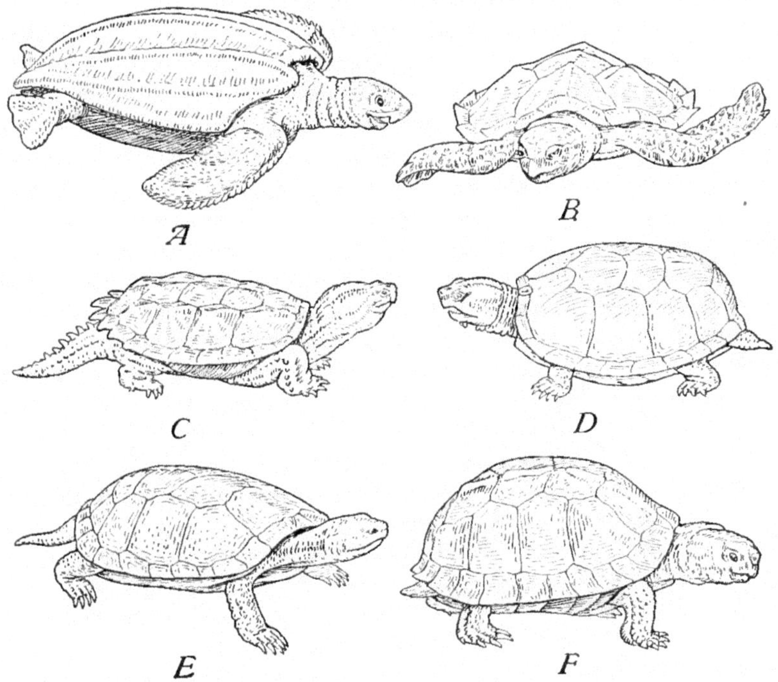

Fig. 133.—Group of Chelonia, I. A, Leatherback Turtle, *Dermochelys* (*Sphargis*) *coriacea;* B, Hawksbill Turtle, *Chelone imbricata;* C, *Chelydra serpentina* (Snapping Turtle); D, Pennsylvania Mud Turtle, *Cinosternum pennsylvanicum;* E, European Pond-Tortoise, *Emys orbicularis;* F, Carolina Box-Tortoise, *Cistudo* (*Terrapene*) *carolina.* (Redrawn after Lydekker.)

putting them in a separate order. Recent discoveries, however, have tended to confirm the conviction that they are an early aberrant offshoot of a primitive land chelonian stock.

Sub-Order II. Thecophora (True Turtles)

The true turtles are subdivided into two assemblages: Cryptodira and Pleurodira.

Division 1. Cryptodira

In these turtles the carapace is covered with horny scutes: the neck is retractile, bending chiefly in a vertical plane; and the pelvis is not fused with the shell. By far the majority of our common turtles and tortoises belong to this division.

Family 1. Chelydridæ (*Snapping Turtles*)—The **common snapper** (*Chelydra serpentina*) and the alligator snapper (*Macrochelys temmincki*), both North American species, are the only living representatives of this primitive family. The common snapper (Fig. 133, C) is our most generalized modern turtle. Its head, body and tail are rather evenly balanced, and the limbs are proportionally heavy and typically reptilian. There is also less complete boxing in of the movable parts than in most other species. In the tail of *Chelydra* are found not only the rows of plates and scutes that are homologous with those in the armature, but at least five rows that have disappeared from or are merely vestigial in the latter. Hence the ancestral condition of the armature is probably more nearly duplicated in the basal portion of the tail of *Chelydra* than anywhere else. The snapper is a slow and clumsy creature, exceedingly sullen and ill-tempered in captivity. When irritated it snaps blindly with widely open mouth, and seizes indiscriminately any object within reach. It is decidedly aquatic in habit and is not fond of basking in the open. More often it is found in shallow, warm pools partly buried in the mud. At times it goes on journeys cross-country from one body of water to another. The snapper makes its nest in loose gravelly or sandy soil at no great distance from the water's edge, though it may wander some distance inland before selecting a suitable nesting place. In excavating the nest a shallow, funnel-like depression is first made; then a crude tunnel is scraped out and enlarged at the bottom into a chamber. All of the digging is done with the hind feet, which are armed with heavy claws. About thirty to forty spherical eggs with tough elastic shells are laid layer on layer with pads of sand packed between; and a layer of sand is packed in and smoothed over the top. *Chelydra* is carnivorous, feeding on fish, frogs, young ducks and all other aquatic animals that come its way. Active prey is caught by stealth. The dull, mud-colored body renders it inconspicuous and aids it in slipping up close to an unwary frog or fish. If the snapper ever approaches a prospective victim so as to be able to snap its jaws upon it, the victim is doomed, for once closed the jaws are like a steel trap.

While the ordinary snapper may reach a weight of twenty pounds the **alligator snapper** grows to twice that weight or more and is proportionately more ferocious. It is said that a large specimen is capable of biting off a piece of board over an inch in thickness.

Family 2. Dermatemydæ.—This is a small group of mostly Central American tortoises, with a strictly aquatic habitat. They are primitive in having a row of scutes between the marginals and plastrals, called inframarginals. This row is represented by the merest vestiges in other families of Chelonia.

Family 3. Cinosternidæ (*Skunk or Musk Turtles*).—This is another group that is primarily aquatic, but not so exclusively so as the first two families described. They are small turtles (Fig. 133, D) that show in their structures evidence of having reverted to an aquatic abode after a prolonged ancestry upon the land. Their box-like shell is not the type of armature that is characteristic of the really aquatic turtles. The commonest representative of the group is *Aromochelys odorata*, a name redolent of the peculiar sickening odor that it gives off from the inguinal glands when disturbed. The "stink pot," as it is commonly called, lives at the bottom of ponds, crawling over the mud, but seldom swimming freely in the water. Its heavy shell makes it a sort of chelonian diver. In warm weather it is often seen floating at the surface supported upon a mass of floating pond-scum. They rarely bask openly above the water. On land they are slow and clumsy of gait, but in spite of this they wander about at night through the grass and shore herbage, hunting for worms and slugs. I have also found them in the daytime rooting about in the moss for insects or grubs, using their snouts for the purpose and snuffing like little pigs. Sometimes they stay out of water so long that they become light in weight from desiccation. When caught they make a great show of fierceness, hissing and opening the jaws widely, looking almost as formidable as a small snapper; but this is either a mere "bluff" or due to fright, for when given the opportunity to bite they do not take advantage of it. Their appetite is insatiable and indiscriminate; anything that could by any stretch of courtesy be described as edible meets with their approval. *Aromochelys* is a curious mixture of a primitive and specialized turtle. It is very aquatic at certain times and decidedly terrestrial at others. It pretends to be fierce, but is gentle; it is omnivorous. It makes the crudest nest of any of the species that the writer has studied. On one occasion a female was observed to dig a shallow hole about two inches wide and about as deep. Two china-like eggs were laid in the nest and covered up loosely with débris. Sometimes the nest is constructed with somewhat greater care, but it is less elaborate than in other species studied.

Family 4. Platysternidæ.—This family is represented by one species native to Borneo, Siam, and Southern China. *Platysternum* is an extremely flat type, with unusually large head and hooked beak.

Family 5. Testudinidæ (the common *pond tortoises*).—This is much the largest family of chelonians and is represented in North America by *Graptemys geographica* (the map tortoise), *Chrysemys picta* (the painted tortoise), *Nannemys gutatta* (the spotted tortoise), *Terrapene carolina* (the box tortoise) and, as an aberrant derivative of North American chelonians, the giant land tortoises of the Galapagos and other oceanic islands. In habits they range from aquatic to purely terrestrial forms. Some are purely carnivorous, other purely herbivorous.

Perhaps the commonest example of our pond tortoises is *Chrysemys picta* (eastern variety) or *C. marginata* (western variety). These rather small tortoises are found in ponds or sluggish streams. They are most frequently seen when basking in the sun along the shore or upon floating logs. They are excellent swimmers and somewhat difficult to catch. They feed upon dead fish and other carrion in the water, tearing up the flesh with their long, sharp claws and sharp-edged beaks. The nest is made with a narrow neck and a flask-shaped chamber at the bottom. It is situated in moist sand along the shores of still waters. Four to eight oval eggs are laid; these are placed in the flask-like enlargement and are covered up neatly with sand, which is pounded down with the knuckles of the hind feet. *Chrysemys* is a bright, intelligent little tortoise, showing little sullenness when captured, and no disposition to snap or to take alarm. They soon learn to come to one who habitually feeds them and will eat from the hand.

Terrapene carolina (Fig. 133, F) is the common land terrapin of the Southern and Eastern States. Structurally they differ little from some of the pond tortoises, but they have acquired exclusively terrestrial habits. If put in the water they soon drown. They are, like the pond tortoises and unlike the giant land tortoises, largely carnivorous. In captivity they become very tame and are often used as pets. There are records of individuals having lived in captivity for fifty years or more. They bask in the heat of the sun most of the day, but at dusk they become active, hunting for slugs and worms, which form their chief diet. At night they retire to their burrows. Their nesting habits are much like those of the pond tortoises.

The *true land tortoises* range from forms of moderate size, like *Testudo græca,* the common European species, to the giant land tortoises of the oceanic islands (Fig. 134, C). These creatures do not differ materially from others except in size, a character which may have been the result of the easy conditions of life on oceanic islands or it may be merely one of the effects of senescence. They are herbivorous and devour quantities of young plant shoots and other succulent vegetation. In the Galapagos Islands there is a different species of land tortoise for almost every island. It is believed that the first individual or pair of these animals reached the Galapagos land mass when it was a single small continent, that subsidence of that part of the earth's crust left only the high places above water, and that these are the present islands. Isolation of the tortoises on the different islands is supposed to have been the principal agency in establishing different species on the various islands. The largest specimens of land tortoises weigh over five hundred pounds and are over four feet in length of shell. They are said to exhibit remarkable longevity, some having a record of about one hundred and fifty years.

Family 6. Chelonidæ (*Sea Turtles*).—This group is best known for the so-called tortoise-shell, a product derived from the horny scutes with which the carapace is covered. They are large turtles with paddle-like limbs, small head, short neck and rudimentary tail. They come ashore only to lay their eggs in the beach sand of the tropical sea-shores. At that time they are captured in large numbers and brought to the metropolitan markets, where their flesh meets with a ready sale as a material for soup. When they are out of the water they are very clumsy and are easily caught. All that the hunter has to do to capture his prey is to turn it over on its back, where it is safe until such time as it is convenient to load it into boats. Usually they are kept in water-filled inclosures till needed and shipped alive to the market.

Chelone mydas, the *green turtle*, is the largest and best known species of sea turtle, though the "*hawksbill*," *Chelone imbricata* (Fig. 134, B) is a close rival in popularity. *Thallasochelys*, "*the loggerhead turtle*," though it is of no commercial value on account of its rank flesh, is of

FIG. 134.—Group of Chelonia, II. A, Snake-necked Tortoise, *Hydromedusa maximiliani;* B, Soft-shelled Tortoise, *Aspidonectes spinifer;* C, Giant Land Tortoise or Elephant Tortoise, *Testudo elephantina*. D, The "Matamata," *Chelys fimbriata*. (All redrawn, A and C after Lydekker, B and D, after Gadow.)

considerable interest on account of the fact that it exhibits a remarkable diversity of scute and plate number and arrangement. This is probably an evidence of primitiveness, and may approach the ancestral condition.

Division 2. Pleurodira

The *Pleurodira* play the same rôle in the southern hemisphere that is played by the Testudinidæ in the northern regions. They are less diversified, however, than the northern tortoises in that they are all aquatic. They differ from our tortoises mainly in that the neck, instead of being withdrawn within the carapace between the shoulders, is bent laterally and tucked under the edge of the shell on one side. The pelvic girdle, unlike that of our tortoises, is fused to the tail and to the carapace.

The genus *Chelodina* will serve as an example of these southern tortoises. The carapace is much like that of *Chrysemys*, but the plastron has a novel feature in the form of a small median scute, the interplastral, which is believed to be a vestige of an ancestral row of scutes that has been lost by most turtles. They are good swimmers and feed exclusively upon aquatic animals such as frogs and water insects. The long neck undulates from side to side like that of a snake. When basking they tuck the head away under the shell in the manner described. There seems to be no striking difference between these tortoises and our own with respect to breeding and nest-making habits. The Snake-Necked Turtle, *Hydromedusa maximiliani* (Fig. 134, A) is another familiar example of this sub-order.

Trionychidæ (Soft-shelled Tortoises)

The distinguishing character of these tortoises is their lack of the scaly or chitinous armature. They also lack parts of the bony armature possessed by other groups. All over the body there is a reduction of the scaly elements; on the feet the scales are reduced to soft folds of skin. The "soft shells" are, however, not to be pitied for their defenseless state, for they make up for their loss by their greatly increased intelligence and rapidity of locomotion. *Aspidonectes spinifer* (Fig. 134, B) is the common "soft shell" of the Mississippi basin and is familiar to most residents of that region. Of all our tortoises they are the most exclusively aquatic, coming inshore only for nesting purposes, and seldom basking except upon floating logs and upon low river banks very close to the water's edge. They always turn around after crawling out of the water, so as to have the head turned toward the water, ready to scramble into the river again at the slighest suggestion of danger. They have need to be wary, for they are excellent food for both man and beast. I have frequently seen young specimens lying in shallow water with only the proboscis-like snout and the dorsally placed eyes protruding above the surface. The body is usually covered over with a film of mud which has been thrown up by rocking the body from side to side and allowing the sediment to settle. When thus camouflaged they are reasonably safe from their enemies. But so swift and alert are the adults that it is unlikely that they would be caught by any of the creatures that inhabit their native waters. Even man with all his equipment for catching animals has the greatest difficulty in securing these tortoises. When one happens to be caught, however, it "keeps its wits about it," as my assistant once said, and is ever on the alert to escape. The captor must be equally wary, for the long neck and strong jaws have an unerring aim quite in contrast with the blind, furious lunge of the "Snapper." The food of the "soft shell" consists chiefly of crayfish and insect larvæ, which they swallow whole without rending in pieces. The nest of this species is a rather deep, neatly made, flask-shaped cavity dug in clean, moist sand. The female comes ashore with the greatest caution, usually very early in the morning, and while making the nest stretches the head on high on the lookout for danger. There are from 15 to 25 spherical tough-shelled eggs, placed in several layers, with sand pads between. The completed nest is covered over so neatly that no trace of it is to be seen from the surface. All of the activities of this species of tortoise appear to the writer to indicate a considerably higher order of intelligence that that shown by any other chelonian.

ORDER CROCODILIA

This order is characterized by its well-proportioned body form, with long tail and well-developed fore and hind limbs; fixed quadrate bone; teeth fixed separately in alveoli. These characters apply not so well to ancestral groups of crocodiles, such as the Pseudosuchia and Eosuchia, as they do to the modern types, the Eusuchia. It is believed that the true crocodiles have been derived from a generalized diapsidian stock as far back as the middle of the Triassic, and we find true fossil crocodiles during the late Jurassic and a continuous line of them up to the present.

Structural Characters.—The *exoskeleton* is composed of squarish corneous thickenings, with narrow channels of flexible skin separating the islands of hard horn. On the back and down the tail the scales are supported by bony cores, and the principal scale rows are keeled, giving a ridged effect to the middle of the back and tail. The hide is used extensively in commerce.

The *tongue* is flat and thick and incapable of protrusion. The *lungs* are large and better developed than in other reptiles. The *teeth* are large and formidable and very irregularly arranged. Though the mouth is provided with very powerful muscles for closing the jaws, those for opening them are very weak, so that a man can easily close with his hands and keep closed the jaws of a large specimen. The *heart* (Fig. 135) and *vascular system* is more advanced in the crocodiles than in any other living reptiles, for the ventricle is almost completely divided by a septum into a right and a left chamber, leaving only a small foramen between. Thus there is practically a complete separation of venous and arterial blood, as in the warm-blooded vertebrates. Though both right and left aortic arches are functional, the left arch is relatively somewhat reduced.

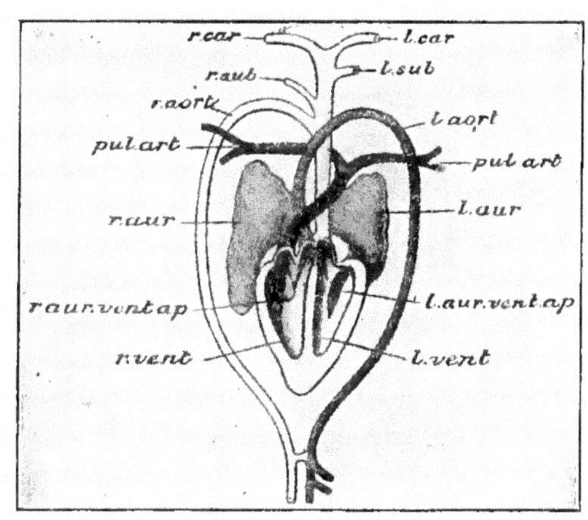

Fig. 135.—Heart of Crocodile with the principal arteries (diagrammatic). The arrows show the direction of arterial and venous currents. *l. aort,* left aortic arch; *l. aur,* left auricle; *l. aur. vent. ap,* left auriculo-ventricular aperture; *l. car,* left carotid; *l. sub,* left subclavian; *l. vent,* left ventricle; *pul. art,* pulmonary artery; *r. aort,* right aortic arch; *r. aur,* right auricle; *r. aur. vent. ap,* right auriculo-ventricular aperture; *r. car,* right carotid; *r. sub,* right subclavian; *r. vent,* right ventricle. (From Parker and Haswell, after Hertwig.)

The *brain* (Fig. 136) is decidedly advanced in structure for a reptilian brain, the large cerebral hemispheres being especially noteworthy. The *tympanic membrane* is sunk in a pit, a tendency that is carried much further in the birds and mammals. It will thus be seen that the crocodiles have followed part way several of the evolutionary paths that have been carried out fully by the birds.

The **geographic distribution** of the crocodiles is wide, but confined chiefly to the tropical regions. They are found over a large part of Africa, in India, Southern China, Malaysia, South and Central America, and along the Gulf of Mexico in North America. Formerly they occurred in Europe and Northern Asia.

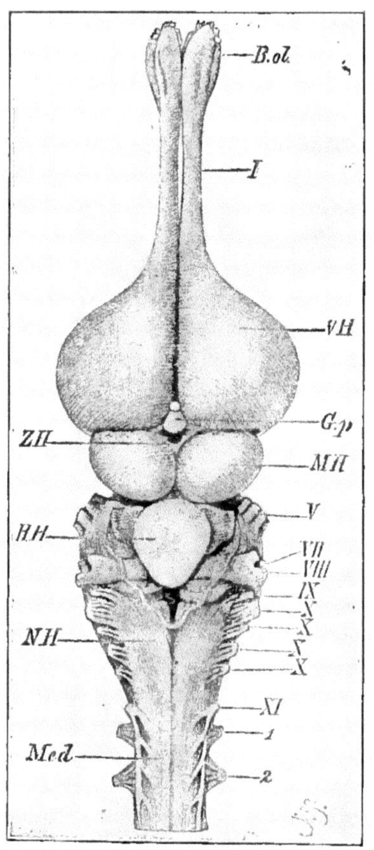

Fig. 136.—Brain of Alligator, from above. *B. ol*, olfactory bulb; *G. p*, epiphysis; *H. H*, cerebellum; *Med*, spinal cord; *M. H*, optic lobes; *N. H*, medulla oblongata; *V. H*, cerebral hemispheres; *I–XI*, cerebral or cranial nerves; *1*, *2*, first and second spinal nerves. (From Wiedersheim.)

Habits.—The crocodiles, alligators and gavials are all fierce predaceous creatures, most of them being enemies of both man and beast wherever they grow. The older they become the more wily and dangerous they live and the more apt to become man-hunters. Their rusty, bark-like backs give them the appearance of partly sunken logs and many an unwary creature attempting to gain support upon such a "log" has suffered a rude awakening. The eggs are laid in the sand much after the fashion of turtles. They reach a great age, probably often breaking over the century mark.

Living Crocodilia belong to two families: Gavialidæ and Crocodilidæ.

Family Gavialidæ.—There is but one living species of gavials, *Gavialis gangeticus* (Fig. 137, C), confined to the Ganges and other large rivers of India. They reach a length of over twenty feet, but are less dangerous to man than are the true crocodiles, although they are believed to be ever on the alert to capture man. As a matter of fact it is stated by competent authorities that they never attack man, but feed entirely upon fish. They differ from the other Crocodilia in that they have an extremely long, narrow snout, which resembles that of a gar-pike. Little is known as to the habits of the gavials.

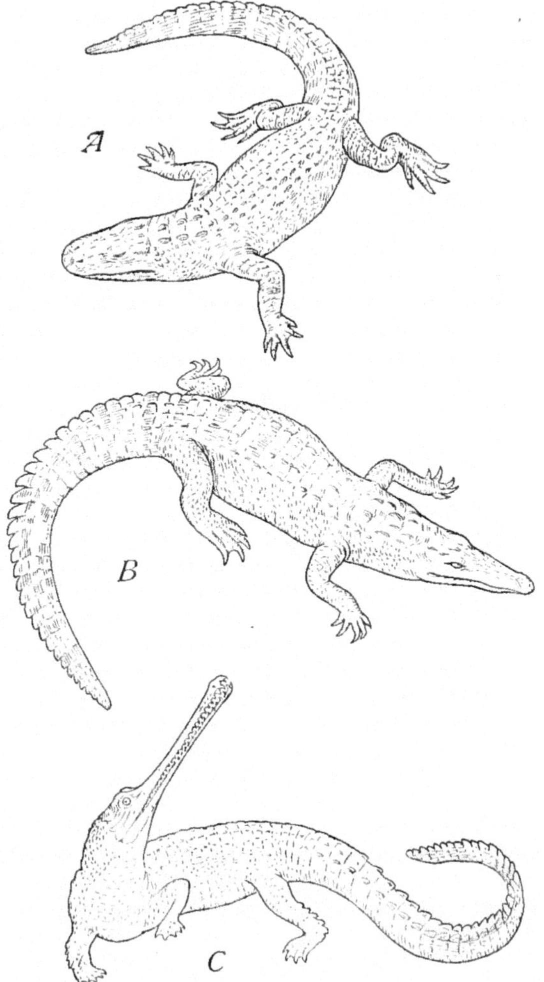

FIG. 137.—Group of Crocodilia. A, *Alligator mississippiensis;* B, *Crocodilus americanus;* C, *Gavialis gangeticus.* (Redrawn, A and B, after Ditmars, C, after Lydekker).

Family Crocodilidæ.—This group includes the old world crocodiles and old and new world alligators.

The common *American alligator* (Fig. 137, A), *Alligator mississippiensis*, occurs largely in the southeastern States, living in the smaller streams and ponds. They usually lie in shallow water with only the eyes and the nostrils exposed. When basking on the shore and disturbed by enemies they take to the water and quickly seek the bottom, where they bury themselves in the mud, from whence it is difficult to dislodge them. They are not as large as the largest crocodiles, reaching a length hardly over twelve feet. The female digs a large nest in the humus and dead leaves, which are piled up into a mound and then hollowed out into a receptacle not unlike a huge bird's nest. The eggs are about three inches in length and of an oval shape and are laid to the number of twenty to thirty to a nest.

The most typical crocodile is the classic *Crocodilus niloticus* (Fig. 137, B), the *Nile crocodile*, which is believed to be the "leviathan" of the Book of Job. The armor is exceedingly heavy and impenetrable to any weapons but bullets. The crocodile makes a long tunnel-like burrow thirty to forty feet in length, with an opening below the water level, used as an entrance, and with a large chamber at the inner end well above the water level. The nest is large and flask-shaped like that of some tortoises, but with a flat bottom grooved around the periphery, causing the eggs to lie in a circular ring. The mother lies over the covered-up nest and takes considerable care of the young after they have hatched.

ORDER SAURIA (SQUAMATA) LIZARDS AND SNAKES

The lizards and snakes to-day are playing much the same rôle for the reptiles that is played by the frogs and toads for the Amphibia. They represent climax conditions and exhibit very pronounced adaptive radiation, being in both groups terrestrial, fossorial, arboreal, amphibious, and aquatic.

They are characterized by the possession of: a movable quadrate bone, which enables the mouth to open more widely; a transverse cloacal aperture; and double copulatory organs.

Contrary to the generally accepted idea that they have arisen from some ancestral prosaurian like *Sphenodon*, they are now traced back to the early lizard-like group represented by *Varanops* (Fig. 123, A) of the Permo-Carboniferous. This ancestral type has been called by some authors *Proterosauria* and was probably ancestral to all of the Sauria or Squamata, past and present.

DIVISION I. LACERTILIA (LIZARDS)

It now appears that the lizards have stolen the laurels of *Sphenodon*, the reputed prototype of all the reptiles; for the earliest known reptile of the Permo-Carboniferous was a very generalized and decidedly lizard-like creature. Some of the modern lizards have departed very little from that type. Perhaps this is the secret of their success in outlasting the great reptilian orders that have come and gone; in that the generalized types that do not go to excesses of specialization are able to weather the age-long vicissitudes of world change, adapting themselves to new conditions and always plastic enough to adjust themselves to a new environment.

Characters of Lacertilia.—It is not so simple a matter as one might think to set up distinctions between lizards and snakes. One might think that the presence of legs in the lizards and their absence in snakes would readily separate the two groups; but there are limbless lizards and there are snakes with at least rudimentary legs. The vast majority of lizards, however, have well-developed legs, only a few degraded burrowing forms being limbless. The lizards also have no elastic ligament between the two halves of the lower jaw as in the snakes. The ventral scales are usually smaller than the dorsal.

The Lacertilia may be divided into three sub-orders: Geckones, Lacertæ and Chamæleontes.

Sub-order 1. Geckones.—The geckos (Fig. 138, A) are primitive lizards with the following peculiarities: four-footed; amphicœlous vertebræ; no bony temporal arches; dilated clavicles; separate parietals; eyes with movable lids; tongue broad, fleshy, protrusible and nicked on the end.

The geckos are practically cosmopolitan within the warm temperate countries. In the United States they are confined to our southwestern Pacific regions. They are wonderful climbers. By means of adhesive pads on the toes they are able to ascend the smoothest surfaces such as walls, ceilings, or even window-panes. Adhesion is accomplished by the vacuum-cup principle, but the "cup" consists of a complicated system of lamellæ. They feed on all sorts of small animals, especially insects and spiders. They are absolutely harmless to man in spite of an undeserved reputation for venomousness. Their chief defense consists of an extremely loosely articulated tail, which comes off with great readiness when seized. When cornered and in grave danger they wag the tail over the body, appearing to offer it for seizure. The enemy is usually satisfied and the tailless gecko proceeds to regenerate another tail, which fortunately it is able to do very readily.

Sub-order 2. Lacertæ.—To this group belong the great majority of modern lizards. As many as eighteen families of Lacertæ are distinguished by the authorities, but in a volume of the present scope it would be unprofitable to deal with more than a few of the most significant types.

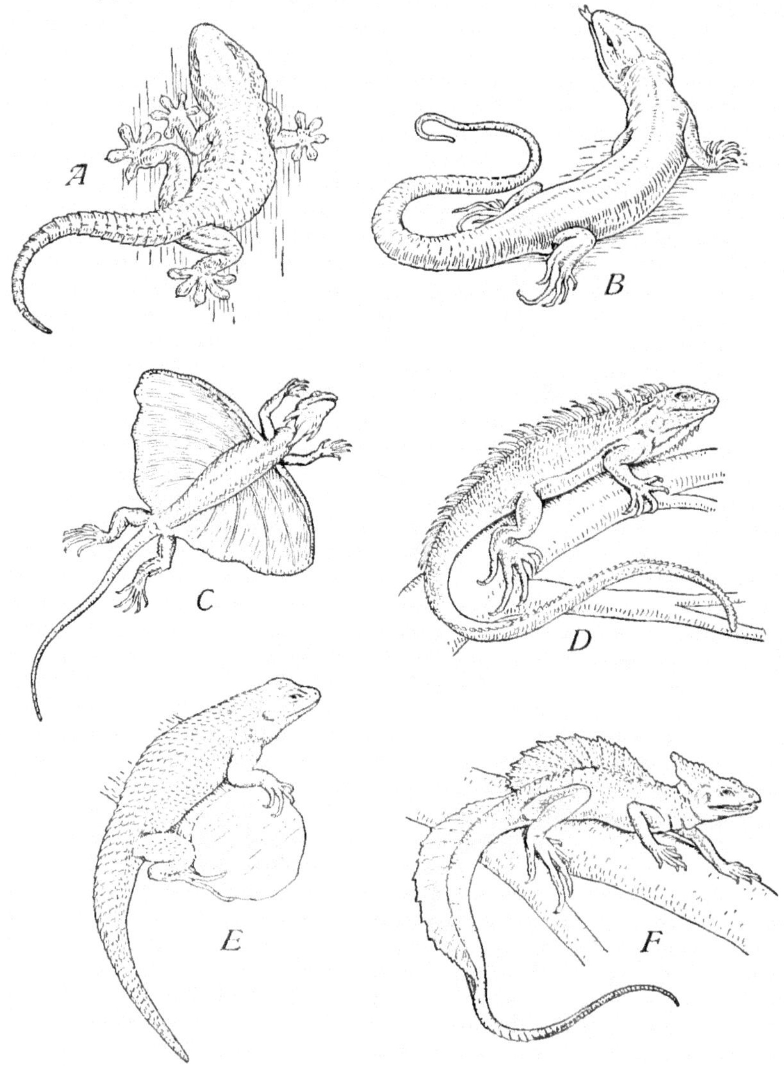

FIG. 138.—Group of Lacertilia, I. A, Wall Gecko, *Tarentola mauritanica;* B, *Lacerta viridis;* C, *Draco volans* (Flying Dragon); D, *Iguana tuberculata,* E, *Sceloporus spinosus;* F, Helmeted Basilisk, *Basiliscus americanus.* (All redrawn, A, B, and F, after Lydekker); C and D, after Gadow; E, after Ditmars.)

The members of this sub-order are distinguished from the other sub-orders of lizards by the fact that the vertebræ are procœlous and solid, and that the ventral portions of the clavicles are not dilated.

It is proposed to describe the characters of a few well-known species with particular reference to their special adaptive features: a cursorial type, an arboreal type, a volant type, an aquatic type, a fossorial type and an ant-eating type.

Lacerta viridis (Fig. 138, B) the common European "wall lizard" is an excellent example of generalized lizard. It is a small type with long slender proportions, is a beautiful green above and yellow below. It runs very swiftly upon the ground and over rocks and hides in thickets and under any available shelter. From some such generalized type as this have radiated all of the more specialized types.

Sceloporus spinosus (Fig. 138, E), one of the commonest American lizards, is a good example of an arboreal type, though it also has a strong liking for the ground if there are thickets available. It is a rusty-colored lizard, harmonizing wonderfully with the bark of the mesquite and other trees which it haunts. During the heat of the day it lies basking on the trunk or exposed branches of trees, and retires to holes in trees or among the roots at night. In the winter it hibernates in shallow holes in the ground or under stones or other shelters. During the cool of the day they are actively in search of food, which consists mainly of tree-inhabiting insects. In the breeding season the male takes on a steely blue sheen about the throat and head. The courtship and mating activities are rather striking. The male stands in front of the female with his brilliant throat inflated and thus displayed to the utmost; then raises himself up and down on the fore legs with a quick rhythm. This the female seems to watch as though fascinated and is soon won. The nest is dug in loose soil in the form of a fairly deep tunnel in a sloping bank. Excavation of the nest is accomplished with the hind feet as in tortoises. The eggs, which are much like tortoise eggs in appearance, number a dozen or more, and when laid are in a stage equivalent to about a 72-hour chick.

Draco volans (Fig. 138, C), the flying dragon, is the best example of the volant type of lizard. The body is dorso-ventrally depressed and the skin is stretched out into two fan-shaped, folding membranes, which are supported on five or six of the greatly elongated ribs. On the neck are three hooks which probably enable the animal to secure a hold when alighting from a flight. The wings are mere parachutes and do not in any sense serve as propellers. Only short soaring leaps from limb to limb or between adjacent trees can be accomplished.

FIG. 139.—Group of Lacertilia, II. A, Horned Toad, *Phrynosoma cornutum;* B, Gila monster, *Heloderma horridum;* C, *Anguis fragilis* (the Glass Snake or Blind-Worm); D, Cape Monitor, *Varanus albigularis;* E, Galapagos Sea-Lizard, *Amblyrhynchus cristatus;* F, *Moloch horridus.* (All redrawn, A, after Gadow; others after Lydekker.)

When the animal is at rest the "wings" are folded against the sides. The flying dragons are natives of Indo-Malayan countries. Not much is known about their habits, but it is said that they live among gorgeous flowers whose colors they closely approximate. Doubtless this camouflage aids the lizard in securing insect food.

Phrynosoma cornutum (Fig. 139, A), the horned toad, is chosen as a desert type. Of course this animal is not a toad at all but a short, flat, spiny lizard, with a very reduced tail, a character that evidently suggested the name "toad" for it. They live in the semi-arid regions of the southwestern States and in Mexico. The only water they seem to take is in the form of dewdrops, and they are capable of living for a long time without any water, growing flatter and lighter as desiccation progresses. Their chief food appears to be ants, though other small insects are not unwelcome. They are fond of basking in the hottest sun during the day, but when night approaches they bury themselves in the sand while still warm from the sun, leaving only the top of the head and the horns exposed. The nostrils are provided with valves to prevent the inhalation of the fine sand. They are colored a dull sandy gray, and this, together with their rugose appearance, makes them very inconspicuous against the usual desert background. One curious habit which the writer had heard of with considerable skepticism and only believed when he saw it with his own eyes, is that of squirting a tiny stream of blood out of the eye, when cornered and in danger. The blood is expelled from the inner corner of the eye and can be shot to a distance of two feet or more. What advantage is gained by this curious habit no one seems to know. The horned toad is a docile little creature and is easily tamed. Of all animals that the writer has experimented with, they are the most readily hypnotized by turning them on the back and pressing gently but firmly against the ventral surface.

Anguis fragilis (Fig. 139, C), the slow-worm or blind-worm, is also called in some sections of the country the "glass snake." These lizards are true fossorial or burrowing types. They are limbless forms, representing the climax of degeneration among the Lacertilia. There is a current legend of the Southern States that this creature can be shattered by a blow into a number of pieces and that these pieces get together again into an entire animal, which then goes on its way rejoicing. The truth underlying the legend is that, like other lizards, the tail is quite brittle and readily knocked off by a blow from a stick. Both animal and tail wriggle about vigorously after such violent treatment, but only the body is able to resume the journey.

Basiliscus americanus (Fig. 138, F), the American basilisk, may be chosen as an amphibious type. It is a large, conspicuous lizard about a yard in length. It is characterized by a very pronounced dorsal crest, which looks like a fin, a secondary sexual character limited to the males. They lie on the branches of trees overhanging the water and at the slightest danger drop off into the water and swim rapidly ashore, using the fin only as a rudder.

Amblyrhynchus cristatus (Fig. 139, E), the sea lizard, is as near an approach to a true aquatic type as the Lacertilia afford. These rather large, heavy-bodied lizards inhabit certain rocky shores on the Galapagos Islands. They are great swimmers, using the flattened, finned tail as a propeller. They habitually feed upon the seaweeds that abound beyond the breakers, and they have to weather the waves in order to secure their food. Often they prefer the really dangerous breakers to their enemies on land, and seek shelter in the sea.

Moloch horridus (Fig. 139, F) is one of the strangest of lizards. Its integument is remarkable for its heavy spines. This animal has been described as a lizard ant-eater and its peculiarities are considered to be primarily adaptations for the ant-eating life. It certainly looks to be well protected to withstand the attacks of ants. One peculiar feature of the integument has attracted considerable attention; for the skin is said to be hygroscopic, capable of absorbing moisture from the air. This strange lizard rivals in bizarre appearance the most fanciful monsters of long ago. Only its small size redeems it from utter frightfulness of aspect.

The only venomous lizard is the *gila monster* (Fig. 139, B), *Heloderma horridum,* a large, heavy-bodied lizard of the arid lands of our southwest and Mexico. It has fang-like recurved teeth, which are so grooved as to form ducts for the poisonous secretion of the labial glands. The Gila is conspicuously marked with contrasting black and orange patches and is often cited as an example of warning coloration, a common phenomenon among venomous reptiles.

The largest living lizard is the *monitor* (Fig. 139, D), *Varanus salvator,* a species that reaches a length of seven feet or more. Apart from its great size the Monitor is a very generalized lizard, differing very little from the primitive lizard-like reptile, *Varanops,* which lived in Permian times. In Southern China and the Malaysian region, where this lizard has its home, it is hunted by dogs and used for food.

The largest American lizard is *Iguana tuberculata* (Fig. 138, D), a native of South and Central America. The *Iguana* reaches a length of five or six feet. Its habits are much like those of the Basilisk.

Sub-order 3. Chamæleontes (Chameleons).—The chameleons are the most highly specialized of the lizards. The body is laterally compressed, the tail prehensile, the toes are parted in the middle into two groups used for grasping, a group of three being opposed by a group of two. Most of them are African or Madagascan, though one species (*Chamæleon vulgaris*) extends into Southern Europe.

As an example of extreme arboreal specialization the group is of unusual interest. Two characters of chameleons have become notorious: their ability to change color and their habit of "shooting" insects with their tongues. Accounts of their color versatility are exaggerated, but the fact remains that they are probably the most effective color changers known, having a range from very light gray to leaf green; and the change can be made in a few seconds. The tongue is capable of "shooting" a fly at a distance of seven inches and the aim is unerring. Probably the aim is improved by the curiously modified eyelids which are grown together with the exception of a mere pin-hole in the center. Apparently the tongue aims at the exact point of focus of the two eyes. Several signs of racial senescence are displayed by chameleons: their high compressed bodies, their lack of scales, and the specialized eyes, feet and tails.

An excellent account of the activities of the chameleon is given by Gadow, accompanied by a composite illustration (Fig. 140).

> "It is most interesting to watch them stalking their prey. Suppose we have introduced some butterflies into their roomy cage, which is furnished with living plants and plenty of twigs. The Chameleons, hitherto quite motionless, perhaps basking with flattened out bodies so as to catch as many of the sun's rays as possible, become at once lively. One of them makes for a butterfly which has settled in the furthest upper corner of the cage. With unusually fast motions the Chameleon stilts along and across the branches and all seems to go well, until he discovers that the end of the branch is still 8 inches from the prey, and he knows perfectly well that 7 inches are the utmost limit to a shot with his tongue. He pauses to think, perhaps with two limbs in the air, but stability is secured by a judicious turn of the tail. After he has solved the puzzle, he retraces his steps to the base

of the branch, climbs up the main stem, creeps along the next branch above, and when arrived at the 7 inch distance he shoots the butterfly with unerring aim. The capacity of the mouth

Fig. 140.—Chamæleons. A–D, *C. vulgaris*, showing various attitudes and changes of color; *C. pumulis* in upper right hand corner. (From Gadow.)

and throat is astonishing. A full grown Chameleon will catch, chew, and swallow the largest moth."

While we may object to the statement that a Chameleon "pauses to think" or "knows perfectly well," we cannot but admire the vividness of the verbal picture here presented.

Division II. Ophidia (Snakes)

Snakes may be defined as Sauria or Squamata that have the right and left halves of the lower jaw connected with an elastic ligament, which enables the mouth to stretch much more widely than it otherwise could; they are limbless or at best have rudimentary limbs under the skin, as in the pythons. They represent a more advanced stage of specialization than do the lizards and are a much more modern develment than any of the other living reptilian groups. In certain respects the snakes are degenerate. As in the eel-like fishes and amphibians the greatly elongated body is accompanied by loss of limbs.

The majority of the snakes have the quadrate very loosely articulated with the squamosal; which aids in increasing the gape of the jaws and enables snakes to swallow objects greater in diameter than their own bodies. This is well shown in the illustration (Fig. 141, A) representing a python swallowing a large bird.

The vertebral column consists sometimes of nearly three hundred vertebræ, which are little if at all specialized in the different regions of the body. The skin is covered with scales devoid of bony cores. The ventral scales are usually broad, band-like and erectile, and are used as an accessory to locomotion; for they point backwards and thus give a good friction surface against the ground or trunks of trees. The outer skin is shed several times a year all in one piece. The eyes have no lids, but each eye is covered with a watch-glass-shaped membrane which is transparent and is shed when the rest of the skin is moulted. This explains why the snake is blind shortly before and during the moult; for the dead eye-membrane is opaque. The ear is peculiar in that the columella has a fibrous pad at the outer end, which plays against the quadrate; so that when the quadrate is pulled away from the skull in swallowing, the columella must be so dislocated as to produce a tremendous roaring sensation, if the auditory organ is at all sensitive. Fortunately, perhaps, the snake has not a keen sense of hearing. The tongue is very slender and forked and is used as a tactile organ; for it is thrust out with a flickering motion against any object that requires investigation.

The internal organs are greatly elongated, having the relations that those of a lizard would have if it were stretched out to more than twice the normal length. The copulatory organ is usually covered with re-

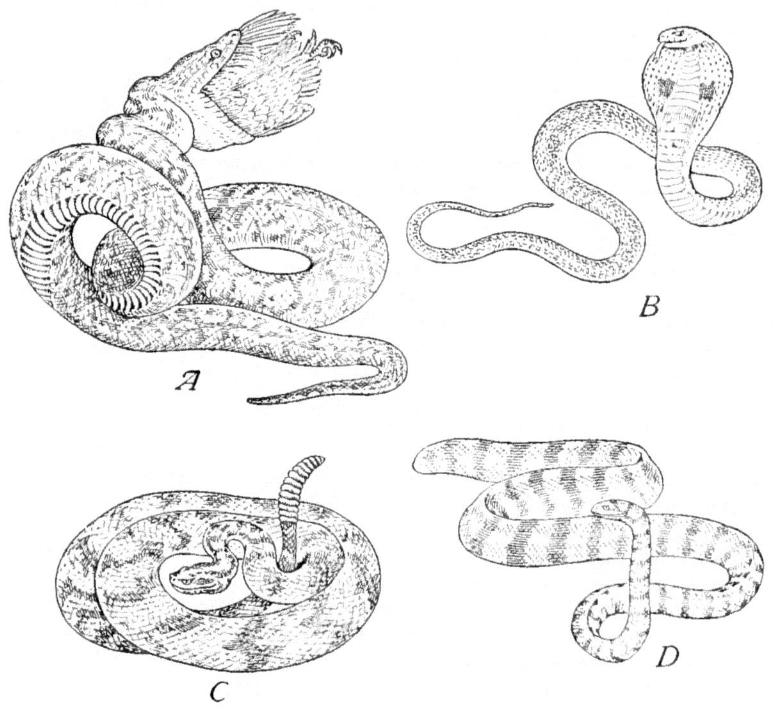

FIG. 141.—Group of Ophidia. A, African Python (*Python seba*) swallowing a bird; B, Cobra, *Naja tripudians;* C, *Crotalis durissus* (Rattlesnake); D, Banded Sea-Snake, *Platurus laticaudatus.* (Redrawn after Lydekker.)

curved hooks and spines, which ensures prolonged copulation in spite of writhing of the two bodies. Though the majority of snakes lay eggs, many of them, including the common garter snakes, are viviparous.

According to Gadow, "Snakes are intelligent creatures; some become quite affectionate in captivity, but most of them are of a morose disposition, and they do not care for company." So far as the average man is concerned this feeling is mutual; for the first human reflex is to kill a snake on sight. Whether this is the result of tradition or is a residual instinct dating back to the arboreal period of man's ancestry, we cannot say. This much should be said for the snakes, however, that most of them deserve nothing but kindly treatment, since they are far more beneficial than many animals that have a much better reputation. It would appear that the few venomous snakes have given a bad name to the whole group.

Snake Venom.—Many snakes, but a small percentage of the whole group, are more or less venomous, but as a rule they are much less deadly than they are supposed to be. Unfortunately there is no simple criterion for distinguishing the poisonous snakes from the non-poisonous. One merely has to acquaint himself with the habitat and appearance of the various snakes native to the country in which he resides or in which he is sojourning. The poison is secreted in a pair of enlarged labial glands, homologous with the parotid glands of the mammals. A duct leads from these glands to the hollows of the paired tubular fangs. The strike of the snake presses upon the gland and causes the poison to exude from the tip of the fang into the deepest part of the wound. Fortunately for us there are only five kinds of venomous snakes in the United States: coral snakes, water moccasin, copperhead, rattle-snakes, and opisthoglyphs.

There are two species of *coral snakes* both belonging to the genus *Elaps;* both are native to the Southern States. They are extremely conspicuous owing to the vivid contrasting bands of red, black and yellow, another example of the so-called warning coloration. The coral snakes are said to be extremely poisonous, but their biting equipment is so constructed that they cannot open the mouth wide enough to bite a human being; so they may be set aside as harmless, so far as man is concerned.

The **water moccasin** (*Ancistrodon piscivorus*), the so-called "cotton mouth," is a large, heavy, aquatic species that reaches a length of five or six feet. It is really a species of rattle-snake without a rattle. This snake has the reputation of being by far the most venomous of all North American snakes, but it is very unusual for a human being to be bitten by it, and fatal cases are exceedingly rare.

The **copperhead** (*Ancistrodon contortrix*) ranges from Massachusetts to Florida and west to Texas. It also is a species of rattler without any rattle.

The true **rattle-snakes** comprise a number of species belonging to the genus *Crotalus* (Fig. 141, C). Of these the Texas rattler is much the largest and that of Canada the smallest. The largest known specimens reach a length of seven feet and are stockily proportioned. The bite is serious but seldom fatal. The rattle of the "rattler" is a curious structure, made by leaving the end of the moulted skin attached to the tip of the tail, each moult adding a new ring to the rattle. The rattling sound, which is more like a shrill hiss, is made by quivering the tail, a movement of excitement or fear rather than a purposeful warning signal. Nevertheless it is a sound that, even when heard for the first time, causes one to "bring up all standing" and watch one's step. Give a rattle-snake half a chance and he will run away without attempting to attack.

Adaptive Radiation Among the Snakes

Although somewhat limited in their adaptive versatility by the lack of limbs, the snakes show quite a wide range of specialization for the various life zones. The more generalized types are the common ground snakes that have holes in the ground merely as retreats in time of danger or for hibernation.

A great many arboreal types have been developed, as the structure of the snake is peculiarly well adapted for that type of climbing, for which we have no other name than "serpentine." The *Boidæ* (*boa-constrictors*) are typical examples of arboreal snakes (Fig. 141, A). These large, rapacious creatures secure their prey by dropping upon it out of trees and crush it to death within their powerful coils. The largest of these snakes are upwards of twenty feet long, about six inches in diameter, and capable of crushing a tiger or a stag. They are unable, however, to eat such large prey, their limit being rabbits and fairly large birds, which they are able to swallow whole without difficulty. There are several types that are more highly specialized for arboreal life than the Boidæ. Among these are the members of the family *Colubrinæ*, which are characterized by their great length and slenderness, and by the great flexibility of the prehensile tail.

Another adaptive type is that which is native of the arid regions and which has adopted the burrowing habit to protect itself against the extremes of temperature so characteristic of desert regions. A specialized burrowing type is represented by the genus *Typhleps*, of which there are about one hundred species. They dig typical burrows in the ground in which they spend much of their time.

The *marine snakes* are fine examples of a purely aquatic type, that never, except for breeding purposes, come out of the water. The species *Platurus laticaudatus* (Fig. 141, D) illustrates the structural and functional adaptations for marine life. They are laterally compressed in the tail region, with dorsal and ventral fin folds; their mode of swimming is precisely like that of eels. They are decidedly venomous. They are viviparous, the female coming ashore to give birth to her young among the rocks. The new-born young are about two feet long and much less specialized for aquatic life than the adults.

The **Cobra** (Fig. 141, B), *Naja tripudians*, is perhaps the king of all the snakes, and with a description of its habits we shall bring this brief account of a not too pleasant topic to a close. The writer finds it impossible to wax even moderately enthusiastic about snakes. The cobra is a native of India, China and Malaysia. Very large specimens reach a length of six feet; but it is not for their size that the cobras are so noteworthy, but for their striking appearance, their venomousness and their sacredness. They are distinguished by the huge hood or neck swelling, upon which appears a color pattern resembling a death's head or a pair of spectacles, depending on the strength of one's imagination. They are an almost invariable accompaniment of the typical Indian conjurer, who charms them and makes them dance to his weird music. The dance is done by erecting the head with inflated hood and by waving it back and forth to the rhythm of the music. The cobra is by nature docile and has no inclination to bite; but when it does strike it is a serious matter, and the number of victims of cobra bite every year is appalling. Some of the natives possess snake stones, a sort of porous material that appears to have the property of absorbing the poison. The owner of such a stone is deemed by his acquaintances to possess a priceless talisman. In India the cobra is considered a sacred animal and, on that account, no systematic campaign of extermination has been started against it.

In concluding this chapter on reptiles it may be said that no account of development has been given, for the reason that reptilian and avian embryology are so similar that the account given for the bird at the end of the next chapter will do duty for the reptilian type of development also.

CHAPTER VIII

CLASS V. AVES (BIRDS)

The propriety of giving class value to the birds is open to serious question. Fundamentally birds are flying reptiles, highly specialized for aërial locomotion. The close affinities of birds to reptiles was in the mind of Huxley when he combined the two divisions under the name *Sauropsida.*

There is never the slightest difficulty in distinguishing a bird from any other animal. The presence of *feathers* is of itself a differentiating character. All birds, moreover, are *bipeds* and have the fore limbs modified as wings, which in some cases are rudimentary, in others, secondarily specialized as flippers for swimming under water. The absence of teeth and their replacement by the horny bill is a nearly universal avian character. The tail is greatly reduced or foreshortened, much as it is in man. There are not, except in certain domestic races of fowls, more than four toes, of which one is the hallux or great toe. Now none of these characters, except the possession of feathers, really demarks the birds from the reptiles; for some group of reptiles, living or extinct, is characterized by bipedality, by wings, by beak, by lack of teeth, by reduced tail, or by four toes. Is a bird then merely a reptile with feathers? In a sense, yes; if feathers be taken as an index of a complex of structural and functional adaptations for flight. The bird, therefore, may be thought of as essentially a heavier-than-air flying machine, a monoplane with propeller planes, a type of motor mechanism that man has failed to duplicate. One of the most effective ways of presenting an account of the bird's characteristic structural and functional peculiarities is to compare it in considerable detail with an aëroplane.

The Bird an Automatic Aëroplane

The essential features of a heavier-than-air flying machine are:— 1, Planes or wings; 2, great and sustained power, including fuel, engine, propeller; 3, minimum weight consistent with maximum rigidity of framework; 4, steering and balancing devices, including rudder, ailerons, stabilizers. Let us consider the ways in which the bird meets these requirements.

1. Planes or Wings.—The wing of the bird (Fig. 142) is a complex of several structural elements consisting of: a framework of bones,

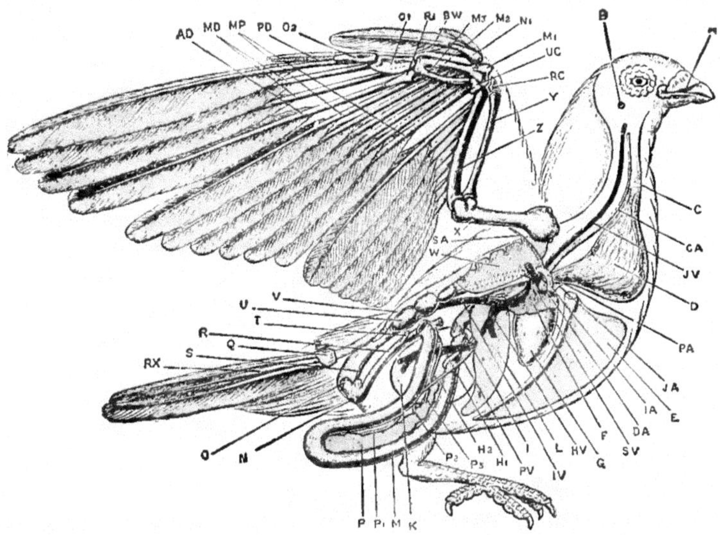

FIG. 142.—Anatomy of the pigeon. *A*, nostril; *AD*, ad-digital primary feather; *B*, external auditory meatus; *BW*, bastard wing; *C*, œsophagus; *CA*, right carotid artery; *D*, crop; *DA*, aorta; *E*, keel of sternum; *F*, right auricle; *G*, right ventricle; *HV*, hepatic vein; *H1*, left bile-duct; *H2*, right bile-duct; *I*, distal end of stomach; *IA*, right innominate artery; *IV*, posterior vena cava; *JA*, left innominate artery; *JV*, right jugular vein; *K*, gizzard; *L*, liver; *M*, duodenum; *MD*, mid-digital primary feathers; *MP*, metacarpal primaries; *M1*, preaxial metacarpal; *M2*, middle metacarpal; *M3*, postaxial metacarpal; *N*, cloacal aperture; *N1*, preaxial digit; *O*, bursa Fabricii; *O1*, proximal phalanx of middle digit; *O2*, distal phalanx of middle digit; *P*, pancreas; *PA*, right pectoral artery; *PD*, predigital primary; *PV*, portal vein; *P1*, first pancreatic duct; *P2*, second pancreatic duct; *P3*, third pancreatic duct; *O*, pygostyle; *R*, rectum; *RC*, radial carpal bone; *RX*, rectrices; *R1*, ulnar digit; *S*, ureter; *SA*, right sub-clavian artery; *SV*, right anterior vena cava; *T*, rectal diverticulum; *U*, kidney; *UC*, ulnar carpal bone; *V*, pelvis; *W*, lung; *X*, humerus; *Y*, radius; *Z*, ulna. (From Hegner, after Marshall and Hurst.)

muscles, nerves, blood vessels, and feathers. The bony framework is that of a modified fore limb of which the human arm is a good prototype. The humerus is large and has heavy ridges for the attachment of the huge pectoral flight musculature. The radius and ulna are largely unmodified, though the ulna is larger than the radius and has a larger than usual head for muscle attachment. The wrist, hand, and finger bones are highly modified both through loss of whole bony units and by the fusion of the remaining bones into strong complexes. The thumb or pollex is reduced to a small rudiment, the index finger is the largest, the second finger fairly well developed, but there is no trace of the third and fourth fingers. The phalangeal part of the fore limb is reduced essentially to a one-fingered condition.

Of the **wing muscles** those of the upper arm are very large and powerful, those of the lower arm much reduced, and those of the hand atrophied. The only movements of the wings are those of elevating, depressing, extending and flexing. The real flight muscles are the chest muscles or pectorals, massive groups of fine-grained striated fibers, which are inserted upon the keel of the sternum. These muscle masses, which are capable of prolonged exertion without fatigue, correspond to the cylinders of the aëroplane motor.

The *wing feathers* are the main factors in giving large planing surface to the wing. A feather (Fig. 143) from the morphological standpoint, is no more nor less then an elaborately subdivided scale, rolled up into a cylinder proximally and expanded into a flat vane at the distal end. The quill is residue of the embryonic rolled-up stage. The vane is composed of a number of subdivisions called barbs, each of which is redivided into minute barbules which are hooked to the barbules of adjacent barbs so as to give stability to the whole vane and to make the feather as a whole a coherent, springy plane. A single row of large flight feathers grows out from the back of the arm and hand bones, each partly overlapping its neighbor. Several rows of so-called coverts overlie these like shingle rows. The overlapping arrangement of all the feathers contributes greatly to make the wing a fairly rigid, but sufficiently flexible plane, which is better adapted for the purpose than the perfectly rigid planes of man-made machines. The wing differs also from the plane in that it is jointed and capable of being folded away when not in use, or of regulating its exposed surface by flexures.

2. Power.—The secret of great and sustained power lies in the capacity to convert chemical energy into mechanical motion through rapid and complete combustion of fuel. In the aëroplane, gasolene is the fuel, the electric spark is the combustion agent and oxygen the combustor; in birds carbohydrates, etc., constitute the fuel, the nerve impulse is the combustion agent and oxygen the combustor; the wing muscles, especially the pectorals, are in flying birds extremely massive, which means that a great excess of energy is always available; the nervous system is highly efficient; and the supply of oxygen is ensured by the extraordinary development and unique structure of the lungs and air passages, as well as by the adequate blood supply and its circulation. The *lungs* proper are not unduly large, but their capacity is greatly increased by the addition of large *air-sacs*, that branch off from the lungs. These air-sacs fill all of the cœlomic spaces and even send fine branches into the hollows of the bones. By this scheme two functions are subserved: that of sending oxygen directly to many tissues, and that of lessening the weight of the body. The lungs moreover differ from those of reptiles or mammals in that a through draft of air is made possible through a system of *excurrent bronchi*, passages that carry used air out of the lung alveoli without interfering with the fresh air that enters through the *incurrent bronchi*. Thus the bird's oxygen supply is much better provided for than that of any other vertebrate, and in some respects approximates that possessed by the flying insects. Adequate oxidation is further provided for by the large heart (Fig. 142) and by voluminous blood vessels, both of which are proportionately more generous in their blood-carrying capacity than those of other vertebrates.

The **high temperature** of the bird is another important element in its power plant. Obviously, the higher the temperature, the more rapid the combustion. The bird's temperature is considerably higher than that of mammals, as anyone knows who has felt the skin of a live fowl. In the best fliers it runs up to 110° or 112° F., even when the birds are at rest. Two elements are concerned in maintaining the characteristic avian temperature: a vaso-motor system, similar to that of mammals, and an unusually effective non-conductive coat of feathers, which prevents surface loss of heat; and no known material does this more effectively than the feather coat of a bird, especially when the feathers are arranged as they are in nature. With this equipment the bird is able to endure the intense cold of the upper atmospheric strata without undue loss of heat and without the least danger of freezing.

The **alimentary system** is also proportionately effective. It must be, for it is the fuel refinery. Crude power materials are taken into the crop or storage tank, are gradually fed into the grinding mill (gizzard) and passed into the stomach proper, and subsequently into the intestines, in such a condition that digestion, or the refining of the fuel, is rapid and complete. Much might be said of the efficiency of the excretory apparatus, but this may be assumed.

The **mechanics of propulsion** is difficult of explanation because of its extreme complexity; but this much may be said: the wing stroke is practically like the arm stroke in swimming. It must do two things: prevent the body from falling, and give a forward impulse. The stroke must therefore be downward and backward; but a forward and upward stroke, like the recovery stroke in swimming, alternates with the power stroke. The possibility of effective and rapid propulsion depends on the relatively frictionless character of the recovery stroke. This is accomplished by bringing back the wing edgewise to the resistance of the air. Many birds make progress by planing up and down the air currents with nearly rigid wings. In this phase of flight man has equaled, if not surpassed, the bird.

3. Lightness and Rigidity.—Many elements combine to make the bird a model of mechanical perfection in this respect. The skeleton (Fig. 144) exhibits instances of the use of nearly all of the recognized architectural principles designed for getting the most strength and rigidity out of the least material. The T and I beam principles are used in many of the bones, the most striking example being the sternum, an ideal T beam. Many of the bones are broadened and flattened; there is much overlapping, as in the uncinate processes of the ribs; and there is very extensive fusion of adjacent bones, with resultant increase of rigidity. The vertebral column, with the exception of the cervical region, is practically rigid, extensive fusions having taken place between the vertebræ themselves, and between the latter and the bones of the pelvis. The bones of the skull are almost paper-thin, but are so fused into a unit as to make a practically sutureless brain-box. A large number of bones are lost, especially in the wings and legs, and those that remain are filled with air instead of with bone-marrow. Thus the skeleton of the birds is, among vertebrates, much the lightest for its size, yet the strongest, as it must be to withstand the racking strains incident to flight.

In a sense the bird is also partially a balloon in that quantities of hot air are carried, not only in the extensive air-sac system, but also inclosed between the body and the feathers and among the innumerable feather interstices. Nearly half of the contour volume of a bird is air-filled.

4. Steering and Balancing Devices.—The tail and its feathers (rectrices) is a rudder which may be used as well for vertical as for lateral steering. Elevating the tail produces an upward slant, depression a downward turning. Tilting from side to side gives lateral steerage. Expanding the feathers like a fan, or closing them together, increases or decreases the effectiveness of the rudder. Balancing devices are used especially in soaring, when irregular wind currents strike the outspread wings and tend to capsize the vessel. To equalize irregularities of air pressure on the two wings the bird may decrease the surface of the wing by partially flexing it at elbow or shoulder, or by twisting the tip of the wing so as to spill off the excess wind. Part of the stabilizing equipment consists of the flexible ends of the feathers which bend upward and spill off the air, much after the fashion of the ailerons on an aëroplane. In the bird no elaborate stabilizer is necessary, for each individual is an automaton, with an effective system of balancing reflexes ever on the alert.

Any more extensive discussion of the flight adaptations of the bird would lead us into a technical exposition quite out of place in the present volume. Enough has been presented to impress the reader with the fact that almost all of the characters that distinguish a bird from a reptile are fundamentally elements belonging to its flying equipment. Unless therefore these characters were evolved in connection with flight they are meaningless; for no other set of conditions could have called forth this peculiar combination of characters.

Birds and Reptiles Compared

Apart from its flight adaptations the bird has been shown to be extraordinarily reptile-like. The avian egg is essentially like that of the reptile, both in size and in envelopes. The developmental history, though much more rapid, as the result of higher temperatures, is essentially reptilian. This speeding up of the developmental rate has evidently been an important element in the evolution of the bird. Like the reptile the bird's jaw consists of several bones and articulates with the quadrate. The skull bones are not materially different from those of the reptile; while the vertebræ, in their variable number especially in the cervical region, and their lack of epiphyses, are reptilian. The hind limbs and pelvic girdle are strikingly like those of some of the dinosaurs. The circulatory system is somewhat different from that of any living reptile, but is the logical development of tendencies observable in the latter. The only aortic arch is the right (but there is a reduction of the left arch in many reptiles); the heart is completely four chambered (but the crocodile heart is nearly so); the red blood corpuscles are nucleated as in the reptile.

Formal List of Characters of a Typical Bird

External Characters.—Body short and spindle-shaped; head, neck and trunk clearly defined; tail short and broad; horny beak with patch of swollen skin at base, called *cere;* two slit-like oblique nostrils between beak and cere; eye with upper and lower lids and a complete third eyelid or *nictitating membrane;* auditory aperture behind the eye, without external ear, leading to the tympanum; wings already described; legs covered with scales and armed with claws.

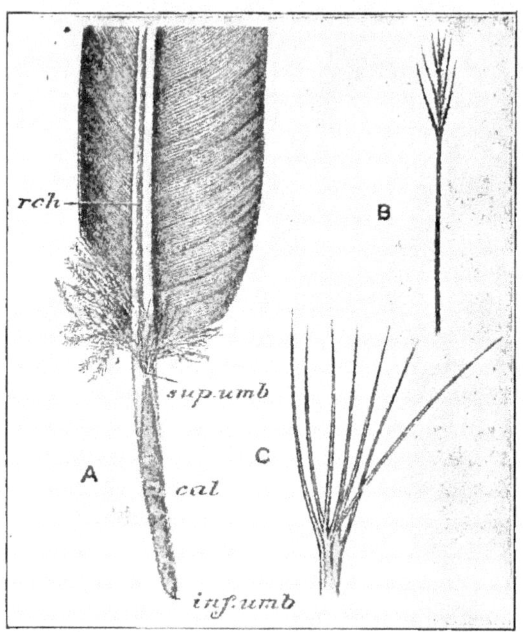

Fig. 143.—Feathers of pigeon. *A*, part of a tail feather; B, filoplume; C, nestling down. *cal*, calamus; *inf. umb*, inferior umbilicus; *rch*, rachis; *sup. umb*, superior umbilicus. (From Parker and Haswell.)

Feathers (Fig. 143.)—A feather is a modified scale, that arises from a dermal papilla and is at first covered with an epidermal sheath. A typical feather consists of a stiff axial rod or stem, of which the basal portion is hollow and forms the quill or *calamus;* the distal part is filled with pith and is called the *rachis.* The rachis supports the *vane* or flat part of the feather, which is composed of parallel *barbs*, each barb divided into numerous *barbules* along either side, that hook themselves to barbules of adjacent barbs and thus help to make a coherent plane out of a series of separate parts. Three principal types of feathers are to be distinguished: a, *contour feathers,* which include the flight feathers; b, *down feathers,* possessing a soft shaft and a vane without barbules; c, *filoplumes,* with slender, hair-like shaft and few or no barbs. Feathers are arranged in tracts, called *pterylæ,* with naked spaces between, called *apteria.* Moulting of feathers occurs periodically, old feathers being dropped and new ones, sometimes of different color, growing out of the old follicles.

The Skin is dry and practically without glands. The only skin gland is a single *oil gland* on the tail; even this is absent in some species.

The Skeleton (Fig. 144).—Most of the skeletal peculiarities have been already discussed. The sternum is keeled except ostriches, etc.; ribs have *uncinate processes,* except Screamers; skull is rounded, has large orbits and the facial bones are extended out upon the beak; quadrate is movable and articulates with the squamosal; a single occipital condyle; no teeth, except extinct forms; cervical vertebræ have saddle-shaped articular surfaces, giving the neck great flexibility and rendering the beak an unusually versatile implement; trunk vertebræ mostly fused; three or four free caudal vertebræ with terminal pygostyle: two cervical and three to nine thoracic ribs, the latter attached to the sternum; pectoral girdle consists of paired blade-like scapulæ, paired coracoids, that are united to the sternum, and free clavicles, fused in the middle to make the "wish-bone"; the pelvic girdle is a solid bone, consisting of the fused *ischia, ilia,* and *pubes,* and the pelvis is firmly fused with the sacral vertebræ; the wing skeleton has been sufficiently described; the leg skeleton consists of a large *femur,* a slender *fibula* and the long, stout *tibio-tarsus,* composed of the fused tibia and proximal tarsal bones; the ankle joint is between the tibio-tarsus and the tarso-metatarsus; foot has four digits, with hallux usually directed backward.

Digestive System.—Mouth hard and narrow; tongue hard and often of great functional value; œsophagus with enlargement, called the crop; stomach with *proventriculus* that secretes gastric juice, and a muscular *gizzard* or gastric mill; intestine U-shaped, composed of *duodenum, ileum* and *rectum;* between ileum and rectum are two *cæca;* rectum opens into a cloaca. There are two bile ducts but no gall-bladder; a *pancreas* empties into the duodenum.

Circulatory System.—The heart is large and four chambered; right auricle receives venous blood, left receives blood from the lungs; the

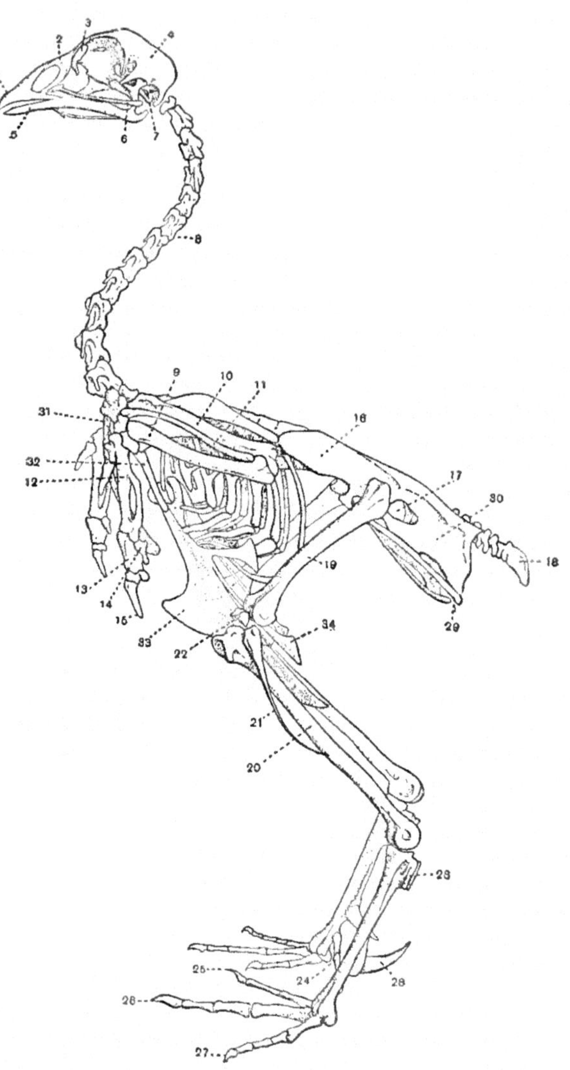

FIG. 144.—Skeleton of a Bird (Common Fowl.) *1*, premaxilla; *2*, nasal; *3*, lachrymal; *4*, frontal; *5*, mandible; *6*, lower temporal arcade in region of quadratojugal; 7, tympanic cavity; *8*, cervical vertebræ; *9*, ulna; *10*, humerus; *11*, radius; *12*, carpometacarpus; *13*, first phalanx of second digit; *14*, third digit; *15*, second digit; *16*, ilium; *17*, ilio-ischiatic foramen; *18*, pygostyle; *19*, femur; *20*, tibio-tarsus; *21*, fibula; *22*, patella; *23*, tarso-metatarsus; *24*, first toe; *25*, second toe; *26*, third toe; *27*, fourth toe; *28*, spur; *29*, pubis; *30*, ischium; *31*, clavicle; *32*, coracoid; *33*, keel of sternum; *34*, xiphoid. The forked bone in front of 7 is the quadrate. (From Shipley and McBride.)

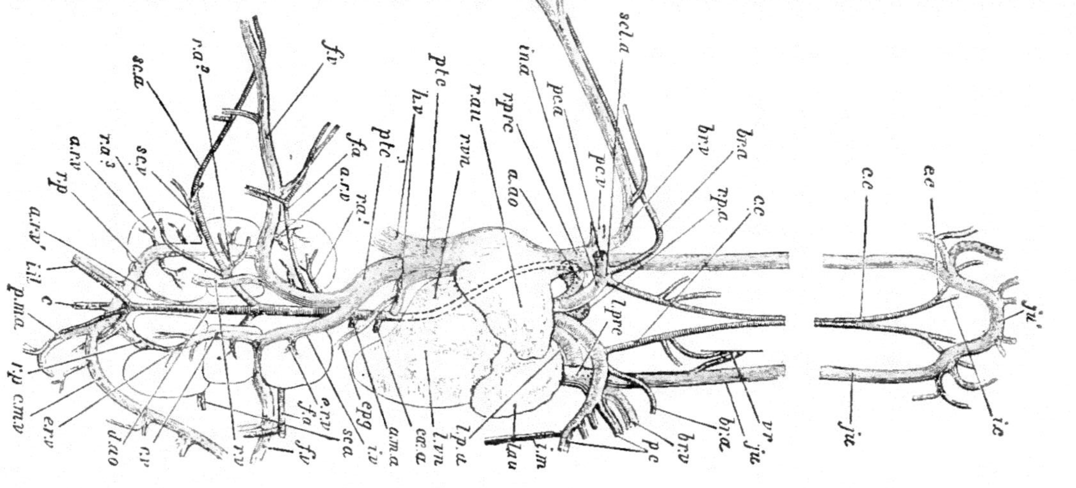

Fig. 145.—Circulation of Bird (Pigeon), showing the heart and chief blood vessels, ventral aspect. *a. ao*, arch of aorta; *a. m. a*, anterior mesenteric artery; *a. r. v*, afferent renal veins; *a. r. v'*, vein bringing blood from pelvis into renal portal system; *br. a*, branchial artery; *br. v*, branchial vein; *c*, caudal artery and vein; *c. c*, common carotid artery; *c. m. v*, cocygeo-mesenteric vein, displaced to the right; *cœ. a*, cœliac artery; *d. as*, dorsal aorta; *e. c*, external carotid artery; *epg*, epigastric vein; *e. r. v*, efferent renal vein; *f. a*, femoral artery; *f. v*, femoral vein; *h. v*, hepatic vein; *i. c*, internal carotid artery; *i. il*, internal iliac artery and vein; *i. m*, internal mammary artery and vein; *in. a*, innominate artery; *i. v*, iliac vein; *ju*, jugular vein; *ju'*, anastomosis of jugular veins; *l. au*, left auricle; *l. p. a*, left pulmonary artery; *l. pre*, left precaval vein; *l. vn*, left ventricle; *pc*, left pectoral arteries and veins; *pc. a*, right pectoral artery; *pc. v*, right pectoral vein; *p. m. a*, posterior mesenteric artery; *ptc*, postcaval vein; ra^1, ra^2, ra^3; renal arteries; *r. au*, right auricle; *r. p*, renal portal vein, on left side of figure, supposed to be dissected so as to show its passage through the right kidney; *rp. a*, right pulmonary artery; *r. pr. v*, right precaval vein; *r. v*, renal vein; *r. vn*, right ventricle; *sc. a*, sciatic artery; *sc. v*, sciatic vein; *scl. a*, subclavian artery; *v. r*, vertebral artery and vein. (From Parker.)

right aortic arch carries all of the arterial blood to the system. Further details of the circulatory system are best understood from the illustration (Fig. 145).

Respiratory System.—Large lungs each with nine thin-walled air-sacs. Air enters bronchi, passes to air-sacs and thence in a warmed condition into the alveoli of the lungs and makes its exit through the

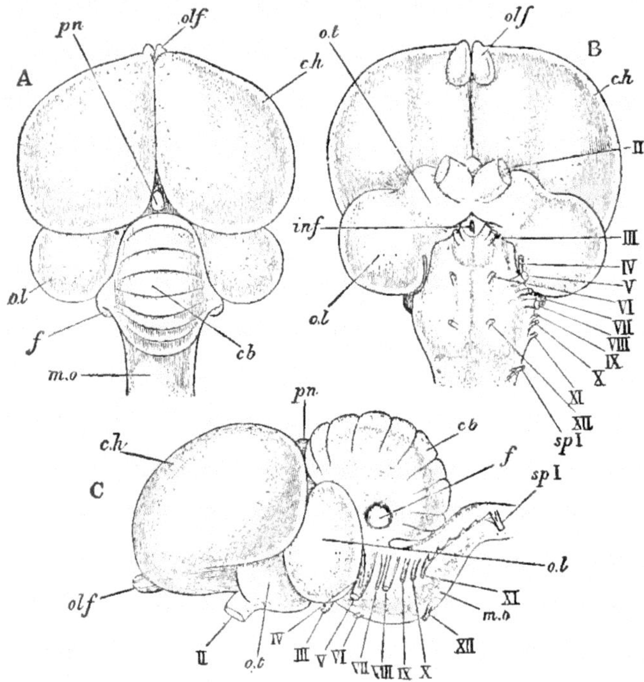

FIG. 146.—Brain of Bird (Pigeon). A, dorsal; V, ventral; C, left lateral view; *cb*, cerebellum; *f*, flocculus; *inf.* infundibulum; *m. o*, medulla oblongata; *o. l*, optic lobes; *o. t*, optic tracts; *pn*, pineal body; *II–XII*, cerebral or cranial nerves; *sp. I*, first spinal nerve. (After Parker.)

excurrent bronchi. A complete change of air occurs at each inspiration and expiration. The trachea and the larger bronchi are kept open by means of rings of cartilage; the trachea is enlarged, just before it divides, into a *syrinx* or voice box, a structure limited to birds and that is in no way homologous with the larynx of mammals; the mechanics of voice production in the bird depends upon forcing air through a flexible valve, which is set in vibration.

Excretory System.—Paired tri-lobed kidneys empty by means of ureters directly into the *cloaca*. Fæces and urine are given off mixed. The bird kidney is a *metanephros*.

Reproductive System.—Testes are oval and are situated dorsally along the back. Each testis has a *vas deferens* leading to a *seminal vesicle*, where sperm is stored. In copulation sperm is simply transferred from the cloaca of the male to that of the female; for there is no penis in most birds. The *left ovary* only is functional, the right becoming vestigial early in development. The single *oviduct* is large and complex, and is provided with albuminous and shell glands. The egg, which is what is popularly called the yolk, is fertilized before it descends very far into the oviduct, soon becomes wrapped round with layers of albumen and, before it is laid, is covered by a shell which is secreted by glands in the lower part of the oviduct. Eggs are incubated outside of the body usually by means of the body heat of one or both parents.

Nervous System (Fig. 146).—The brain is very short and broad. The *cerebrum* is large but not convoluted; the *cerebellum* is very large and complex; *optic lobes* are well developed; *olfactory lobes*, rudimentary, indicating poor sense of smell.

Sense Organs.—The olfactory epithelium is poorly developed; sense of taste is almost as poorly developed as the olfactory. The inner ear, especially the *cochlea*, is more complex than in reptiles. The *eye* is the bird's main dependence; it is large and highly organized, probably keener than that of any other animal. *Sclerotic plates* cover the eye-ball. A fan-shaped *pecten* (absent in Apteryges) of unknown function is suspended in the vitreous humor.

The Origin of Birds

About the origin of birds palæontology says but little. Only one link definitely connecting the true birds with their reptilian ancestry has been discovered. This one link is the bird-reptile *Archæopteryx*, a form distinctly intermediate between the bird and the reptile, about which we shall have more to say in other connections. The evolution of modern avian characters from those seen in *Archæopteryx* is easy to imagine, for all of the avian characters are foreshadowed in this creature, which though more reptile-like than any other bird, is really not a reptile but a bird. What we need to find is a pro-avian ancestor of the birds, some true reptile that exhibits unquestioned tendencies in the direction of avian traits. The pterosaurs might seem at first thought to be the ideal group from which to derive the birds, but unfortunately these highly specialized flying reptiles are anatomically too different from birds to offer any hope of using them as a connection between the birds and the reptiles. The pterosaurs have arrived at their flying mechanism in an entirely different fashion.

The bipedal dinosaurs have been chosen by some authorities as the group offering the strongest suggestion of avian affinities. It is argued that the birds took their origin from some rather generalized offshoot of the Triassic bipedal dinosaurs, which developed flight, and, after a long period of transition, gave rise to *Archæopteryx* and other primitive birds. This is possibly the best clue as to the pro-avian ancestry of birds, but this is at best far from a satisfying phylogeny. In lieu of a definite ancestral group of reptiles from which to derive the birds the problem has become somewhat less concrete and concerns itself with an attempt to explain the origin of the flying habit. Three distinct theories of the origin of flight are held at the present time: that of the cursorial, that of the arboreal, and that of the diving origin of flight.

The theory of the **cursorial origin of flight** was advanced by Nopcsa, a Hungarian palæontologist. This author considers that there

Fig. 147.—Restoration of a hypothetical pro-avis, supposed cursorial ancestor of birds. (From Lull, after Nopcsa.)

is a very fundamental distinction between flight based on membranous planes, like those in bats and pterodactyls, and planes made up of feathers; for the former involves marked adaptations of the hind limbs, whereas the latter involves only the fore limbs and leaves the hind limbs unchanged. The hind limbs of birds are essentially homologous with those of the cursorial dinosaurs. It is therefore argued that the origin of flight involved changes in the fore limbs only and that the beginnings of flight occurred while running efficiency was at its height. The conclusion is that the first birds arose from some long-tailed reptile (Fig. 147) that sped over the earth on its strong hind legs and stretched out its fore limbs for the sake of maintaining balance and probably flapped these limbs to aid the speed of flight. These flapping fore limbs, or pro-wings, developed more surface, partly by flattening out and partly by the backward growth of the scales of the posterior margin. Similar large scales are supposed to have developed laterally on the tail. The evolution of these specialized flight-scales into feathers is thought to have been a mere matter of a continued increase in size and numbers, accompanied by regional specialization; for in reality a feather is morphologically no more nor less than a specialized scale. The gradual modification of the remaining body scales into feathers would be the logical sequence of events, and the long list of flight adaptations would appear as correlated variations. The first steps in flying would be prolonged leaps, aided by the flapping pro-wings; then short soaring flights would be made, followed by longer flights accomplished by energetic flapping of the wings alternating with periods of soaring. While rather plausible in some ways the theory of the cursorial origin of flight has not gained any general acceptance.

The theory of the **arboreal origin of flight** has met with more widespread approval. Two phases of this general theory have been advanced: the pair-wing theory, and the four-wing theory.

The "**pair-wing**" **theory** is derived directly from a study of the characters of *Archæopteryx* (Fig. 148, I). The long clawed, prehensile, probably climbing wing-fingers of this ancestral bird point toward an arboreal habitat. It is believed to have been not a true flyer, but merely a soarer or glider, capable of only short passages from limb to limb, or from tree to tree. The lack of any foundation for a flight musculature argues against the possibility that the creature could have taken any long flights in which propulsion by means of wings would be necessary.

The "**four-wing**" **theory** of Beebe is the most recent theory dealing with the origin of flight. This author made the remarkable discovery that vestigial flight feathers occur on the thighs of a number of species of modern birds. Traces of similar feathers were found on the thighs of *Archæopteryx*. These discoveries led to the conclusion that the first flyers had wings on both arms and legs (Fig. 148, J)

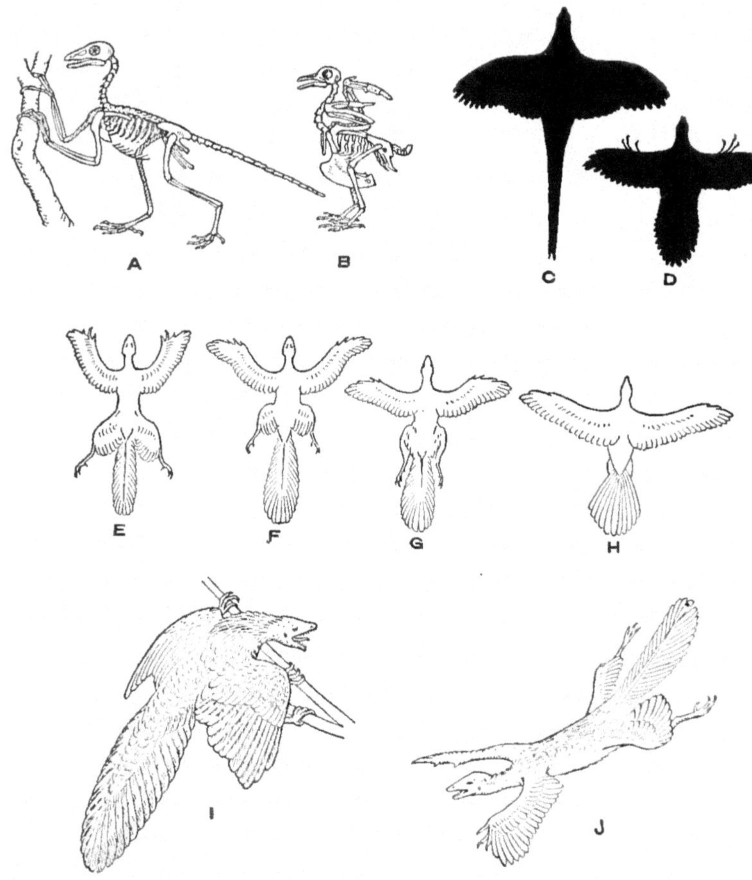

Fig. 148.—Group of figures to illustrate theories of the origin of flight. A, reconstruction of skeleton of *Archæopteryx* compared with that of a pigeon, B; C, D, silhouettes of pheasant (left) and *Archæopteryx* (right) to illustrate two wing theory of origin of flight; E, F, G, H, four stages in the hypothetical evolution of the two winged from the four winged bird. I, restoration of *Archæopteryx*, after Heilmann. J, Tetrapteryx, the hypothetical four-winged ancestral bird of Beebe. (Redrawn after Osborn's "Origin and Evolution of Life.")

and used both sets in parachuting from trees to the ground and from tree to tree. Later the wings of the legs degenerated as the tail feathers took up the duty of acting as a posterior plane, and the arm-wings increased in size and effectiveness as motor organs, as shown in Fig. 148, E, F, G, H.

Gregory's compromise theory of the origin of flight is perhaps more nearly acceptable than any of those hitherto given, and is herewith presented in his own words:

"The pro-aves were surely quick runners, both on the ground and in the trees, but it is not clear whether the upright position was first attained upon the ground or in the trees. The very early acquired habit of perching upright on the branches, as shown by the consolidated instep bones, grasping first digit and strong claws of *Archæopteryx*. Their slender arms ended in three long fingers provided with large claws which were at first doubtless used in climbing. These active pro-aves contrasted widely in habits with their sluggish remote reptilian forebears. In pursuit of their prey they jumped lightly from branch to branch and finally from tree to tree, partly sustained by the folds of skin on their arms and legs and later by the long scale-feathers of the pectoral and pelvic 'wings' and tail. That they held the arms and legs perfectly still throughout the gliding leap appears doubtful, for all recent animals that do that have never attained true flight. I cannot avoid the impression that a vigorous downward flap of the arms even before they become efficient wings, would assist in the 'take-off' for the leap, and that another flap just before landing would check the speed and assist in the landing."

Diving Origin of Flight.—So far as the writer is aware, no one has proposed a theory of flight involving the idea that flight may have originated in connection with soaring over the water and diving after fish. Yet there are certain considerations that strongly support such a conception. According to this view the pro-aves used the fore limbs, together with their membranes and elongated scales (possibly also the similar structures of the legs), as planes to aid in diving. The value of such accessories is obvious; the dive being more definitely directed, the descent being made flatter so as to carry the diver farther out from shore, and the force of the plunge being eased up sufficiently to avoid shock. If the wings were flapped more or less a longer glide out over the water could be made, and possible circling movements could be made over the water while searching for fish. It would appear therefore that the use of the pro-wings as planes in diving would serve as useful a function as in running or leaping from bough to bough.

We would then have to suppose that some of the archaic diving birds, such as the penguins, underwent a specialization of the primitive wings, using them for under-water "flying"; that others, such as the grebes, never developed them into fully effective organs of flight; while still others, such as the loons, became good flyers though still retaining their diving propensities. According to Dr. Coues, the loon practically flies under the water, using the wings as well as the feet as propellers. The strong flying sea-birds would then be derived from ancestral diving types that had gradually perfected their flight; while land-birds of all sorts would be derivatives of the sea-birds. There are, in fact, many evidences that the sea is the ancestral home of the birds and that they have invaded the land in comparatively recent times. If one turns to page 289, where the orders of carinate birds are listed, he will note that Brigade I (largely archaic birds) consist almost entirely of water birds, while Brigade II (largely modern types) consists exclusively of land birds, with the arboreal birds confined to the more highly specialized sub-orders. If this classification represents an approximation to the phylogenetic order, the arboreal birds, instead of being the most primitive (as the theory of arboreal origin of flight maintains), are a modern product, and life in the trees is a modern habit.

Archæopteryx, of course, seems to militate against the diving origin of flight, for it is assumed to be a climbing arboreal bird. But might not climbing be equally appropriate as an aid in scaling cliffs after diving and swimming in the water? Moreover, the teeth of *Archæopteryx* would be of great service in seizing fish. On the whole, then, the existence of *Archæopteryx* is no more a barrier to the acceptance of the diving than to the cursorial origin of flight; while other considerations appear to make the former more probable than the latter.

Archaic Birds (Archæornithes)

It is customary to divide all birds into two sub-classes: *Archæornithes*, consisting of but one species (*Archæopteryx lithographica*); and *Neornithes*, including all other birds living and extinct.

It is highly probable that the period of avian evolution began not later than the Triassic; hence the birds are the latest of the vertebrates to have made their appearance in the world. The earliest actual birds whose fossil remains have been found are not materially different from modern birds. The only avian creature that is not fully a bird is *Archæopteryx*.

Archæopteryx (The Lizard-Tailed Bird)

Much has already been said about this reptile-like bird, two specimens of which have been recovered from the Upper Jurassic slates of Bavaria (Fig. 149). It was a creature about the size of a crow, but with smaller wings. It had a number of reptilian characters the most important of which are:—

1. There is no true bill, but the reptile-like jaws were armed with conical teeth in distinct sockets (Fig. 150).
2. The hand-wing had three fingers, long and probably prehensile, armed with curved claws. These were probably used both for seizing prey and for clinging to trunks and limbs of trees.
3. The sternum had no keel, or a very poorly developed one at best; hence the bird could not have had strong flying muscles.

Fig. 149.—*Archæopteryx lithographica.* Berlin specimen. *c*, carpal, *cl*, furcula; *h*, humerus; *r*, radius; *sc*, scapula; *u*, ulna; *I–IV*, digits. (From Parker and Haswell.)

4. The centra of the neck and back vertebræ were biconcave, as in many primitive reptiles.
5. The fibula and tibia were not coalesced.
6. The tail was long and lizard-like, composed of about 21 free post-sacral vertebræ (Fig. 148, A). At least the first 12 vertebræ bore paired flight feathers with well-defined shafts.
7. There were structures that have been interpreted as abdominal ribs.

Two additional characters that are not reptilian nor fully avian should be mentioned.

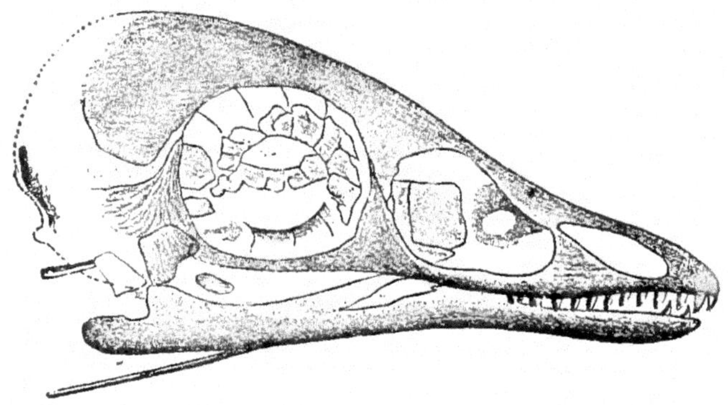

Fig. 150.—Skull of *Archæopteryx*, showing teeth and sclerotic plates. (From Headley, after Dames.)

1. The leg was rather weak as compared with that of most modern birds, a fact that militates against the theory of the cursorial origin of flight. The pelvic girdle is much smaller than in birds of the present, and is not fused with the sacral vertebræ.
2. The feathers of the wing were entirely typical both in form and in arrangement, but were rather small for the size of the body. The general contour feathers were evidently less abundant than in a modern full-fledged bird. A theoretical reconstruction of the plumage is shown in Fig. 148, I.

Modern Birds (Neornithes)

On account of the fact that *Archæopteryx* differs fundamentally from all other birds both living and extinct, it is placed in a separate sub-class: *Archæornithes.* All other birds are placed in the sub-class *Neornithes,* of which three divisions are distinguished:

Division 1. Neornithes Odontolcæ (toothed diving birds), represented by *Hesperornis* and *Baptornis.*

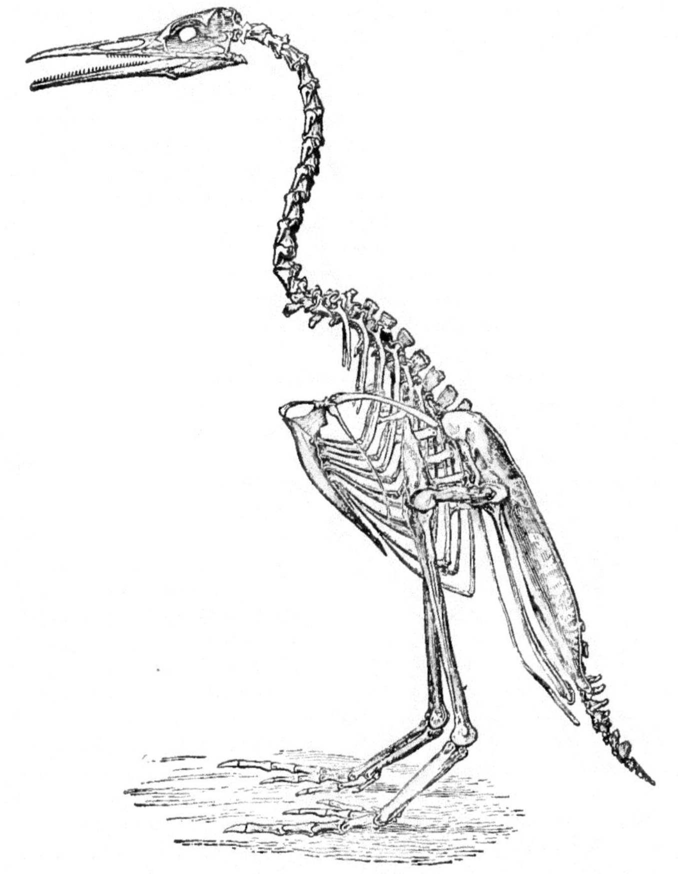

FIG. 151.—*Hesperornis regalis.* Restored skeleton. (From Parker and Haswell, after Marsh.)

Division 2. Neornithes Ratitæ (running birds), exemplified by the ostrich, and represented by both extinct and living species.

Division 3. Neornithes Carinatæ (keeled or flying birds), modern birds mainly, with a few extinct species.

The Toothed Diving Birds (Neornithes Odontolcæ)

The oldest avian remains next to those of *Archæopteryx* belong to Cretaceous times, and occur in strata that are characteristically

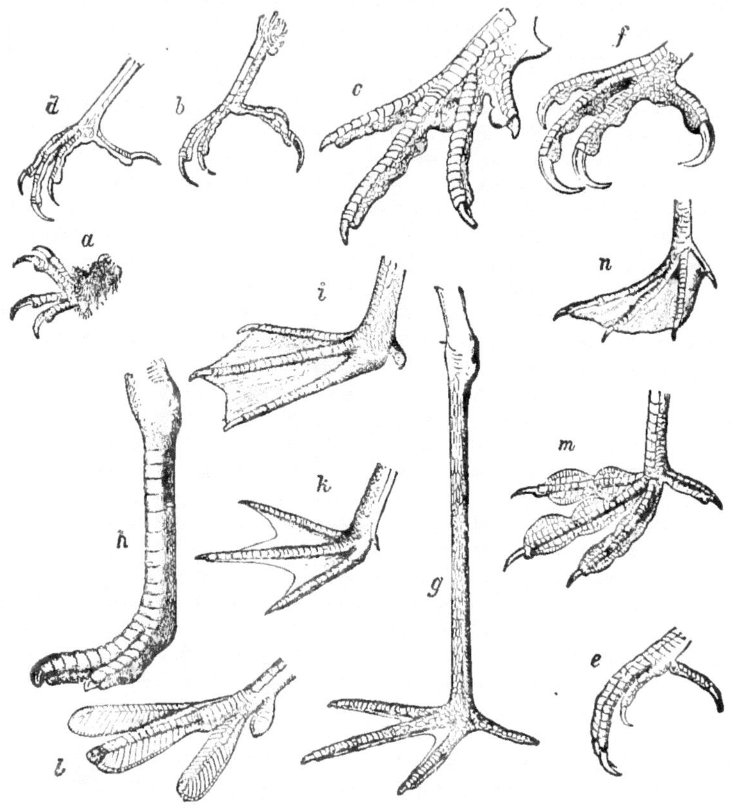

Fig. 152.—The most important forms of birds' feet. *a*, clinging foot of a swift, *Cypselus*; *b*, climbing foot of woodpecker, *Picus*; *c*, scratching foot of pheasant, *Phasianus*; *d*, perching foot of ouzel, *Turdus*; *e*, foot of kingfisher, *Alcedo*; *f*, seizing foot of falcon, *Falco*; *g*, wading foot of stork, *Mycteria*; *h*, running foot of ostrich, *Struthio*; *i*, swimming foot of duck, *Mergus*; *k*, wading foot of avocet, *Recurvirostra*; *l*, diving foot of grebe, *Podiceps*; *m*, wading foot of coot, *Fulica*; *n*, swimming foot of tropic-bird, *Phaëton*. (From Hegner, after Sedgwick's Zoölogy: b, c, d, f, n, from règne animal.)

marine; for the other fossils in these strata are essentially sea types. The bird fossils referred to evidently belonged to a type that was primarily a sea diver, as is evidenced by the rudimentary wings and flat sternum, and by the fact that the well-developed legs were set far back as in modern penguins. This species, *Hesperornis* (Fig. 151), was a large bird about four feet in length. It had a large head and its jaws were provided with true teeth imbedded in sockets of the maxillary and dentary bones. In general appearance it must have resembled the modern loons except for the wings which were very much reduced, consisting merely of a long, slender humerus, without any fore-arm or hand.

Fragmentary remains of another toothed diving bird, *Baptornis*, have also been referred to this division.

BIRDS OF TO-DAY

The Present Status of Birds.—The birds of to-day rank with the teleost fishes as a climax group. They appear to be at the height of their evolution and have undergone a very elaborate adaptive radiation, being specialized for life in the trees, for life on and under the ground (in caves), for life in waters shallow and deep, and for life in the air. They have many specialized types of diet: carnivorous, insectivorous, herbivorous, and graminivorous.

Modern birds show also many signs of racial senescence, especially in their extreme specializations of beaks and of feet, in their over-elaborate integumentary structures, and in their riotous coloration. The various types of beaks and feet are well shown in Figs. 152 and 153, and explained in the legends.

Birds exhibit very pronounced *sex-dimorphism*, the males usually being more highly colored and with more elaborate plumage, wattles, spurs and other excrescences; while the females are usually colored more like the background and are in other ways much less specialized. Birds differ from all other vertebrates in that the female is the heterozygous sex, yielding two kinds of eggs, male-producing and female-producing; whereas in other vertebrates it is the male that is heterozygous and produces male and female sperms.

The Running Birds (Neornithes Ratitæ)

This division of modern birds is comparatively small in number of species; but they make up for their small numbers by their large size. They are characterized by: absence of keel to the sternum; greatly reduced wings incapable of flight; coracoid and scapula fused together;

horny sheath of the bill in several separate pieces; large penis; tail functionless; no oil gland on the tail.

If feathers, wings, air-sacs, hollow bones, high temperature, etc., are, as we have maintained, adaptations for flight, we have no alter-

Fig. 153.—The most important forms of birds' beaks. *a*, flamingo, *Phœnicopterus; b*, spoonbill, *Platalea; c*, yellow bunting, *Emberiza; d*, thrush, *Turdus; e*, falcon, *Falco; f*, duck, *Mergus; g*, pelican, *Pelicanus; h*, avocet, *Recurvirostra; i*, black skimmer, *Rhynchops; k*, pigeon, *Columba; l*, shoebill, *Balæniceps; m*, stork, *Anastomus; n*, aracari, *Pteroglossus; o*, stork, *Mycteria; p*, bird of paradise, *Falcinellus; q*, swift, *Cypselus*. (From Hegner, after Sedgwick's Zoölogy: a, b, c, d, k, after Naumann; g, i, m, o, after règne animal; l, after Brehm.)

native but to conclude that these and other flightless birds have been derived by reverse adaptation from ancestors that were able to fly. Possibly, however, the Ratitæ represent several independent offshoots from a primitive avian stock, in which the powers of flight had not fully developed, and in which wings ceased to evolve when the powers of running served a more important function. The Ratitæ are a very old group, comparatively speaking, for their fossil remains have been found in Cretaceous rocks. Six families of Ratitæ are distinguished, four living and two extinct.

The Ostriches or Camel-birds (*Struthioniformes*).—These largest of living birds are more highly specialized as runners than are any others. The foot is a hoof-like running appendage with only two toes, with heavy claws on the short stout toes. Beneath, the foot is heavily padded with calluses. The beak is short and broad but is split back far enough to give a wide gape to the mouth. The head is comparatively small; the neck is very long and flexible. The plumes of commerce are homologous with the flight and steering feathers of the flying birds, but the barbs are not attached to one another as in the flat vane of the typical feather.

There is some difference of opinion as to how many species of ostriches exist. Some authorities recognize only one species, *Struthio camelus* (Fig. 154, D); others distinguish two additional species which they call *S. australis*, and *S. molybdophanes*. It seems advisable to treat these doubtful "species" as varieties and to deal with only one species of ostrich.

The **ostrich** lives in arid or desert country, thriving in the Sahara Desert and similar environment complexes. It is able to make good progress in the sand, for its foot is very much like that of the camel. On hard soil it is probably the swiftest runner known, being able to outdistance a good horse easily. It has, however, the unfortunate habit of running in a circle, and thus may be caught by men on horseback who know how to short-cut across the circle and thus to intercept it. Its stride is said to be fully twenty-five feet in length and when at full speed the wings are stretched out as balancers and probably partially lift the weight off the ground after the manner of the hypothetical pro-avian cursorial ancestor of the birds.

A single cock has a following of several hens, which lay their eggs in a common nest, a shallow excavation in the sand or dry soil, covered up with débris. The eggs are not left, as is popularly supposed, to be incubated by the sun's heat, but are brooded by the cock. Brooding of eggs is necessary, for the eggs would be chilled and doubtless killed by the low nocturnal temperatures characteristic of deserts and arid regions.

When cornered the ostrich fights viciously, delivering a sidewise kick that would compare favorably with that of a mule. They also bite and peck with the strong beak, but the feet are their main dependence. In captivity they are quite tractable and they are extensively cultivated on farms for the sake of their valuable plumage.

Two stupid traits are popularly attributed to the ostrich: first, that he hides his head in the sand in order to conceal himself from his enemies; second, that he eats tin cans, railroad spikes, and similar non-nutritious articles. The first is a slander on this alert, wary, and decidedly intelligent creature; for competent observers report exactly the opposite behaviour, in that when hiding it crouches low among the grasses or underbrush and only raises the top of the head and eyes above the shelter. The second is only partially true, and there is method even in this apparent show of madness; for when the bird is in captivity it sometimes is forced to use various unusual articles for abrasive purposes, in lieu of gravel or more suitable gizzard-filling material.

The Rheas (*Rheiformes*).—The rheas are much like the ostriches in general appearance and in habits, but are smaller and less highly specialized for running. They have three toes furnished with rather heavy, but typical, claws. The wings are better developed and the feathers less plume-like than in the ostrich. The head, neck, and thighs are feathered. The rheas are popularly confused with the ostrich; in fact *Rhea americana* (Fig. 154, A) is called the "American ostrich." This species lives upon the pampas of Argentine, southern Brazil, Bolivia, and Paraguay. They are swift runners, with a habit of doubling upon their pursuers and occasionally lying down in the long grass with only the head protruding. Often they lie in this position until almost trodden upon, apparently relying implicitly on the efficacy of their concealment. When running at full speed they materially aid their progress by vigorously flapping their wings. Mating and nesting habits are almost identical with those of the ostrich.

The Emeus and Cassowaries (*Casuariiformes*). These large birds are characterized by: rudimentary wings; long, limp, bifurcated contour feathers; no plumes; three toes with typical claws; legs proportionately shorter than in the two preceding families.

There are several species of cassowaries (Fig. 154, C), native to Australia and to several islands of the Malay Archipelago. They live in wooded country, keeping to the densest parts. They are swift and apparently reckless runners for they go at breakneck speed through the heavy underbrush, over logs and other obstacles six

FIG. 154.—Group of Ratite Birds. A, Rhea, *Rhea americana;* B, the Kiwi, *Apteryx australis;* C, Cassowary, *Casuarius uniappendiculatus;* D, Ostrich, *Struthio camelus;* E, Emeu, *Dromæus novæ-hollandiæ.* (Redrawn after Evans.)

feet or more high. Rivers are no obstacles, for they are excellent swimmers.

The plumage is much like long, soft fur and is used for weaving rugs and ornaments. The head is quite a striking object, generally blue in color, with flesh-colored wattles and an orange stripe down the middle of the back of the neck. A black shield or casque with green sides adorns the top of the head. This color description gives the suggestion that such a head would be decidedly conspicuous, but our modern knowledge of camouflage would lead us to believe that in the dense woods such a combination of colors might be practically invisible. A nest is made of leaves and grass and a few large green eggs are deposited therein. As in the other Ratitæ, the cock broods the eggs.

The *emeu* (Fig. 154, E) is a native of Australia and is not unlike the cassowary in habits and habitat, except that it lives in woods that are less dense. They are purely monogamous, differing in this respect from other ratite birds. The male incubates the eggs that are laid to the number of a dozen or more in a hollow, scraped out of the surface soil. The flesh is palatable and the subcutaneous fat is used for oil.

The Kiwis (*Apterygiformes*).—The kiwis (Fig. 154, B) are frequently called "New Zealand wingless birds." They are the smallest of the modern ratite birds, unless we include the tinamous, whose ratite affinities are in question. The beak is long and slender; the neck and the legs are comparatively short; the wings are more rudimentary than those of any living bird and are completely concealed beneath the long, hair-like plumage; there are four toes, but the hallux is quite short. Five species of the genus *Apteryx* are distinguished. These are distributed on the various islands of the New Zealand group, where they occupy wooded, hilly country. These strange birds live a nocturnal life, hiding in burrows of their own making during the day. The burrows are dug out by scratching movements of their strong feet. They can run much more swiftly than one would expect them to do, considering the comparatively short legs. Their stride measures at least a yard long and involves leaving the ground at every step. When they are cornered they strike viciously with the feet, raising the leg as high as the breast and delivering a downward blow. Their food consists mainly of earthworms, which are best secured at night. The bird seizes the worm with the long beak and gently pulls it out of its hole, using a curious wriggling motion. The name "kiwi" was suggested by their loud, whistling note. The nest, if such it may be called, is an enlarged chamber at the end of the tunnel-like burrow and is made by the female. The male, however, with true ratite chivalry, assumes the main responsibility of incubating the two large eggs.

Extinct Ratitæ

The Moas (*Dinornithiformes*).—When British explorers first occupied New Zealand nearly seventy years ago the skeletons of gigantic wingless birds were found scattered about the plains. These skeletal remains were in such a good state of preservation that it seems probable that there were living moas less than five hundred years ago. It may well be that the last of these birds were exterminated by the Maoris. *Dinornis* was in general appearance not unlike the ostrich, but was very much more heavily built in the legs and had either no wing bones at all or at best the merest rudiments of wings. The birds were somewhat taller than the ostrich, with head and neck much like those of the latter.

The Elephant Birds (*Æpyornithes*).—These birds probably were living in Madagascar less than two centuries ago. They are believed to have furnished the factual foundation for the mythical "Rocs" of Sinbad the Sailor. They were out of accord with these birds of oriental fiction in that they were incapable of flight and were much less gigantic in size, being only about seven feet in height, though of massive build. The eggs were surprisingly large in size, some of those which are still used by the natives as receptacles, measuring thirteen by nine inches and having a capacity of two gallons. This is the largest egg on record, though doubtless some of the extinct giant reptiles had larger ones. No doubt this fine bird was hunted out of existence by the native tribes of Madagascar. Possibly the collecting of their eggs was more destructive to the species than was the slaughter of adults.

Appendix to the Ratitæ

The Tinamous (*Crypturiformes*).—The systematic relations of this interesting group of birds is in dispute, some authorities placing them near the ostriches among the ratite birds, and others classing them as an aberrant family of the order Galliformes, among the carinate birds. By considering them as an appendix to the Ratitæ we shall avoid doing violence to the opinions of either faction.

The home of the tinamous (Fig. 157, A), of which there are about forty species, is the New World, their range being from the extreme lower end of South America to Mexico. Though they bear a strong superficial resemblance to gallinaceous birds, especially the partridges, it is believed by some authorities that they are more fundamentally related to the Ratitæ, though they are able to fly. Their wings are short and rounded, but the keel of the sternum is well developed and the pectoral musculature is large in size. The tail feathers are reduced in size, even rudimentary in some species. They are strong, swift runners and are reluctant to resort to real flight; nevertheless when they do fly they make a fairly good job of it for short distances. With a great whirring of wings and extraordinary effort they rise to fifty or sixty yards above the ground and then with expanded wings glide slowly down to the ground, covering distances of about a thousand yards, which may be repeated several times if necessary.

It seems likely that the tinamous represent a condition intermediate between the flying birds proper and the running or flightless birds, for they possess characters that relate them to both groups. Some authorities believe that the ostriches and their kin have probably been derived from an early group of rather weak fliers that is now represented by the modern tinamous.

Keeled or Flying Birds (Neornithes Carinatæ)

Nearly twelve thousand species of modern birds belong to this great division, as compared with a dozen or so species of all other living birds. The study of birds has grown into the highly specialized science of Ornithology, and a very large number of both professional and amateur naturalists and bird lovers have been engaged in adding to the already voluminous annals of bird lore and pseudo-lore. A vast literature dealing with the habits, distribution, migrations, and adaptations of birds has accumulated, much of which is worthless, because exaggerated, inaccurate, and superficial. But the authentic literature on all phases of bird life is so voluminous that no one but a specialist can hope to keep abreast of it.

The classification of the carinate birds, though elaborate, is in a fairly satisfactory condition. Only in a few minor points is there radical disagreement among authorities. Among the most acceptable classifications of the Carinatæ is that of Knowlton in his "Birds of the World," and that of Evans in the volume on "Birds" in the Cambridge Natural History, an outline of which is as follows:

Brigade I. (Largely archaic types)

LEGION I. COLYMBIMORPHÆ (Diver-like Birds)
Order 1. Ichthyornithiformes
" 2. Colymbiformes
" 3. Sphenisciformes
" 4. Procellariiformes
LEGION II. PELARGOMORPHÆ (Stork-like Birds)
Order 5. Ciconiiformes
" 6. Anseriformes
" 7. Falconiformes

Brigade II. (Largely modern types)

LEGION III. ALECTOROMORPHÆ (Fowl-like Birds)
Order 8. Tinamiformes
" 9. Galliformes
" 10. Gruiformes
" 11. Charadriiformes
LEGION IV. CORACIOMORPHÆ (Crow-like Birds)
Order 12. Cuculiformes
" 13. Coraciiformes
" 14. Passeriformes

Knowlton's classification differs from that of Evans in only a few major particulars:— 1, the tinamous are placed among the Ratitæ in close association with the ostriches, instead of next to the Galliformes; 2, the penguins are placed among the flightless birds, immediately following the Ratitæ; the genus *Ichthyornis* is placed in the same order as the toothed diving birds *Hesperornis* and *Baptornis* instead of among the flying or carinate birds; 4, there is no brigading of the birds into archaic and modern brigades, and there is no grouping of orders into legions, a proceeding that is less likely to lead into false phylogenetic implications than the somewhat artificial grouping of Evans.

The present writer prefers to follow neither classification rigidly but to use a combination of the two methods. The limitations of the present volume forbid any but the most cursory treatment of this immense and extremely attractive assemblage of modern vertebrates. Our plan will be to give a brief characterization of each order, to indicate the various groups that comprize it, and to select a few

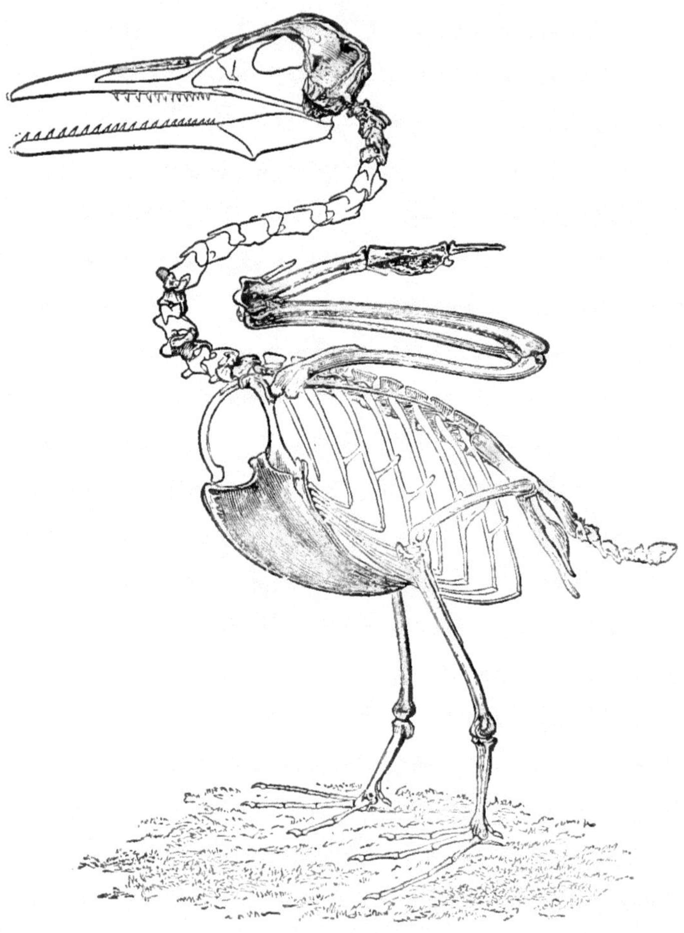

Fig. 155.—*Ichthyornis victor*. Restored skeleton. (From Parker and Haswell; after Marsh.)

of the most interesting or significant species for illustrative types. In many cases the species singled out for description are of no more intrinsic interest or importance than many others that might have been chosen. As a rule when the choice has lain between a native and a foreign type the native type has been given the preference.

Toothed Flying Birds (Ichthyornithiformes)

This order is represented by the single extinct genus *Ichthyornis* (Fig. 155), several species of which have been found in Cretaceous strata. These birds were evidently rather gull-like divers, if one may judge by structure, but differed from all modern carinate birds in that they had true teeth in sockets. Were it not that they are distinctly keeled or flying birds they might appropriately have been placed, as Knowlton places them, in the order with the toothed diving birds of even earlier times.

The Penguins (Sphenisciformes)

These curious, highly specialized, marine diving birds (Fig. 156, B), have a wide distribution among the Antarctic Seas. They are really flightless birds and might on that account be excluded from the Carinatæ, but they have well-developed wings and a fairly good keel to the sternum, the wings being used for "flying" through the water instead of through the air; for the wings and not the feet are the chief organs of locomotion; a unique character among diving birds. The legs of the penguin are set so far back on the trunk that in the water they are used primarily as a rudder, and on land their terminal position makes the bird practically sit upright on the tail. The wings are modified into flippers not unlike those of the whale; they are quite devoid of flight feathers and the bony framework is quite stiff and inflexible. The swimming stroke, when under the water, consists of alternating rotary sweeps of the two flipper-like wings, which drive the pointed body through the water at a fine speed. Penguins live on fish, mollusks, and crustaceans. They are markedly gregarious, especially during the breeding season, thousands of them being congregated upon the narrow confines of rocky islets and points of land along the sea shores. From various elevations they are constantly diving into the icy water after their food, emerging wet and glistening, but capable of almost instantly drying their plumage by vigorous shaking of the muscular skin. The penguins and the screamers are the only birds that have the skin completely covered with feathers. In the penguins the feathers are lance-shaped and have flattened shafts; they overlap one another in the most perfect fashion so as to shed effectively all water from the skin. Certain burrowing species of the Falkland Islands differ rather sharply from the others, in that they lay their eggs in rather shallow burrows. The penguins are considered to be so radically different in structure from both flying birds and ratite birds that they might well be placed in a separate division coördinate in rank with the Ratitæ and the Carinatæ.

The Loons and Grebes (Colymbiformes)

This archaic and quite isolated group of diving birds is placed first among the modern flying birds because they possess a more generalized structure than any other. The **loon** or great northern diver (Fig. 156, A), is the example of the order most familiar to dwellers in the Northern States. Its weird, laughing cry is one of the outstanding features of our northern woodland life. The ability of the loon to dodge a bullet by diving is proverbial, even if not true. The coloration of this striking bird is a study in contrasting blacks and whites, with a checkered pattern on the back, white breast, black head, and white and black bands on the neck. On the land the loon is quite clumsy and makes poor progress in walking. It really never seems to come ashore except for nesting purposes, when it deposits its two large eggs in some slight depression not far from the water's edge. Fortunately, the eggs are of a brownish mottled color and are so nearly in harmony with the background that they are very difficult to detect.

Another species of loon, the Pacific Loon, has been studied by Coues, who gives the following realistic description of its behavior in the water:

> "Now two or three would ride lightly over the surface, with neck gracefully curved, propelled with idle strokes of their paddles to this side and that, one leg, often the other, stretched at ease almost horizontally backward, while their flashing eyes, first directed upward with sidelong glance, then peering into the depths below, sought for some attractive morsel. In an instant, with a peculiar motion, impossible to describe, they would disappear beneath the surface, leaving a little foam and bubbles to mark where they went down, and I could follow their course under the water; see them shoot with marvelous swiftness through the liquid element, as, urged by the powerful strokes of the webbed

FIG. 156.—Archaic Carinate Birds. A, Loon, or Great Northern Diver, *Colymbus gracialis;* B, Rock-hopper Penguin, *Eudyptes chrysocoma;* C, Albatross, *Diomedea exulans;* D, White Stork, *Ciconia alba;* E, Red-breast Goose, *Bernicla ruficollis;* F, Red Kite, *Milvus ictinius*. (D and D after Lydekker, the rest after Evans.)

feet and beats of the half-opened wings, they flew rather than swam; see them dart out the arrow-like bill, transfix an unlucky fish, and lightly rise to the surface again."

Loons are almost as efficient as flyers as they are as divers and swimmers; in this respect they are much more generalized than are the penguins.

The grebes are somewhat more like penguins than are the loons, though they too are good flyers. They are much smaller than loons and have a much wider distribution, being practically cosmopolitan in their range. The European little grebe, or "dabchick," is an interesting little fellow about nine inches in length. It has attracted a good deal of attention on account of its unique habit of taking its young one under its wing when diving into the water to escape from its enemies on the land or in the air. The American Eared Grebe is characterized by the presence of conspicuous tufts of feathers on the sides of its head that look like ears. The great crested grebe and the pied-billed grebe, which is an American dabchick, are two other well-known species.

The Petrels and Albatrosses (Procellariiformes)

These sea birds are characterized by powers of flight more marked than any other group. Because of their ability to travel great distances with the greatest ease they have attained a world-wide distribution.

The **petrels** are birds of moderate size, with extremely long, narrow wings and hooked bill. They soar about over the waves and dive into the sea after fish, their main food. The stormy petrel is considered by mariners as a prophet of rough weather when it hovers about ships at sea.

The **albatross** (Fig. 156, C), is one of the largest of flying birds, considerably larger than a goose. The following vivid word picture of Professor Hutton will serve to acquaint the reader with one of our noblest birds:

"With outstretched, motionless wings he sails over the surface of the sea, now rising high in the air, now with a bold sweep, and wings inclined at an angle with the horizon, descending until the tip of the lower one all but touches the crests of the waves as he skims over them. Suddenly he sees something floating on the water and prepares to alight; but how changed he now is from the

> noble bird but a moment before all grace and symmetry. He raises his wings, his head goes back, and his back goes in; down drop two enormous webbed feet straddled out to their full extent, and with a hoarse croak, between the cry of a Raven and that of a sheep, he falls 'souse' into the water. Here he is at home again, breasting the waves like a cork. Presently he stretches out his neck, and with great exertion of his wings runs along the top of the water for seventy or eighty yards, until, at last, having got sufficient impetus, he tucks up his legs, and is once more fairly launched into the air."

Several less well-known types of bird are also classed in this order: fulmars, shearwaters, and diving petrels.

The Stork-like Birds (Ciconiiformes)

This rather mixed assemblage of water birds includes: tropic birds, gannets, cormorants, darters, frigate birds, pelicans, bitterns, herons, ibises, storks, spoonbills, flamingoes, etc. So extensive and varied a group is it that it is difficult to characterize it as a whole. It is subdivided into eleven families, most of which are birds capable of sustained flight; many of them are wading rather than swimming birds.

The **tropic birds** are true denizens of the tropic oceans, flying hundred of miles from land, and taking shelter and rest amid the rigging of ships when opportunity affords. **Gannets** are also sea birds, but frequent the colder regions, coming ashore during stormy weather. **Cormorants** are rather large sea-coast birds with pronounced fish-eating proclivities; they are not exclusively marine, but frequent inland lakes during the breeding season. **Darters** or **snake birds** are not marine, but frequent inlets of the sea and fresh-water lakes; they are not very strong flyers, but excel as divers, leaving scarcely a ripple upon the surface when they go down after fish.

The **frigate bird** or man-of-war bird is a true sea-bird seldom coming ashore except to nest. The long wings and extremely long tail are distinctive features. **Pelicans** are familiar large birds of the tropics. The bill is very large and the lower jaw is provided with a capacious pouch in which a large supply of captured fish can be stored. The stubby tail and short legs are familiar attributes of this interesting bird. **Herons, ibises** and **storks** (Fig. 156, D) have a strong general resemblance; their long legs, collapsible necks, and slow flapping wing-stroke are known to every child. The **spoonbill** is a stork-like bird with a spoon-shaped bill with which it captures insect prey, larvæ, fish, frogs, etc.; they are largely tropical in distribution. **Flamingoes** are large, extremely long-legged, long-necked birds, with wonderful pink plumage. They are good flyers, but are more characteristically waders. In the breeding season they are decidedly gregarious, building extensive colonies of tall, chimney-like nests of mud, that are hollowed out at the top to receive the eggs. These nests, which look like a lot of tree-stumps, are made high partly to keep the eggs out of reach of the water, for they are built on low ground, and partly because they are of a convenient height for these long-legged creatures to sit down upon. One might imagine the rather precarious situation involved in an attempt of these stilted birds to sit down on a nest built at the level of the ground. In certain respects the flamingoes are transitional between the storks and the geese.

The Goose-like Birds (Anseriformes)

This order is divided into two quite well-defined sub-orders consisting of the screamers and the Anseres proper.

The **screamers** are quite unlike the goose tribe in general appearance and in habits, and it is only on the basis of skull and skeletal characters that they are classed as Anseriformes. They are about the size of turkeys and have a fowl-like head and bill. They are highly unique in two respects; the ribs are entirely devoid of uncinate processes, which are possessed by all other living birds; and they share with the penguins and ratites the distinction of being the only birds having the entire skin covered with feathers, no apteria or naked areas being present. Some writers consider these two characters so distinctive that they would assign to the screamers rank as a separate order.

The **horned screamer** is the best known species, characterized by the possession of a forward curving brow-horn about five inches in length. It also has on the anterior margin of each wing two sharp claw-like spikes, that could readily do considerable damage to an antagonist. The disproportionately loud screaming note of these strange birds has given them their name.

The remaining members of the order are **Anseres,** familiar types to everyone. The **swans** are large birds emblematic of grace of form and movement. The **geese** proper are the most generalized members of the order, and are intermediate between the swans and the ducks in their characters, especially in the length of the neck. Some of the ducks are among the most brilliant in plumage among birds. Few handsomer vertebrates exist than the male *mandarin duck.* The *eider ducks* are natives of the north and are the most widely known and highly prized members of the duck family. The *mergansers* or *fish-ducks* differ from the true ducks in having more slender bodies, long compressed bills, and grebe-like necks, and in having the edges of the bill serrated so as to give the impression that they have teeth. On account of their fish-eating habits they are not nearly so desirable for food as are most of the ducks and geese, which are largely grameniferous.

Falcon-like Birds or Birds of Prey (Falconiformes)

Just as the great carnivores among mammals are designated as the "kings of beasts," so the great birds of prey (eagles, hawks, falcons, etc.) are "kings among birds." The members of this order are characterized by hooked, raptorial beak, strong talons, large crop and predaceous habits. So much are the eagles objects of human admiration that they have been chosen as emblems of empire; even our own naturally peaceful commonwealth is proud to be represented by the king of American eagles. The order Falconiformes falls into three subdivisions, represented respectively by: the American vultures; the secretary bird of Africa; and the falcons, eagles, hawks, buzzards, Old World vultures, etc.

The **American vultures** are large birds with wonderful powers of flight, though somewhat sluggish in habit. The common **turkey buzzard** is the most conspicuous example of this division and is a familiar part of the scenery in most of our Southern States. They are economically of considerable importance on account of their effective work as scavengers, and on this account there are laws protecting them from marksmen. In spite of their value as sanitary agents they are generally looked down upon because of their disgusting feeding habits and because they have a trick of vomiting upon their adversaries. If the truth were known it would probably be found that the buzzard is a victim of chronic dyspepsia due to the unwholesome character of its food, and that it accepts with gratitude any offerings of fresh meat that may come its way. It is said to eat carrion because its beak is not strong enough to enable it to kill living prey. Perhaps the poor buzzard is more to be pitied than censured. The *Andean condor*, and the *California condor*, and the *king vulture* are other familiar members of the present group.

The **secretary bird** (*Gypogeranus*) is perhaps the strangest of all birds of prey. It is a long-legged bird, rather more like a crane in proportions than like the other members of its order; it stands about four feet in height on long slender legs, upon which it places more reliance for speeding than upon its wings. It is especially fond of snakes, though it accepts lizards, frogs and insects. Its method of attacking a snake is unique. The snake is incited to strike, and when it does the bird side-steps and receives the blow on the edge of its stiffly extended wing. The force of the blow seems to stun the snake momentarily, and the bird pounces on it and grasps it by the neck with its powerful talons.

The remaining subdivision includes the following types: falcons, gyrfalcons, duck-hawks, kestrels, falconets, carrion-buzzards, numerous types of hawks, caracaras, true eagles, hawk-eagles, harpies, harriers, and Old World vultures. Several other less known types might be mentioned, but these will suffice.

The **golden eagle** may well be allowed to represent the entire collection. This characteristic American bird is nearly a yard long and has a wing-spread of nearly seven feet. It is proverbial for its courage, but one is somewhat taken aback by what Major Bendire says about it:—"Notwithstanding the many sensational stories of the fierceness and prowess of the golden eagle, especially in defense of its eyrie, from my own observations I must confess, if not an arrant coward, it certainly is the most indifferent bird, in respect to the care of its eggs and young, I have ever seen." This disclosure might possibly make us doubt the wisdom of our selection of a national emblem, but, as though to compensate its faults in some respects, it is given credit for being "a clean, trim-looking, handsome bird, keen-sighted, rather shy and wary at times, even in thinly settled parts of the country, swift of flight, strong and powerful of body, and more than a match for any animal of similar size." To say the very least the bird is *efficient*, and in this respect not so bad an emblem. Let him who will find a better bird!

The Fowl-like Birds (Galliformes)

This order is a large and cosmopolitan one and is divided into four sub-orders, three of which are small and the other contains all of the numerous types of game birds.

The **Magagascar mesite** is the sole representative of the first sub-order (*Mesœnatides*). It is decidedly aberrant, having a head and bill more like that of a rail than like that of a game bird. So anomalous is this bird that various authorities have classed it respectively with the rails, with the cranes, and even with the song-birds.

The **Hemipodes** or **bustard quails** (*Turnices*), representing the second sub-order, are in outward appearance not unlike small quails or partridges, but differ so fundamentally from the latter in skeletal structure that they are placed in a separate division.

The **Gallinaceous game birds** (*Galli*) comprise a large assemblage of more or less familiar types, most of which need no description. Apart from the game birds proper the Galli include two families of unfamiliar birds, represented by the *brush turkeys* (Megapodes), of Australia and New Guinea, and the *curassows* and *guans* (*Cracidæ*) of tropical America.

The true **Gallinaceous game birds** consist of wild turkeys, guinea-fowls, grouse, partridges, quails, ptarmigans, prairie-hens, bob-whites, pheasants, jungle-fowls and pea-fowls. The most highly specialized types of Galli are characterized by most gorgeous plumage, notable example being the males of the golden and Lady Amherst pheasants, which are native to South China and Eastern Thibet. As described by Mr. Ogilvie-Grant, the male of the golden pheasant has the top of the head, crest and rump brilliant golden yellow, the square-tipped feathers of the back and neck brilliant orange, tipped and banded with steel blue, while the throat and sides of the head are pale rust color, the shoulders and remainder of the under parts crimson-scarlet, and the middle tail feathers black with rounded spots of pale brown; the tail is twenty-seven out of a total length of forty inches. If, as we have maintained, extravagance of coloration is one of the criteria of racial senescence, this is one of the most senescent of the birds. The **pea-fowls** are almost as wonderfully colored as the finest of the pheasants, but they are too familiar to require description. Their native home is in oriental countries, but they have been domesticated and widely distributed by man.

The **jungle-fowls** deserve special mention because it is from them that our domestic poultry have been derived. Four distinct species of jungle-fowl are known, all of them native to the dense jungles of the Indo-Malayan region. Of these it is believed that the red jungle-fowl (*Gallus gallus*) has given rise to all of the domestic breeds of poultry. The breed known as the black-breasted game has retained more completely than any of the others the characters of the wild ancestor. The most extreme deviations from the primitive characters of the species are seen in the *Japanese tosa fowl*, in which the tail feathers have been known to reach a length of fifteen feet, and the *cochins*, with their short, plump appearance and feathered shanks.

The **hoactzin,** representing the fourth sub-order (*Opisthocomi*), is one of the most curious of birds. In the adult condition it is not unlike a small type of pheasant, but it has certain anatomical characters that set it apart from all other birds:—the breast bone is wider behind than in front; the keel of the sternum is confined to the posterior part; the crop is extremely large and muscular, invading the space usually taken up with pectoral muscles and the anterior part of the sternum; and the bones of the shoulder girdle are fused completely to one another and to the sternum. The most remarkable features of the hoactzin, however, concern the young bird, which, when newly hatched, has a well-developed clawed thumb and index finger on the wing, reminding one of the condition in *Archæopteryx*. By means of these wing-digits and the feet which are extraordinarily large and strong for a young bird, these youngsters are able to clamber about among the branches and hunt for their own food. They are really practically quadrupedal in the use of both pairs of limbs in climbing. It is believed by some writers that the juvenile characters of the hoactzin are reminiscences of an Archæopteryx-like ancestry. Inasmuch, however, as they belong to one of the more highly specialized groups of birds, and inasmuch as no other known type of bird exhibits similar juvenile characters, it seems more likely that these characters are adaptive, juvenile specializations. In view of all of the peculiarities of the hoactzins it is difficult satisfactorily to classify them in any order; but the rather strong resemblance of the adult to the pheasants seems to place them among the Galliformes.

The Crane-like Birds (Gruiformes)

The majority of the members of this order are waders, but some, such as the bustards and the wekas, are decidedly terrestrial. The group does not hold together as well as some of the others, and probably should be divided into two orders. Seven families are distinguished, represented by the following types:—rails, gallinules, and coots; bustards; the kagu; sun-bitterns; and finfoots.

The common **sandhill crane** is probably the most abundant and conspicuous example of the larger Gruiformes in America. Coues, much impressed by their appearance in migration flight, writes of them as follows:—

> "Such ponderous bodies, moving with slow-beating wings, give a great idea of momentum from mere weight—of force of motion without swiftness; for they plod along heavily, seeming to need every inch of their ample wings to sustain themselves. One would think they must soon alight fatigued with such exertion, but the raucous cries continue, and the birds fly on for miles along the tortuous stream, in Indian file, under some trusty leader, who croaks his hoarse orders, implicitly obeyed."

The **great bustard** is the largest European bird, being about forty-five inches long and weighing nearly thirty pounds. In general appearance it is not unlike a goose, but has a head and bill more like that of a crane. **Sun-bitterns** are rather small birds something like a rail and a heron, but with rather short legs, a very thin neck and large head with long pointed bill. When at rest the head is sunk down on the body so as to give it the appearance of being practically neckless. The **finfoot** tribe consists of birds about whose relationship there is a good deal of controversy; some authorities placing them among the grebes, on account of the grebe-like head and bill. The **rails** are rather ordinary birds, so far as appearance goes; but they are of interest because they are believed to be intermediate between the two orders Galliformes and Charadiiformes. In general appearance they remind one of both the quail and the plover.

The Plover-like Birds (Charadriiformes)

This order is considerably more homogeneous than the last, but it is difficult to select a good popular name for the group; for the gulls and pigeons are in truth not very "plover-like." The plovers are the most generalized members of the order and may well give to it their name. There are four sub-orders: *Limicolæ* (typical plover-like, shore-feeding birds), *Lari* (the gulls and terns), *Pterocles* (the sand grouse), and *Columbæ* (the pigeons).

The **Limicolæ** are marsh and shore birds, with fairly long neck, long slender bill, legs moderately long and slender, short tail and wings, and plumage streaked and of inconspicuous patterns. They usually nest on the ground and the young are capable of running very soon after hatching. To this sub-order belong: the plovers, snipes, and curlews; the sheath-bills; the crab-plovers; the pratinicoles and coursers; the sand snipes; the thick-knees; and the jacanas. Most of these are birds without any outstanding characteristics that might capture the attention. As an example we may well select the *American woodcock* (Fig. 157, D), a species native to the Mississippi valley. This bird has an unusually long bill, which it uses largely for unearthing earthworms from their burrows. It is said that a woodcock will eat half a pound of worms in a day. It is mainly nocturnal and when flushed in the daytime appears to be dazzled by the light. The *jacanas* are strange-looking tropical birds, characterized by enormously long toes and claws, by means of which they are able to walk about with ease over the lily-pads, after the fashion of a man on snow-shoes.

The **Lari** are the **gulls** and their allies, a group almost too familiar to require description. They are aquatic, mainly oceanic, in habitat, are of medium size and have unusually long, pointed wings. Besides the gulls, terns, noddies and such typically gull-like birds the sub-order includes the auks, the puffins and the murres. The **puffins** or sea *parrots* are the most grotesque members of the entire order. They have a brilliantly colored, laterally compressed bill; and their body form and attitude remind one of that of the penguins. The **great auk,** a recently extinct species, is of considerable interest. Of it Knowlton says that "its sad and untimely fate has invested it with a pathetic, not to say melancholy history." It used to be extremely abundant on the islands north of Scotland and near Newfoundland, but it was slaughtered by the millions, largely for its feathers. The eggs were also collected so that nothing is now left of that fine species but heaps of bones scattered about the lonely islands. The last living specimen was seen in 1844.

The **Pterocles** are the *pigeon grouse* or *sand grouse,* a small group that appears to combine the characters of several orders and whose

FIG. 157.—Representative Carinate Birds. A, Great Tinamou, *Rhynchotus rufescens;* B, Pheasant, *Phasianus colchicus;* C, Land-rail, *Crex pratensis;* D, Woodcock, *Scolopax rusticula;* E, Hornbill, *Rhytidoceros undulatus;* F, Parrot, *Psittacus erithacus*. (Redrawn after Evans.)

systematic relations are not at all certain. Outwardly they appear to be intermediate between the grouse and the pigeon, but structurally they more nearly approach the conditions seen in the pigeons.

The **Columbæ** include the pigeons and doves, and the dodo and solitaire. The **dodo** is a recently extinct, aberrant, not to say grotesque and gigantic pigeon. A funnier looking bird could not readily be imagined, if we may credit the pictorial records of it made by travelers of the seventeenth century. That these apparent caricatures were founded on fact is evidenced by the bones of the bird found in pools. It was a short, plump bird, with an eagle-like beak and ridiculously inadequate plumage, wings, and tail.

The **true pigeons** constitute a very large and widely distributed group. Perhaps the most interesting and significant of the species are the rock pigeon, the passenger pigeon, and the great crowned pigeon.

The **rock pigeon** or **rock dove** (*Columbia livia*) is the species from which nearly all of the fancy breeds of domestic pigeons have been derived; and when fancy breeds are allowed to interbreed freely, the offspring tend to revert to the characters of the wild ancestor. The common mongrel pigeon of the city streets represents fairly closely the characteristics of the wild rock pigeon. The **passenger pigeon** a century ago existed in numbers almost incredibly large. Wilson, a pioneer American ornithologist, estimates that in a single flock seen by him near Frankfort, Kentucky, there were over two billion individuals. In describing similar conditions, Henderson says that "the air was literally filled with pigeons, the light of noonday was obscured as by an eclipse," and that their wings made "a noise like thunder." "Nothing," says Nuttall, "can exceed the waste and desolation of the nocturnal resorts (of these pigeons); the vegetation becomes buried by their excrement to the depth of several inches. The tall trees for thousands of acres are completely killed, and the ground completely strewed with massive branches torn down by the clustering weight of the birds which have rested upon them. The whole region for several years presents a continued scene of desolation, as if swept by the resistless blast of a whirlwind." At the present time it is a question whether there are any passenger pigeons still living. An isolated report comes in now and then that someone has seen a specimen, but there is usually some uncertainty about the identification. The fate of this fine species of bird well illustrates the ruthlessness of man when he begins the process of extermination. The *great crowned pigeon,* a native of the Solomon Islands, represents the climax of the evolution of the pigeon family. It is a noble-looking bird, as much as thirty-four inches in length; with a great, erect, fan-shaped crest of feathers on top of the head which gives it a regal appearance.

The Cuckoo-like Birds (Cuculiformes)

This order is sharply subdivided into two sub-orders: the *Cuculi* (cuckoos and plantain-eaters), and the *Psittaci* (parrots, parrakeets, etc.).

Of the **Cuculi** Knowlton says:—

> "Taking everything into account, the Cuckoos comprise a very remarkable and interesting group of birds, being for the most part birds of shams and pretenses, and ever seeking to convey the impression that they are other than they really are."

We might well call them "camouflage birds," a term that would well characterize these interesting traits. They are certainly great mimics both of the appearance and of the voices of other birds. Some **Cuckoos** place their eggs in the nests of other birds. It is perhaps on account of this peculiar parasitic nesting habit that they are best known. Instead of building a nest of her own, the female lays her eggs on the ground and then carries them in her bill to the nest of other birds. The bird thus imposed upon is likely to react against this intrusion by dumping out the foreign egg, or by building a second story to the nest, thus leaving the cuckoo egg walled up in the basement. Doubtless, however, a sufficiently large number of cuckoo eggs are tolerated by other birds to keep up the normal supply of the various species. This parasitic habit belongs to the Old World cuckoos, for the American cuckoos build their own nests. The **road-runner** is an interesting terrestrial Cuckoo familiar to the inhabitants of the Southwestern United States and Mexico. One sees this long-legged bird pacing along ahead of him on lonely country roads, always keeping a respectful distance ahead, but not offering to leave the road or to fly. The *plantain-eaters* seem to be in some ways intermediate between the cuckoos and the parrots.

The **Psittaci, parrots,** (Fig. 157, F), are a very sharply circumscribed group of brilliant and interesting birds. There are over eighty known genera but they are all unmistakably related. They are usually brilliant in plumage, favoring green, yellow and brilliant red tints, but are occasionally brown or black. They are climbing arboreal birds that use the bill as an aid to climbing, which is a unique use for this organ. Perhaps the most striking characteristic of the parrots is their ability to articulate. Though their native language is one of discordant screams, they can be taught to mimic human language with moderate success, thus showing their cuckoo-like propensities of pretending to be what they are not. Among the parrots are included cockatoos, parrakeets and macaws.

Roller-like Birds (Coraciiformes)

This is one of the largest and most heterogeneous of the avian orders; having affinities with the cuckoo-like birds, on the one hand, and with the sparrow-like birds, on the other. There are seven suborders, most of which are not literally roller-like in appearance.

The **Coraciæ** (**true rollers** and their allies) include: rollers, motmots and todies, kingfishers, bee-eaters, horn-bills, and hoopoes; a rather motley collection of types in itself. The *common roller* (Fig. 158, D) is a native of Southern Europe, outwardly resembling many of the typical passerine birds. The *horn-bills* (Fig. 157, E) are the most remarkable of the Coraciæ; they are large birds with enormous bill, used by the male as a trowel in the operation of walling up the female in a hollow tree. Whether the female is a restless sitter and needs thus to be kept on the job, or whether the wall is for her protection while she is confined at her intimate task, it is difficult to say. She is fed, however, by the male, through a small window just large enough for her bill to be thrust out.

The **Striges** (**owls**), are a well-defined group, formerly classed with the Falconiformes on account of their predaceous habits, but now known to have closer affinities with the goat-suckers. The **great horned owl** (Fig. 158, E) is the finest of its kind, a wise-looking, powerful bird of great size, described as a "veritable tiger among birds." It kills quails, grouse, doves, wild ducks, as well as all sorts of smaller and medium-sized mammals. It hunts at night and hides in hollow trees during the day. The little *American screech owl* is the commonest and most widely distributed of our owls.

The **Caprimulgi** (goat-suckers and their allies) include the oil-bird, the frog-mouths, the goat-suckers or night jars, and the whip-poorwills. They are all much alike, being characterized by rather compact bodies, very short, but extremely wide, bill, and deeply

FIG. 158.—Types of Coraciiformes, showing generalized and specialized species. A, Trogon or Quezal, *Pharomacrus mocinno;* B, Toucan, *Rhamphastus ariel;* C, Hummingbird, *Eulampis jugularis;* D, Common Roller, *Caracias garrulus;* E, Great Horned Owl, *Bubo virginianus.* (All redrawn; A, B, C, after Evans; D and E, after Knowlton.)

cleft mouth fringed with stiff hairs, used to trap insects as they fly through the air. Their flight is swift and practically noiseless. Their mournful nocturnal cries sound like "whip-poor-will," "poor-will" "who-are-you," etc.

Micropodii (**hummingbirds** and **swifts**) are the smallest of birds. Of the hummingbirds (Fig. 158, C) much has been written in praise of their beauty. "Glittering fragments of the rainbow," Audobon calls them; while Knowlton characterizes them as "gems of the feathered race." Small though they be, they are among the most highly specialized of all birds; and therefore, of vertebrates. They seldom alight, but feed while upon the wing, hovering over a flower, poised as though resting, but continuously beating the air with vibrant wings, whose speed of wing-stroke rivals that of the insects. Their tiny eggs and nests are objects of intense curiosity among bird-lovers; some of the nests are of the size of a thimble and the tiny eggs are like pearls. Though of miniature size the hummingbirds are pugnacious and full of courage, a pair of them not hesitating to attack such giant intruders as hawks and large snakes.

The **swifts** are less attractive than their relatives, the hummingbirds, and are often mistaken for swallows. They have the bill short and broad, and the wide gape of mouth like the goat-suckers.

The **Colii** (*colies*) are a small group of somewhat anomalous birds, that have often been placed in the order of passerine birds, but, on account of apparent affinities with the plantain-eaters, they are placed in the present position.

The **Trogones** (*trogons*) are highly specialized tropical birds of comparatively small size, with long tail, short strong bill, and very elaborate plumage. The **quezal** (Fig. 158, A) is one of the most wonderfully colored of the trogons, if not of all birds. Its brilliant plumage of gold, metallic greens and blues, and its gracefully drooping, ethereal plumes give it an almost unearthly beauty.

The **Pici** (*picarian birds*) include both familiar and unfamiliar types such as the jacamars and puff-birds, barbets and honey-guides, toucans, woodpeckers and wrynecks. Of these we must be content to examine only the toucans and woodpeckers. The **toucans** (Fig. 158, B) with the possible exception of the horn-bills, have the most remarkably specialized bill known. As Stejneger says, "The first thing which strikes the observer, when looking at one of the large Toucans, is the enormous size of the bill. It is not only as long as the bird itself, but it does not lack much of equaling the body in bulk; and the observer will most likely make the remark that such an enormous bill must be very heavy. The fact is, however, that the bill is extremely light in comparison with its size, being very thin and filled with light, cellular bony tissue." It is not clear of what value such an enormous bill can be to the bird, for none of its activities appear to be connected with this great structure. In all probablity this great bill is an example of an overspecialized structure, much like the enormous horns of the extinct Irish elk, which are believed to have finally caused the extinction of the species.

The **woodpeckers** and **sapsuckers** are among the most familiar of our native birds, and they are especially known for their habit of riddling the bark and wood of trees in their search for insects and larvæ, and for their noisy drumming while engaged in this task. The finest of the woodpeckers is the great ivory billed woodpecker, which has a length of about twenty inches.

The Passerine Birds (Passeriformes)

This order, consisting largely of perching birds, is for the Neornithes what the order Acanthopterygii is for the Teleostomi; the largest, most varied, most distinctively modern order of the sub-class. Over five thousand species, or nearly half of all known species of birds, are included within this single order. The list of families, thirty-six in number, is too long to recite, but the reader may get an idea of the scope and variety of the order from the following list of representative types:—broad-bills, wagtails, rock-wrens, king-birds, oven-birds, ant-birds, lyre-birds, larks, pipits, fork-tails, thrushes, robins, warblers, gnatcatchers, mocking-birds, water-ousels, wrens, tits, swallows (Fig. 159, A), martins, wax-wings, shrikes, nut-hatches, greenlets, titmice, orioles (Fig. 159, D), birds of paradise, crows, ravens, magpies, starlings, honey-eaters, sun-birds, flower-pickers, creepers, quit-quits, tanagers, weaver-birds, finches, starlings, buntings, etc.

In general, it may be said that the passerine birds are of small or moderate size, of conservative or generalized proportions, and without exaggerations of bill or feet. Some of them, however, have developed a wealth of plumage elaborations, which is taken to be one of the criteria of racial senescence. Garrod and Forbes subdivide the passerine birds into two sub-orders: the *Desmodactyli,* in which the hallux or hind toe is weak and the front toes are more or less united; and the

Eleutherodactyli, in which the hallux is the strongest toe and the other toes are free.

The **Desmodactyli** are the broad-bills, a single unfamiliar type native to oriental countries. They do not differ outwardly from the general standard of passerine birds, and are of interest principally to the systematists.

The **Eleutherodactyli,** or *free-toed Passeriformes,* comprise all of the remaining members of the order, and cannot receive the proportionate amount of attention in the present volume that their importance deserves. For particulars as to the families of passerine birds and the habits of the numerous genera and species, the reader is referred to the many good treatises on birds. We shall merely call attention to a few of the most conspicuous types.

The **birds-of-paradise** (Fig. 159, C) are without question the most elaborately plumaged members of the order, and constitute a striking exception to the general rule, that passerine birds have conservative plumage. The only birds of other orders that compare with the birds-of-paradise in brilliancy are the long-tailed trogons or quezals and the hummingbirds, and none of these types have such elaborate feather structure. On the whole, then, these birds may be said to cap the climax in the evolution of plumage specialization. The *great bird-of-paradise* is perhaps the most beautiful of the numerous species. Apart from the striking color scheme, the most remarkable specializations consist of a pair of dense tufts of delicate, drooping plumes, that vary from two to three feet in length, arching upward and then falling downward in a veritable cascade of glistening light. Anatomically speaking, this marvelously handsome creature is no more nor less than "a glorified crow," for when plucked he is seen to be as plain and common a bird as is his black cousin.

The **lyre-birds** (Fig. 159, B) of South Australia rival the birds-of-paradise in elaborate structure of plumage, but are not at all brilliant. They are moderately large birds, about two and a half feet long, with rather long neck, and with fowl-like head and beak. Their only claim to beauty consists of the remarkable lyre-like tail; the sides of the "lyre" consist of two large strong feathers, that curve outward from their base, then curve inward, and again outward, at the ends, in most graceful lines. Two middle feathers, almost as graceful as the frame-feathers, cross each other and droop out beyond the outer feathers; while the remaining feathers are long, slender and

FIG. 159.—Types of Passerine Birds, showing generalized and specialized forms. A, Swallow, *Hirudo rustica;* B, Lyre-Bird, *Menura superba;* C, Lesser Bird of Paradise, *Paradisea minor;* D, Baltimore Oriole, *Icterus baltimore;* E, House-sparrow, *Passer domesticus.* (All redrawn; C, after Knowlton, the rest after Evans.)

comparatively straight, and simulate the strings of the lyre. In color these birds are for the most part of a soft brown, and they are not conspicuous in their natural haunts.

The **sparrows** (Fig. 159, B) are usually placed last in the systems of classification because they are believed to be the most modern type. They are the most numerous and the most familiar of all birds. They are usually small inconspicuously colored birds, characterized by strong, hard, conical bill, compact form and comparatively short body, tail, and wings. Possibly the most significant event in the history of modern birddom was the invasion of North America by the English Sparrow in 1852. It first landed in Brooklyn and spread from there over the North Eastern Atlantic States. In a half century it had spread over a large part of the continent and is now the most numerous bird species in the world. The English sparrow is a modernist among birds and leads us to discuss the probable future of the bird tribe.

THE FUTURE OF BIRDS

It will have been noted that most of the orders of birds have some very generalized types and some highly specialized types. One could select from nearly every order a representative that is conservatively proportioned and has simple bill and generalized feet. In each order we also find certain types with exaggerated proportions, overspecialized bill or feet and highly colored or elaborate plumage. If we may rely on the uniformity of nature, we may expect the events of the past to repeat themselves, and if they do, these specialized birds will become still more senescent and, unable to reverse the course of specialization, become extinct; it is all well enough to be handsome and brilliant of plumage or unduly long of leg or large of bill, but perhaps the birds thus endowed will pay for it in contributing to the prehistoric fauna of the next geologic age; while the sparrow and his ilk will still dispute with other dominant races the domains of earth and tree and air. It is as much as a bird's life is worth now-a-days to have beautiful or elaborate plumage, for primitive man must have its plumes for the adornment of his primitive mate; and he gets what the mate desires whether he has to hunt the trackless forests for it, or merely pays an exorbitant milliner's bill; a type of bill quite unknown among birds. Safety for the bird of to-day lies in homeliness of aspect, adaptability as to environment and food, and a goodly share of pugnacity and resistance to hardship. Let the modern birds consider the sparrow and his ways. He is plain and homely, eats anything, lives anywhere, builds his nests in strange and unfamiliar places, using new and untried materials. He can whip anything his own size in feathers, but does not needlessly pick a quarrel, and he can put up with either cold or heat, drought or flood; they all look alike to him. Doubtless in the distant future he will dispute for the supremacy of the earth with the mouse, the ant, and super-man.

Man owes much to the passerine birds. They give to him who has a naturalistic bent a keener zest for woodland life. Vast numbers of people find their lives enriched by the study of the haunts and varied activities of the birds. As destroyers of harmful insects the passerine birds are of inestimable value to mankind. It is therefore of the utmost importance that all agencies organized for the prevention of slaughter of the song-birds and other passerine birds, should receive the united support of every zoölogist and lover of nature. Organizations such as Audubon Societies and the various Sportsman's Clubs are doing much to spread propaganda favoring bird protection. The writer of this volume would like to go on record as unreservedly urging the support of all agencies designed for bringing about the enforcement of laws forbidding the cruel and senseless slaughter of migrant passerine birds.

Migration of Birds

"The desire to migrate," says Seebohm, "is a hereditary impulse, to which the descendants of migratory birds are subject—a force almost, if not quite as irresistible as the hereditary impulse to breed in the spring." Migrations follow more or less direct paths between winter homes and breeding quarters. Most birds breed in the north and winter in the south. Migration paths follow coast lines, as a rule, and such locations as islands, capes, inlets and other good landmarks are favorite stopping places. Frequently the same birds stop at the same places several years in succession.

Birds have keen powers of orientation, and a strong homing instinct. This is not, as some appear to believe, due to a sixth sense, but to a highly developed place memory, or ability to recognize after a lapse of time elements in a landscape that have been observed one or more times before. If a bird is taken to an entirely new region and released it has great difficulty in orienting itself and only succeeds in getting home if by chance it happens to discover a familiar landmark. Young birds are much less capable of homing than are older birds, and need to follow a leader until they become familiar with the route. Some birds migrate in flocks of great size, others in small numbers or even in pairs. The speed attained by migrating birds may be as high as a hundred miles an hour, but the majority of them scarcely attain half that speed. Even at the rate of fifty miles an hour birds have been known to travel a distance of nearly two thousand miles in two days; for they take little rest while migrating, and are often entirely exhausted when they reach their destination. It is during the migrating season that ignorant and lawless people take advantage of the large numbers of fatigued birds and shoot them in vast numbers, displaying in so doing a lack of sportsmanship truly lamentable.

Geographic Distribution of Birds

Birds of good flying powers are as nearly cosmopolitan as any animal could be; the albatross and the petrels range the oceans from one extreme to the other. Birds of moderate flying powers may be limited to one continent; many birds, for example, are confined to North America, while others breed in North America and winter in Central or South America. Flightless birds are often confined to single islands, such as New Guinea or New Zealand.

One should carefully distinguish between migration and distribution when dealing with birds; for when we say that a given species breeds in Canada and winters in Central America we do not mean that its area of distribution covers all of the intervening territory, for a large part of this territory is not even passed over by the species in question during the migration flights.

DEVELOPMENT OF BIRDS

The classic type for the study of avian embryology is the common fowl. Usually the study of the development of the chick constitutes a major part of college courses in vertebrate embryology; hence only a brief outline of bird development need be given here. There is so little difference between the development of the reptile and the bird that the following sketch will serve to illustrate the salient features of the embryology of the Sauropsida in general.

The "egg" of the bird (Fig. 160) is a large and complex structure, consisting of the ovum proper, the albuminous layers, shell membranes and shell. The *ovum*, or what is usually referred to as the yolk, is a single food-gorged cell inclosed within a vitelline membrane and with a single nucleus or germinal vesicle. The active protoplasm of the ovum is largely aggregated in a small region situated at the animal pole of the cell, called the *germinal spot*, where lies the nucleus.

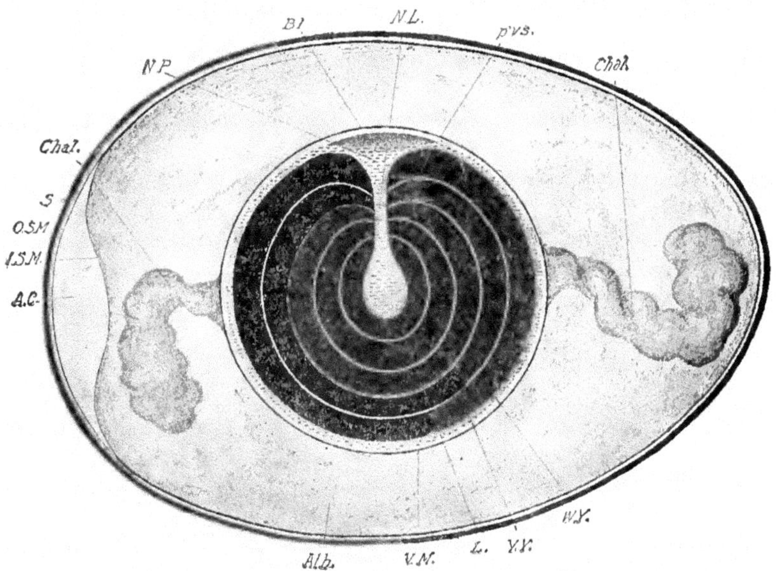

Fig. 160.—Diagram of hen's egg to show envelopes, and general relations of parts. *A. C*, air chamber; *Alb*, albumen; *Bl*, blastoderm; *Chal*, Chalaza; *I. S. M*, inner layer of shell membrane; *L*, latebra; *NL*, neck of latebra; *N. P*, nucleus of Pander; *O. S. M*, outer shell membrane; *p. v. s*, perivitelline space; *s*, shell; *B. M*, vitelline membrane; *W. Y*, white yolk; *Y. Y*, yellow yolk. (From Lillie's "Development of the Chick" [Henry Holt and Company].)

This small mass of hyaline protoplasm is continuous with a thin sheath of protoplasm that surrounds and incloses the entire yolk mass and to a certain extent permeates the body of the yolk.

Immediately surrounding the ovum is a thick viscous layer of albumen that is swathed about the ovum and prolonged on opposite sides into twisted ropes, called *chalazæ*, that fasten the ovum to the shell membranes and suspend it in such a way that it cannot come in contact with the shell. Between the chalazal layer of albumen and

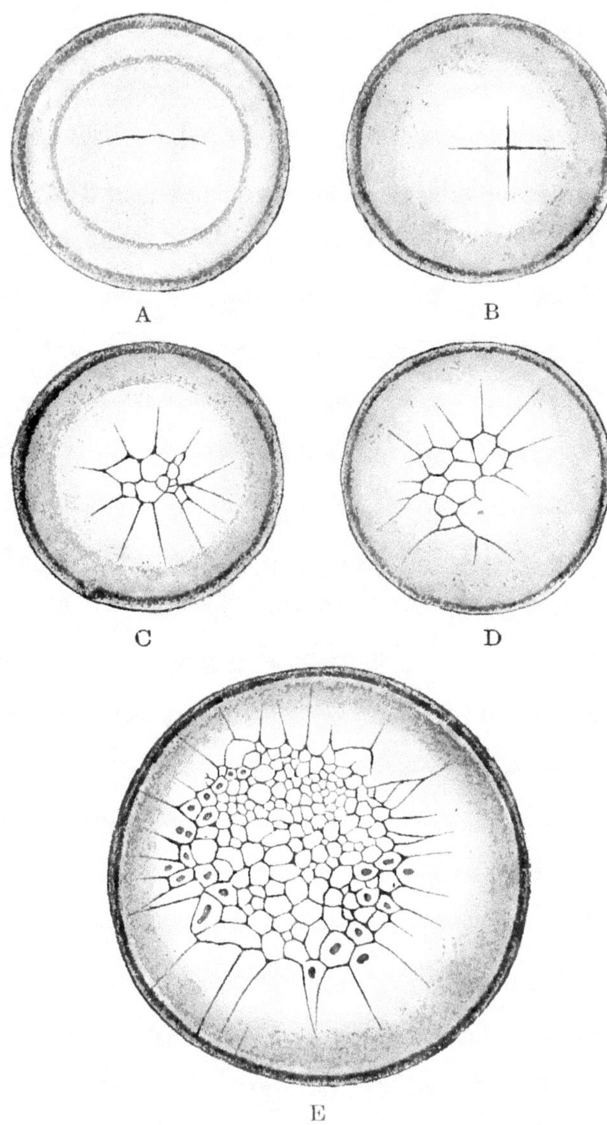

Fig. 161.—Cleavage of hen's egg. A, first cleavage furrow (x 14). The egg came from the lower end of the oviduct. B, Four-celled stage (x 17); from the uterus; C, Ten central and eleven marginal cells (x about 16); D, nine central and sixteen marginal cells (x about 16); E, late cleavage stage (x about 22). (From Lillie's "Development of the Chick," after Kölliker.)

the shell membrane is a second layer of albumen which is quite fluid in consistency. Surrounding the albumen is the double parchment-like shell-membrane, with an air-space between its two layers at the broad end of the egg. The shell proper is a rather complex structure composed of calcium carbonate; it is porous and more or less pigmented.

Cleavage (Fig. 161) is strictly meroblastic, the first cleavage being merely a furrow, and many furrows are formed before any of the cells are furnished with floors or bottom partitions that cut them off from the underlying yolk. Development proceeds beyond the gas-

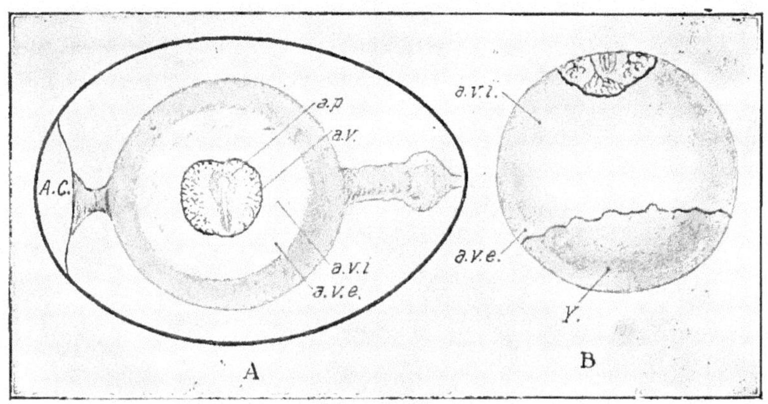

FIG. 162.—Hen's egg at about the twenty-sixth hour of incubation, to show the zones of the blastoderm and the orientation of the embryo with reference to the axis of the shell. B, yolk of hen's egg incubated about 50 hours to show the extent of overgrowth of the blastoderm. A. C, Air chamber; *a. p*, area pellucida; *a. v*, area vasculosa; *a. v. e*, area vitellina externa; *a. v. i*, area vitellina interna; *Y*, uncovered portion of the yolk. (From Lillie, after Duval.)

trula stage before the egg is laid. A newly laid egg shows the embryo as an embryonic disc, a small whitish spot at the animal pole, composed of central transparent area (*area pellucida*) bounded by an opaque ring or germ wall. The pelucid area is two-layered posteriorly, an inrolling of cells having occurred which constitutes the primitive endoderm and is the equivalent of the archenteron invagination of the frog. The *blastopore* is crescentic, as in the frog, and the primitive streak is formed by concrescence of the blastopore. The head forms in front of the *primitive streak*, which constitutes the axis of the trunk.

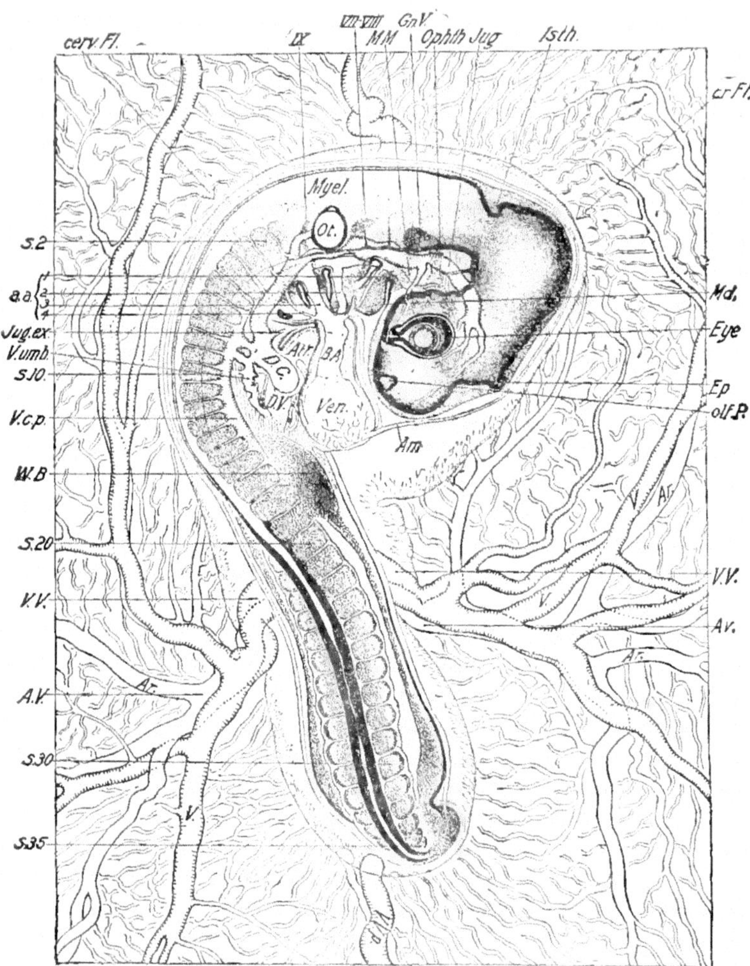

FIG. 163.—Chick embryo of 35 somites, drawn as a transparent object. *a. a. 1, 2, 3, 4,* First, second, third, and fourth aortic arches; *Ar,* artery; *A. V,* vitelline artery; *cerv. Fl,* cervical flexure; *cr. Fl,* cranial flexure; *D. C,* duct of Cuvier; *Ep,* epiphysis; *Gn. V,* ganglion of trigeminus; *Isth,* isthmus; *Jug. ex,* external jugular vein; *Md,* mandibular arch; *M. M,* maxillo-mandibular branch of the trigeminus; *olf. P,* olfactory pit; *Ophth,* ophthalmic branch of trigeminus; *Ot,* otocyst; *V,* vein; *W. b,* wing bud; *V. C. p,* posterior cardinal vein; *V. umb,* umbilical vein; *V. V,* vitelline vein; *V. V. p,* posterior vitelline vein. (From Lillie's "Development of the Chick.")

The *medullary plate* and *medullary groove* forms much as in the frog, beginning at the anterior end and proceeding to close gradually from the anterior toward the posterior.

While the axial parts of the embryo are differentiating the peripheral parts of the blastoderm continue to grow round the yolk, new cells being continually formed at the margin. Finally the whole ovum becomes covered with cells. A considerable part of this sheath of cells is destined to be used only for the formation of embryonic membranes—amnion, allantois and yolk-sac. Only the .parts near the animal pole of the egg are concerned in forming the embryo proper. The embryo is gradually pinched off from the rest of the egg by means of deep grooves that go so deep as finally to leave only a narrow yolk-stalk between embryo and yolk. An extensive vitelline circulation covers the yolk sphere, and through this means the embryo maintains a nutritive connection with the yolk.

Like all other vertebrates the young chick (Fig. 163) develops four pharyngeal clefts (gill-slits), only one of which actually breaks through to the pharynx; this is the eustachian tube of the adult. It has been commonly stated that the bird embryo never exhibits any traces of gill filaments in these gill-slits, but Boyden has recently described not only in the chick but in several reptiles the transitory appearance of tissues, which he believes are undeniably rudimentary branchial filaments.

Embryonic Membranes.—The importance of the amnion and allantois (Fig. 164) as adaptations for land life, and their rôle in the evolution of the terrestrial vertebrates, have been sufficiently dealt with in the chapter on reptiles. In general the mode of origin of these membranes is the same in the bird as in the reptile. The **amnion** begins as a crescentic fold of the extra-embryonic blastoderm in front of the head. This fold, which consists of ectoderm and mesodern only, grows backwards, covering the head like a hood and continues to spread over the body until it meets a smaller, but similar tail fold that has been growing forward. The two folds fuse together and completely inclose the embryo in a sac lined with ectoderm on the inside and mesoderm on the outside. Of course an outer section of the fold is also produced, called the **chorion,** which is lined with ectoderm on the outside and with mesoderm on the inside. Thus two complete membranes shut off the embryo from the albuminous layers. The inner layer of the amnion secretes an abundant watery fluid that

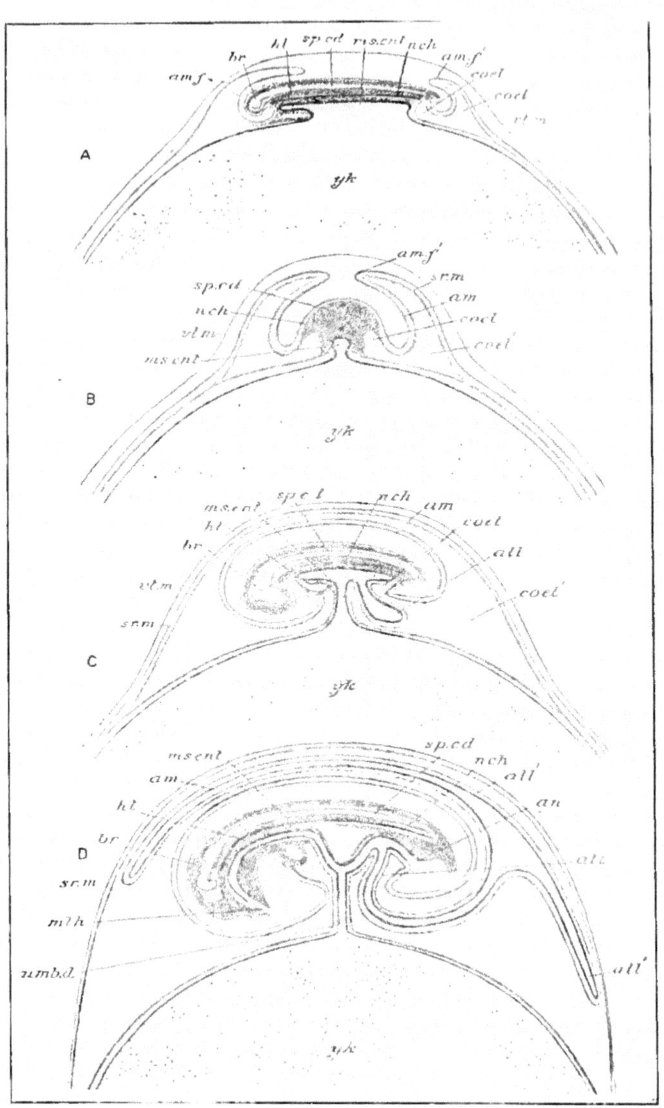

FIG. 164.—Diagram illustrating the development of fœtal membranes in a bird. A, early stage in the formation of the amnion, sagittal section; B, slightly later stage, transverse section; C, stage with completed amnion and commencing allantois; D, stage in which the allantois has begun to envelop the embryo and yolk-sac. The ectoderm is represented by blue, the endoderm by red, and the mesoderm by gray. *all*, allantois; *all'*, the same growing round the embryo and yolk-sac; *am*, amnion; *am. f*, *am. f'*, amniotic fold; *an*, anus; *br*, brain; *coel*, coelom; *coel'*, extra-embryonic cœlom; *h*, heart; *ms. ent*, mesenteron; *mth*, mouth; *nch*, notochord; *sp. cd*, spinal chord; *sr. m'*, serous membrane; *umb. d*, umbilical duct; *vt, m*, vitelline membrane; *yk*, yolk-sac. (From Parker and Haswell.)

bathes the embryo throughout the entire embryonic period and protects it from shocks and injuries due to contacts.

The **allantois** begins as an out-pouching of the hind-gut not far from the yolk-stalk. It pushes outward as a thin-walled sac, lined with endoderm on the inside and with mesoderm on the outside, grows out between the amnion and chorion, and expands into a large umbrella-shaped body until it fills the entire extra-embryonic cœlom, or space between the chorion and amnion. Thus the amnion is covered with the distal part of the allantois and the latter is covered with chorion. In the later stages these three membranes fuse together in a number of places into a single compound membrane. The allantois becomes richly vascular on its outer surface and acts as an embryonic lung, getting oxygen through the porous shell membranes and shell.

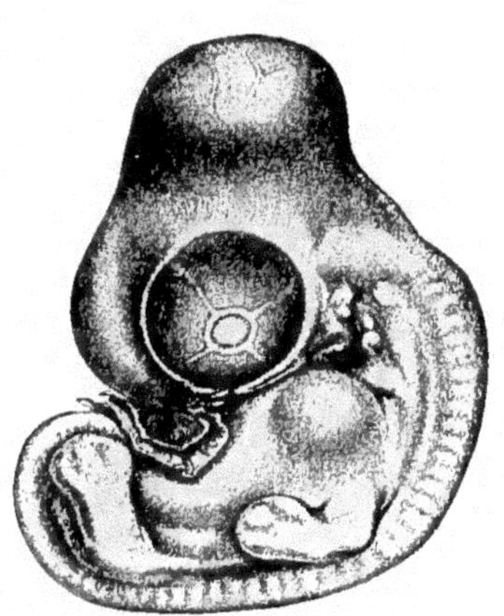

Fig. 165.—Chick embryo at five days' incubation, showing precocious development of head, long tail, wing and leg nearly identical. (From Lillie, after Keibel and Abraham.)

The **yolk-sac** is at first nearly the entire egg, but as development progresses it diminishes in size as the yolk substance is assimilated by the embryo; until finally the tiny sac that remains is drawn into the body cavity of the chick through the umbilicus, and the latter closes.

Changes in Body Form During Development.—It is of interest to note that the tail of the chick of four or five days' incubation is comparatively long and slender, much like that of a lizard at an equivalent stage of development. Also the fore and hind limbs are much alike at that period, as is shown in the illustration (Fig. 165). It is only after about ten days of incubation (Fig. 166) that the tail becomes foreshortened into the typical avian tail and the fore limbs take on the characteristic features of wings. As in most other vertebrate embryos, the head is relatively enormous as compared with the body during a large part of the embryonic period, and it is only in the last stages of incubation that the body becomes larger than the head. The feather rudiments appear about the sixth day and are at first mere papillæ protruding from the skin. At hatching the chick is completely covered with down-feathers, which are the forerunners of the definitive feathers and gradually give place

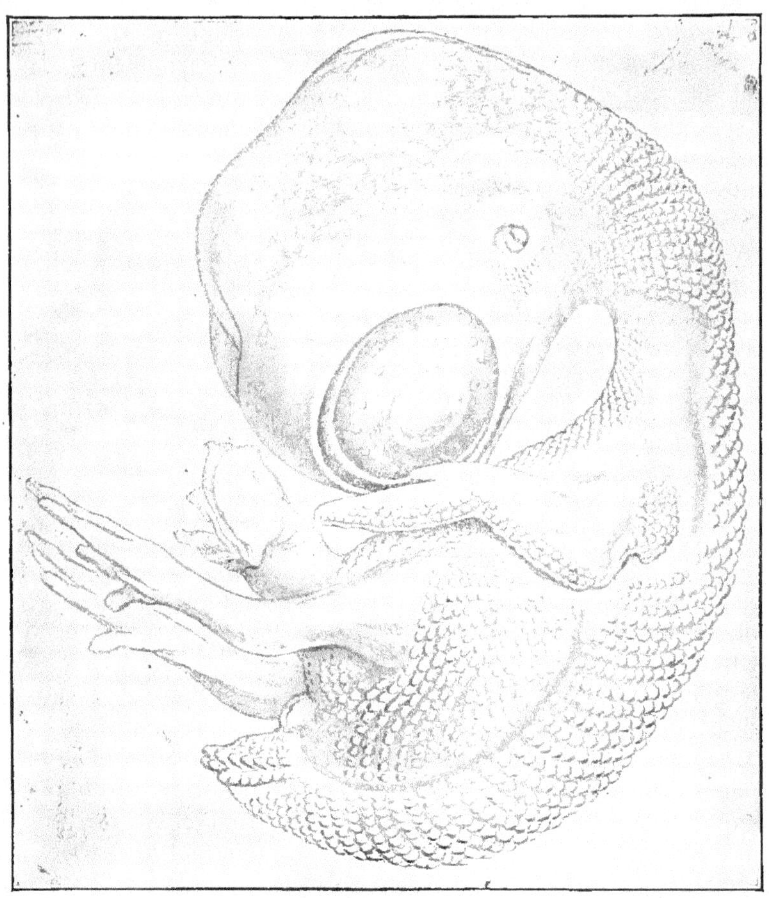

Fig. 166.—Chick embryo at 10 days and 2 hours, showing differentiated wing and legs, shortened tail and feather papillæ. (From Lillie, after Keibel and Abraham.)

to the latter. On about the seventeenth day the amniotic fluid begins to disappear and on the twentieth day it is gone. On the same day the chick, by means of a sharp little egg-tooth on the point of the bill, pecks a hole in the shell and begins to breathe with its lungs. The allantois then gradually shrivels up and its circulation is cut off. On the twenty-first day the chick bursts the shell and emerges. It is quite a capable youngster at hatching, for it can walk, and see, and within a few minutes begins to peck at the ground. This is quite in contrast with the situation in many other birds, whose young are hatched in a naked, blind and entirely helpless condition. As a rule, birds that nest on the ground have precocious young at hatching and are called *Præcoces* or *Nidifugæ;* while birds that nest in trees or in other safer retreats have helpless young and are called *Altrices* or *Nidicolæ.* Intermediate conditions are of course found in many species, especially in those of sea-birds, such as petrels and gulls, whose young are downy at hatching, but stay in the nest for some time.

Nesting Habits of Birds.—Any adequate account of the nesting habits of birds would require a volume in itself, for there are countless different kinds of nests and of materials used. The more primitive nests appear to be crude nests built on the ground, consisting of mere hollows scooped out of the sand or earth after the manner of some of the reptiles. Some birds have no nests at all but merely lay the eggs on rocks; this is probably not a primitive, but a degenerate habit. The members of the higher orders of birds, as a rule, make nests out of grasses or other materials that are suitable for weaving a fabric or basket-like container for eggs. These nests are placed in trees, on cliff-sides, in hollow trees, in burrows under the ground or in caves. Clay or mud nests are common, especially among swallows. Birds that occupy territory inhabited by man are quick to adopt the various materials that man furnishes, such as string, rags, paper and other common waste. The use made of various man-made bird houses illustrates the fact that the bird is decidedly adaptable and not stereotyped in its form of intelligence.

CHAPTER IX

CLASS VI. MAMMALIA

Unfortunately the only vernacular name for the class Mammalia is **mammals,** but the man of the street does not know what a mammal is. He knows birds, reptiles, fishes and has an idea that a frog is an amphibian; but he uses a variety of words to express his idea of a mammal, none of which seems to serve the purpose very well. He sometimes uses the word "beast," but this term does not seem to apply to men, at least not to all men, nor to whales; he uses the word "quadruped," but this term seems scarcely appropriate to bipeds, whales or bats.

It has generally been assumed that mammals represent the apex of organic evolution, or at least that of the chordate phylum. It is, however, not to be granted as axiomatic that the mammals represent a higher level of evolutionary attainment than do the birds; for the birds are a more recent evolutionary product, are more nearly a climax group to-day, and on the whole represent a more highly specialized condition than do the mammals. In only one particular do the mammals exhibit a distinctly higher order of specialization than do the birds; namely, in brain specialization, and particularly in that of the cerebral hemispheres. It has also been said that the mammals surpass the birds in specialization of the teeth and of the feet. This is true in a sense, though the toothless condition of the bird and the replacement of teeth by the bill is really a more highly specialized condition than any in which the teeth still persist; while the wing represents an extreme specialization of the fore limbs more radical than anything in the mammals, except possibly the flippers of whales. It must be admitted, however, that the bird's hind limbs are rather conservative; for the wing has rendered the functioning of the foot of secondary importance. The claim of the class Mammalia to supremacy in taxonomic ranking rests almost entirely upon their superiority of nervous organization. Man as the exemplar of brain specialization adds immeasurably to the claim for supremacy of the class to which he belongs; for there is no dispute as to the supreme status of the human mammal. Without Man the mammals would have at best a disputed claim to highest rank among the vertebrates; with Man included, the mammals reign supreme.

The mammals are to-day as well defined in their structural characters as are the birds, although many of them are exceedingly aberrant. No transitional type at all on a par with *Archæopteryx* exists for the mammals, although the monotremes are intermediate in some particulars between the typical mammals and the reptiles, and the reptilian order of cynodonts shows many mammalian traits.

Distinguishing Characters of Mammals

1. The skin (Fig. 167) is more or less clothed with hair, though in some species there are only a few localized bristles.

2. A muscular diaphragm forms a complete partition between the thoracic and abdominal parts of the body cavity; it functions both in respiration and in parturition.

3. Mammary glands, with or without teats, are always present.

4. Sebaceous glands and sweat glands are always present (Fig. 167).

5. The red blood corpuscles are non-nucleate in the definitive condition and are circular in outline, except in the camels where they are ovoid.

6. The cerebral hemispheres are connected by a heavy commissure of fibers, called the corpus callosum (Fig. 171) which is rudimentary in monotremes and small in marsupials.

7. There is a single aortic arch, the left, which curves over the left bronchus.

8. A larynx, or voice-box, lies at the upper end of the trachea.

9. An epiglottis, a movable cartilaginous plate, covers the glottis.

10. Lips and cheeks are characteristic of all mammals except whales.

11. The mandible (Fig. 169) consists of but one pair of bones, the dentaries, which are firmly fused in the adult; the dentary articulates directly with the squamosal.

12. There is a chain of three bonelets in the middle ear that connects the tympanum with the inner ear. These bonelets are: the *stapes* (believed to be homologous with the columella of reptiles), the *malleus* (believed to be homologous with the articulare of reptiles), and the *incus* (believed to be homologous with the quadrate of reptiles).

13. There is an external fleshy and cartilaginous conchus to the ear.

14. The body of the vertebra is formed of three pieces, one of which makes the centrum and the other two the epiphyses.

15. Cartilaginous disks (intervertebral disks) separate the centra of the vertebræ from one another.

16. The coracoid, except in the monotremes, is represented by a mere vestige fused with the scapula, called the coracoid process.

17. The ribs, except in monotremes and whales, are attached by two heads, the capitulum and the tuberculum.

18. There are characteristically seven cervical vertebræ. (In the manatees there are but 6, and in the sloths there may be 6, 8 or 9.)

19. With few exceptions (whales, edentates), mammals are diphyodont, i. e., have two sets of teeth, a milk and a permanent dentition.

20. The teeth are (a) thecodont (each imbedded in an alveolar pit in the jaw bone); and (b) heterodont (differentiated into incisors, canines and molars); exceptionally homodont or absent.

21. There are two occipital condyles, which are part of the exoccipital bones.

22. Except in monotremes there is no distinct cloaca.

23. Mammals are homothermous (warm-blooded), with a well-developed temperature-regulating apparatus.

24. The heart is completely four chambered, with two auricles and two ventricles and a complete double circulation.

25. With the exception of monotremes, the eggs are minute in size and the young are born alive (viviparous).

26. There are no gills nor gill rudiments at any stage of development.

27. An amnion and an allantois are always present; but the allantois is sometimes vestigial or functionless.

The first 19 characters and 20 (b) distinguish the mammals from all other living vertebrates.

Numbers 21 and 22 distinguish the mammals from modern reptiles and birds (Sauropsida).

Numbers 23 and 24 distinguish the mammals from the reptiles.

Number 25 distinguishes the mammals from the birds.

Numbers 26 and 27 distinguish the mammals from the Amphibia.

The Mammalian Integument.—Under this head we shall discuss briefly hair; skin glands; and claws, hoofs and nails.

The possession of **hair** is as truly diagnostic for mammals as are feathers for birds. Even the apparently naked, glossy-skinned whales have a few bristle-like hairs on the upper lip. Sometimes hairs may be fused into scale-like or horn-like structures as in the scaly ant-eaters and the rhinoceros. Again they may be more or less covered or obscured as in the armor of the armadillos. The hair arises from a slight thickening of the Malpighian layer of the epidermis, which subsequently invaginates so as to form a deep pocket or follicle of the epidermis and a dermal papilla pushes up into the bottom of the invagination after the manner of a pulp cavity in a tooth. Thus the origin and development of the hair is totally different from that of a scale or of a feather. The hair is like nothing else; it is *sui generis*.

There are many kinds of **skin glands** among mammals, but they may all be reduced to two fundamental types: *sudoriparous* or sweat glands and *sebaceous* glands. Generalized sweat and sebaceous glands (Fig. 167) are scattered over nearly the entire skin, while local specializations of both types occur in all mammals. The mammary glands of the monotremes are specialized sweat glands, while those of the Eutheria are specialized sebaceous glands. A great many mammals possess scent glands located in various regions. These serve a variety of uses, principal among which are: to attract the opposite sex; to enable gregarious forms to distinguish their kind; and for defensive purposes, as in the skunk and his tribe.

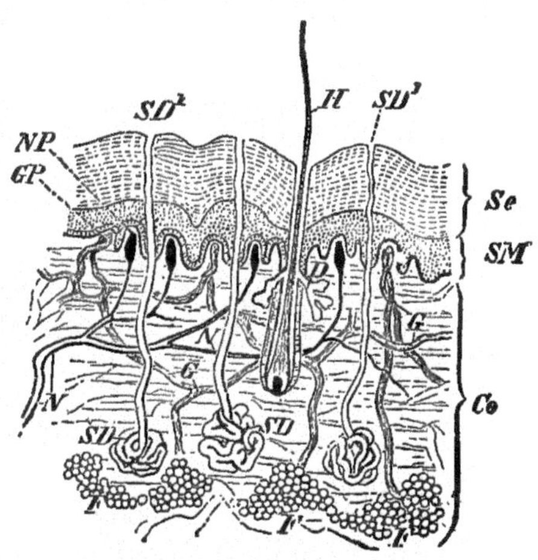

FIG. 167.—Section of human skin. *Co*, dermis; *D*, sebaceous glands; *F*, fat in dermis; *G*, vessels in dermis; *G. P*, vascular papillæ; *H*, Hair; *N*, nerves in dermis; *N. P*, nervous papillæ; *Se*, horny layer of epidermis; *S. D*, sweat gland; *S. D*¹, duct of sweat gland; *SM*, Malpighian layer. (From Wiedersheim.)

Either **claws, hoofs,** or **nails** are present in all mammals except the whales; even in the latter rudiments of claws appear in the fœtus and are subsequently lost. These three distinct types, and the total absence of any such structures, serve to divide the mammals into four great sections: the clawed, the nailed, the hoofed, and the whales, which have no such structures. It is probable that the claw is the most primitive type and that the nail, the hoof, and the naked-fingered type represent a phyletic series of specializations. This idea is more fully discussed in connection with the mammalian orders.

The Mammalian Skull.—The skull of the mammal (Fig. 168) differs from that of other vertebrates in a number of important particulars. It is more compact and contains fewer elements than that of the reptile. The following bones characteristic of the reptilian ancestry have disappeared from the adult cranium: pre- and post-orbitals, pre- and post-frontals, basipterygoids, quadrato-jugals, and

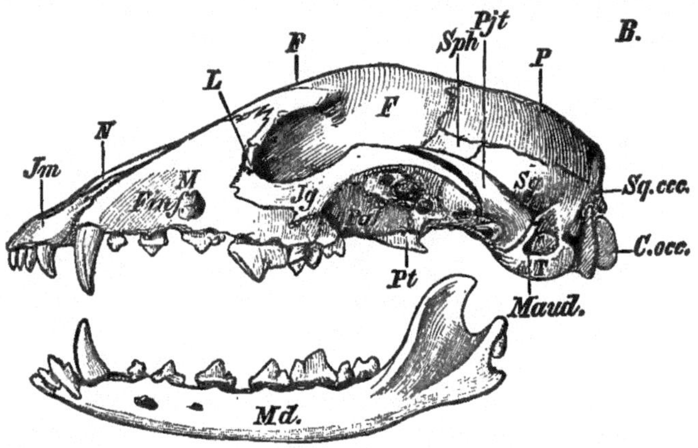

FIG. 168.—Skull of mammal (dog). *C. occ*, occipital condyle; *F*, frontal; *F. inf*, infra-orbital foramen; *Jg*, jugal; *Jm*, premaxilla; *L*, lachrymal; *M*, maxilla; *M. aud*, external auditory meatus; *Md*, mandible; *N*, nasal; *P*, parietal; *Pal*, palatine; *Pjt*, zygomatic process of squamosal; *Pt*, "pterygoid"; *Sph*, alisphenoid; *Sq*, squamosal; *Sq. occ*, supra-occipital; *T*, tympanic. (From Wiedersheim.)

supra-temporals. In the lower jaw, the angulare, splenial, and articulare are gone; but the latter is believed to have been drawn in to form the malleus, one of the ear bonelets. The quadrate has also been drawn in to form the incus bonelet.

Mammalian Dentition.—The teeth of mammals (Fig. 169) are attached only to the dentary, maxillary, and premaxillary bones. They are limited in number, rarely exceeding fifty-four. The *incisors* are generally simple in structure and with a single root; the *canines*, when present, are also simple and with a single root; the remaining teeth (cheek-teeth) are divided into *premolars* and *molars*, and show a wide range of complexity in structure and in number of roots. They range from simple one-cusped teeth like canines to those with a large number of cusps. The primitive types of cheek-teeth are provided with conical tubercles, and are known as *bunodont;* a more highly specialized type of tooth has the tubercles connected by ridges, and is known as *lophodont.* There are usually two sets of teeth, a milk dentition and a permanent dentition, a condition known as *diphyodont* in contradistinction to the condition characteristic of the lower vertebrates, which have but one set of teeth (*monophyodont*). In the lowest mammals there is no second dentition, or only a partial replacement of the first by the second set. Many mammals also have degenerate dentition, involving an entire loss of teeth or merely a loss of incisors, or canines, or some of the molars.

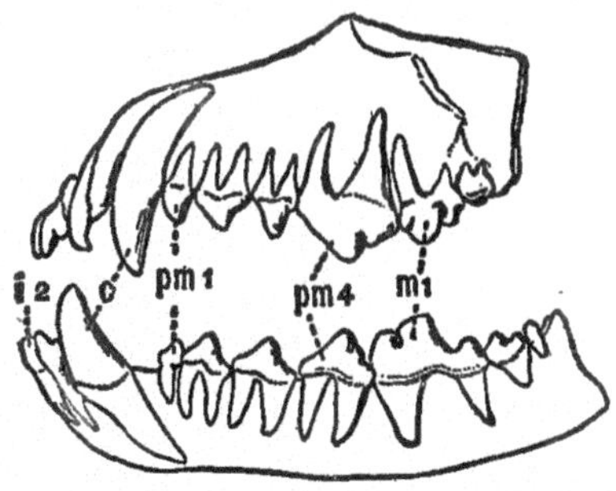

FIG. 169.—Teeth of dog. *i 2*, second incisor; *c*, canine; *pm 1*, *pm 4*, first and fourth premolars; *m 1*, first molar. (From Hegner, after Shipley and MacBride.)

A typical tooth (Fig. 170) consists of three kinds of tissue: enamel, dentine, and cement. The enamel is derived from the epithelium of the mouth cavity and is therefore ectodermal; the other constituents are dermal in origin. The teeth arise as tooth-germs quite independent of the jaws and later become imbedded in sockets of the latter. The dental epithelium is at

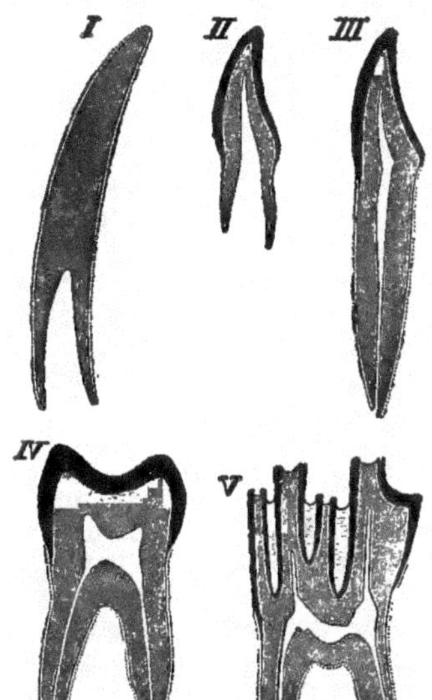

FIG. 170.—Diagrammatic section of various forms of teeth. I, incisor or tusk of elephant with pulp cavity open at base. II, human incisor, during development, with pulp cavity open at base. III, completely formed human incisor, opening of pulp cavity small. IV, human molar with broad crown and two roots. V, molar of ox, enamel deeply folded and depressions filled with cement. Enamel, black; pulp, white; dentine, horizontal lines; cement, dots. (From Hegner, after Flower and Lydekker.)

first invaginated as a continuous fold, covering the jaw from end to end; this fold is known as the enamel organ. At intervals thickenings occur at the bottom of the groove, each of which becomes bell-

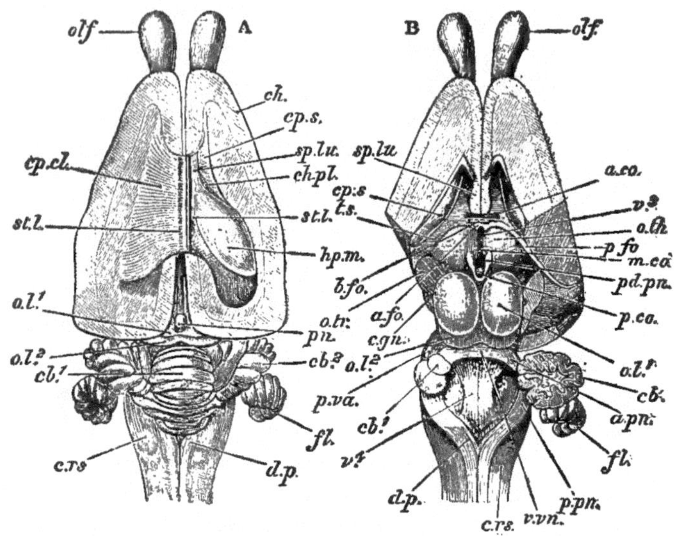

FIG. 171.—Brain of rabbit, especially to show corpus callosum. (Nat. size.) In A the left parencephalon is dissected down to the level of the corpus callosum: on the right the lateral ventricle is exposed. In B the cerebral hemispheres are dissected a little below the level of the anterior genu of the corpus callosum; only the frontal lobe of the left hemisphere is retained; of the right a portion of the temporal lobe also is left; the velum interpositum and pineal body are removed, as well as a greater part of the body of the fornix, and the whole of the left posterior pillar; the cerebellum is removed with the exception of a part of the right lateral lobe. *a. co*, anterior commissure; *a. fo*, anterior pillar of fornix; *a. pn*, anterior peduncles of cerebellum; *b. fo*, body of fornix; cb^1, superior vermis of cerebellum; $c. b^2$, its lateral lobe; *c. gn*, corpus geniculatum; *c. h*, cerebral hemisphere; *c. ph*, choroid plexus; *cp. cl*, corpus callosum; *cp. s*, corpus striatum; *c. rs*, corpus restiforme; *d. p*, dorsal pyramid; *fl*, flocculus; *hp. m*, hippocampus; *m. co*, middle commissure; ol^1, anterior, and ol^2, posterior lobes of corpus quadrigemina; *o. th*, optic thalamus; *o. tr*, optic tract; *p. co*, posterior commissure; *p. fo*, posterior pillar of fornix; *pn*, pineal body; *pd. pn*, peduncle of pineal body; *p. pn*, posterior peduncle of cerebellum; *p. va*, fibres of pons Varolii forming middle peduncles of cerebellum; *sp. lu*, septum lucidum; *st. l*, stria longitudinalis; *ts*, tænia semicircularis; *v. vn*, valve of Vieussens; v^3, third ventricle; v^4, fourth ventricle. (From Parker and Haswell.)

shaped, with a dermal papilla in the hollow of the bell. The top of the bell continues to grow out as the tooth and soon ruptures the gum and protrudes as a naked cusp. Sometimes the enamel becomes folded into deep pockets and gives to the tooth a complex cross-section, as in the ungulates. The tooth remains hollow, a pulp cavity remaining in its center, which contains blood-vessels, nerves, and connective tissue. The dentine is merely a fine quality of bone and has the histological structure of the latter.

The Mammalian Brain.—Although the brains of certain archaic mammals were not much more highly developed than those of some of the reptiles, those of modern mammals, especially those of the more highly specialized groups, show marked advances over the brains of other vertebrates. The mammal brain (Fig. 172) is relatively large, but the cerebral hemispheres show more increase than do other parts. These hemispheres are connected by an elaborate system of commissures, which serve to correlate the two and to make them act as one organ; the *corpus callosum* is the most important of these commissures and it reaches a large size in the highest mammals. The surface of the *cerebrum,* in all but the more primitive mammals, is much infolded into a system of convolutions, which greatly increase the surface without unduly increasing its bulk. It is not strictly true that the degree of complexity of the cerebral convolutions is an index of the grade of intelligence; for the elephant has the most elaborately convoluted cerebrum, but is hardly as intelligent as many other mammals with less convolutions. The optic lobes are four in number, but in size they are small. The *cerebellum* is scarcely as elaborate as in the birds, though better developed than in any reptile.

Urogenital Systems of Mammals.—The kidneys are compact in form and are of the metanephros type. They are usually asymmetrical in position, one lying more anteriorly than the other. The ureters lead directly to the urinary bladder, which is formed out of the remains of the allantois.

The *ovaries* are always paired; never single as in the bird. They are very small in size, since they produce minute eggs with little or no yolk. This small size of ovaries and eggs is in correlation with the habit of uterine gestation. The paired *oviducts* enlarge into paired *uteri,* which in some groups unite into a single median uterus.

The *testes* lie at first in the body cavity, as in reptiles, and occupy positions homologous with those of the ovaries. In most mammals (monotremes, whales, elephants, armadillos, and a few others excepted) the testes descend through the *gubernaculum* into the *scrotum.* The penis of the male mammal is homologous with the clitoris of the female and is a structure quite unique among vertebrates.

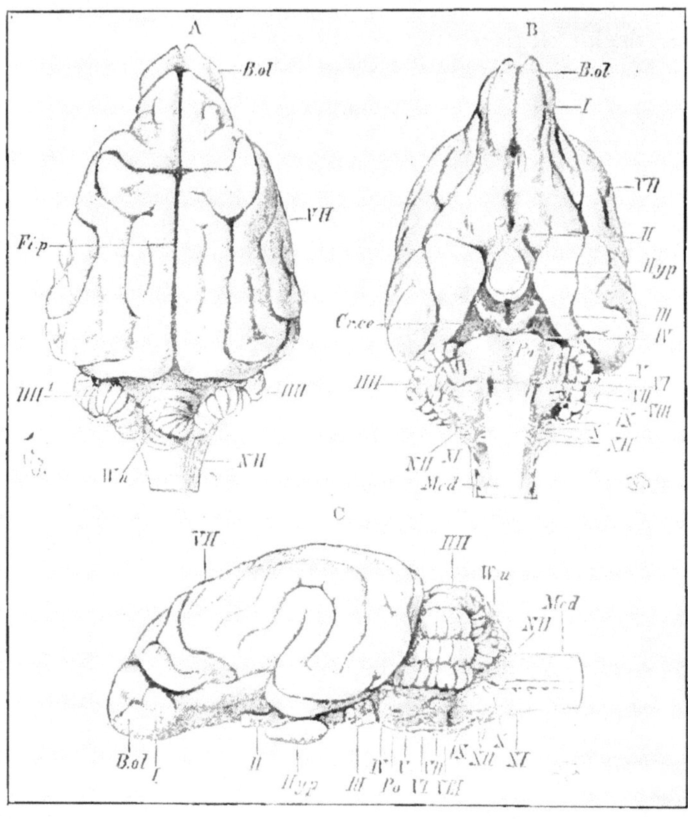

FIG. 172.—Brain of dog. *A*, dorsal, *B*, ventral; *C*, lateral aspect. *B. ol*, olfactory bulb; *Cr. ce*, crura cerebri; *Fi. p*, great horizontal fissure; *HH*, *HH'*, lateral lobes of cerebellum; *Hyp*, hyophysis; *Med*, spinal cord; *NH*, medulla oblongata; *Po*, pons Varolii; *VH*, cerebral hemispheres; *Wu*, middle lobe (vermis) of cerebellum; *I–XII*, cranial or cerebral nerves. (From Wiedersheim.)

THE ORIGIN OF MAMMALS

It has long been held that the mammals are descended from the reptiles, a theory based on the fact that the monotremes, primitive egg-laying mammals, have many distinctly reptilian characters. If mammals descended from any other vertebrate class they must have come from either Amphibia or birds. The latter possibility is out of the question, for the birds are more recent than the mammals. There is, however, some ground for the idea that the mammals may have been derived directly from the Amphibia.

The Theory of Amphibian Ancestry of the Mammals.—This theory was advanced by Huxley and gained considerable vogue until recent discoveries eliminated it from the field. Huxley's argument in brief was as follows: The presence of two occipital condyles separates the mammals from the reptiles, and unites them with the Amphibia. The mammals retain the left aortic arch and lose the right, while birds retain the right arch and lose the left. Reptiles show a tendency to reduce the left arch, which does not look toward a mammalian condition, and therefore discredits the reptile ancestry idea.

This theory is based on the supposition that the condyles and aortic arches of modern reptiles are primitive and were the same in the early reptiles as they are to-day. This is a fallacy, however, for some of the early reptiles, notably the cynodonts, had two condyles like the Amphibia. It is also quite possible that the early reptilio-mammal stock had a reversed symmetry of the aortic arches. Although still advocated by some modern writers, the theory of amphibian ancestry of the mammals confidently may be set aside.

Palæontological Evidence of the Origin of the Mammals.—Within comparatively recent years the fossil evidences of mammalian descent have been vastly strengthened by the discovery in Triassic rocks of South Africa of a large collection of remains of a group of extinct reptiles known as *Cynodontia* (dog-toothed), that have already been dealt with in the chapter on reptiles. There were many types of cynodonts, some of which exhibit mammalian tendencies of one sort, others of another. Certain authorities claim that all of the distinctions between reptiles and mammals, based on bony structures, are transgressed by one or more groups of cynodonts, some groups transgressing with regard to a majority of distinctions, other groups with regard to one or a very few. These creatures were obviously reptiles of a rather generalized type in most respects, but they were evidently making some of the same experiments that the ancestral mammals must have made in order to arrive at the present mammalian status. Whether or not these cynodonts were the actual ancestors of the first mammals it is impossible to say, but there is nothing inherently improbable about such a theory.

The cynodonts (Fig. 173) were mammal-like in a number of ways: *a*, they had a well-defined heterodont dentition, with incisors, canines and molars; *b*, they had two condyles; *c*, the lower jaw was composed primarily of the dentaries, but there were sometimes small vestigial angulare, articulare, and other reptilian bones; *d*, the quadrate was often greatly reduced and must have been functionless as a connection between the mandible and the skull. These and many minor features of the skull and limb skeletons were modified in a mammalian direction, but no single species of cynodont approached very closely a true mammalian condition. Possibly the future has in store for the palæontologists the discovery of the real ancestral mammal.

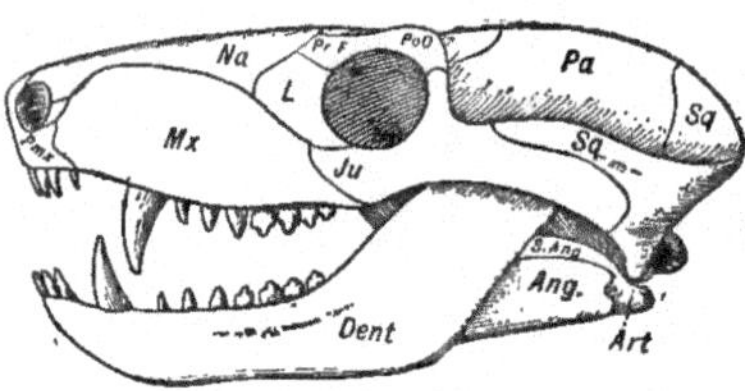

FIG. 173.—Skull of cynodont reptile, *Nythosaurus larvatus*, Trias, South Africa. Note mammal-like tooth differentiation, but complex reptilian lower jaw. *Ang*, angulare; *Art*, articulare; *Dent*, dentary; *Ju*, jugal; *L*, lachrymal; *Mx*, maxillary; *Na*, nasal; *Pa*, parietal; *Pmx*, premaxillary; *PoO*, postorbital; *Pr. F*, prefrontal; *S. Ang*, surangulare; *sq*, squamosal. (From Lull, after Broom.)

Now, the cynodonts belong to the sub-class Synapsida and the order Therapsida, which Williston places very low in the series of reptilian orders, far below the Ichthyosauria, Squamata, Rhynchocephalia, Crocodilia, Dinosauria, and Pterosauria. The only orders of lower rank are Cotylosauria, Chelonia, and Theromorpha. It seems highly probable that the Therapsida were derived from an early very generalized group of Permo-Carboniferous theromorphs, probably the pelecosaurs, of which *Varanosaurus* appears to be the most generalized representative. This type was a long lizard-like reptile with very generalized proportions and with the maxillary teeth somewhat more prominent than the others.

In addition to the mammal-like skull characters referred to above, these South African cynodonts had modifications of the limbs (Fig. 123, E) that appear to have had to do with rapid locomotion, characters that might well have served to introduce the habit of migration and thus to have given these reptiles an advantage over their more sluggish relatives. Migrations would have a tendency to increase the powers of observation and in turn to have served to accelerate brain development. The habit of living in regions with a changeable temperature would doubtless be associated with the development of various temperature-regulating mechanisms that are to-day associated with what we call warm-bloodedness. Probably the gradual development of the homothermous condition paralleled the gradual separation of the right and left ventricles and the resultant complete separation of the arterial and venous blood. One of the consequences of a higher temperature must have been a heightened nervous efficiency, for it is well known that nervous tissues tend to develop more effectively at relatively high temperatures. Furthermore, the habit of uterine gestation would increase the effectiveness of the higher temperatures at the very time when the organism is most responsive; for, as has been experimentally demonstrated, the early stages of development are crucial in determining the character of the nervous system. (In support of this view it may be said that the least highly differentiated brain among modern mammals is that of the monotremes in which there is a less effective temperature-regulating mechanism and in which a constant developmental temperature is impossible because the eggs are allowed to cool periodically while the mother is absent in search of food. The most highly differentiated brain, moreover, is found in Man, who has an exceptionally prolonged period of uterine gestation, and who has learned the uses of clothing and artificial heat as aids in maintaining a constant high body temperature, especially in the young infant prior to the development of its homothermic mechanism; for the human infant is for some time after birth practically cold-blooded in the sense that it is unable to maintain a constant temperature.)

Time, Place and Environment of the Pro-mammals.—A knowledge of the period when the pro-mammals lived should give a clew as to the probable causes of the development of mammalian characters in some reptilian group. The place of origin of the first mammalian experiments appears to have been South Africa, and the time early Permian. The eminent palæontologist and palæogeographer, Schuchert, says: "The evidence is now unmistakable that early in Permian times all of the lands of the southern hemisphere were under the influence of a glacial climate as severe as the polar one of recent times, and that, like the latter, the Permian one also had warmer interglacial periods, for coal beds occur associated with glacial deposits in Australia, South Africa, and Brazil." Now the cynodonts, which we have dealt with as the group of reptiles showing the most pronounced mammalian tendencies, were not Permian animals at all, but lived in Triassic times, a million or so years later. How then could they have given rise to the mammals? The answer is that they themselves probably did not produce the mammals, but that they and the mammals were both derived from a common ancestral stock that lived in the Permian. The mammals represented an offshoot of this ancestral stock that went the entire course in developing mammalian characters, while the cynodonts represent a number of partially successful experiments that fell short in various respects of full mammalian development. Some day it is hoped that the true ancestral mammals will be found in rocks deposited not later than the Middle Triassic and not earlier than the Lower Permian.

MESOZOIC MAMMALS

The first actual relics of mammals proper appear in the Triassic contemporaneously with the cynodont reptiles. It is believed that the beginning of mammalian evolution took place about ten million years ago and that the first mammals were very small creatures about the size of rats or mice. Osborn believes them to have been arboreal forms, probably insectivorous, and obliged to lead a furtive nocturnal life. In the daytime they hid among the trees and thickets and at night ventured forth in search of prey. It may well have been creatures of this sort that were partially or largely responsible for the slaughter of eggs and young of the great Mesozoic reptiles, for they must have lived together during this period. Osborn thinks that these small furry creatures probably resembled the modern tree-shrews, such as *Tupaia* (Fig. 186, B), a species which he believes to be the most nearly prototypic of the modern mammals.

In the study of mammalian evolution particular attention must be paid to the two mechanisms whose contact with the environment is the most intimate; the teeth and the feet. For the evolution of the mammalian orders and families is primarily one of foot and tooth specialization; hence these two characters are of fundamental importance in the classification of the Mammalia.

The **teeth** of reptiles, except the cynodonts, are simple conical bodies, with little or no regional differentiation. The cynodonts, as we have seen, had incisors, canines and primitive molars; the mammals have carried out this differentiation much further. The molars tend to become tuberculate (*bunodont*) in some groups, and flatten out into broad, crushing teeth (*lophodont*) in others. In some orders the incisors are modified into great chisel-edged gnawing teeth; in other groups they become vestigial or are totally lost in the adult. The canines on the whole are the most conservative of the teeth, tending to retain their conical shape; but in some groups they have become specialized into tusks of various kinds, and in other groups they are vestigial or absent.

The **feet** of primitive reptiles are typical five-fingered feet with claws. From this type of foot, it will be recalled, the reptiles underwent an extensive adaptive specialization. Hoofed feet were developed in the heavy herbivorous forms and long raptorial claws developed among the carnivorous groups. A similar adaptive radiation in foot structure occurred among the mammals. It is probable that the first mammals were unguiculate (clawed), a condition very similar to the generalized ancestral foot. From this type were developed the various three-fingered, two-fingered, and one-fingered hoofed types, the curving-clawed predaceous types, the flat-nailed types of the primates, and all of the other specialized types of foot structure.

Fig. 174.—Jaw of primitive mammal, *Dromatherium sylvestre*, Trias, N. Carolina; twice natural size. (From Lull, after Osborn.)

The earliest mammalian remains (Fig. 174) consist of two jaw bones found in the coal beds of North Carolina, a Triassic deposit. The creatures to which these jaws belonged, whether they were true mammals or not, must therefore have been contemporaneous with the South African cynodonts. Except for the fact that the jaw was a one-piece jaw consisting only of the paired dentary bones, it was more like those of the cynodonts than like those of modern mammals. The molar teeth were very generalized in that their number of tubercles was indefinite and the incisors were only slightly flattened.

A somewhat later group of primitive mammals, known as *Triconodonta*, is represented by a few fragmentary remains (Fig. 175) found in Jurassic rocks. These mammals had teeth more perfect in form than those just described, the molars being *trituberculate* with the cusps knife-edged and arranged in a single row like the teeth of a saw. Another group of early mammals of Lower Cretaceous and Jurassic times had teeth in which the tubercles are arranged in a triangular or trigon group, with the main cusp on the inner edge of the tooth. These *Trituberculata*, as they are called, probably had insectivorous habits and may have been the direct ancestors of the insectivores of

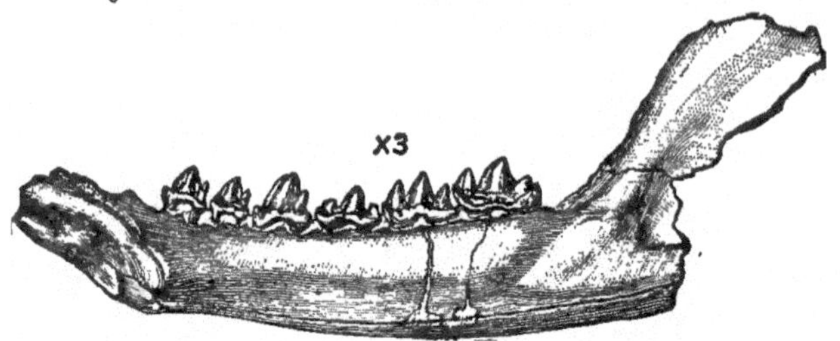

FIG. 175.—Jaw of Triconodont mammal, *Triconodon ferox*, Comanchian, Wyoming. Three times natural size. (From Lull, after Marsh.)

to-day. Still another early group, the *Allotheria* (Fig. 176), found as early as the Jurassic, but lasting over into the Cenozoic, had multi-tuberculate molars and rodent-like incisors. The premolars were in some cases much like the primitive cutting teeth of the Trituberculata.

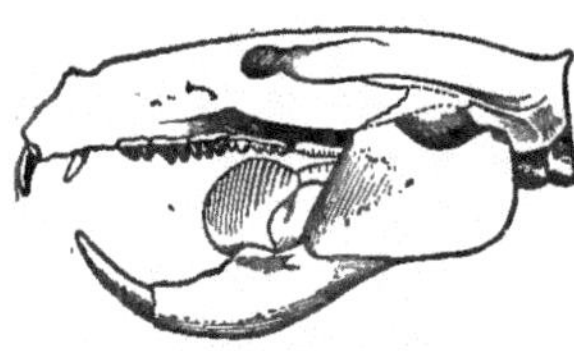

FIG. 176.—Skull of multi-tuberculate mammal (allothere) *Ptilodus gracilis*, Palæocene. (Ft. Union); Wyoming. About natural size. (From Lull, after Gidley.)

All of these mammalian relics indicate that the mammals made a very modest and tentative start in the Mesozoic. If one may judge by the teeth, they had already undergone a limited adaptive radiation into insectivorous, carnivorous, and gnawing types, which foreshadowed the mammalian groups of like habits to-day. The reason for their inconspicuousness during the Mesozoic is not far to seek, for they lived when the reptiles had pre-empted all of the important life ranges. Their very inconspicuousness was their salvation and gave them a chance to live through a trying period and to await the dawning of their great opportunity; this came toward the end of the Cretaceous, when the reptilian dynasties waned and extinction overtook all of the highly specialized dominant types. Perhaps, as has already been suggested, these small, blood-thirsty mammals played an important rôle in hastening the decline of the reptiles by preying on their eggs and young. One may picture the Mesozoic drama in the words of Lull, if we bring "before the mind's eye broad vistas of low-lying well-watered woodland with ever alert furry forms taking such refuge as the trees and shrubbery or occasional hiding holes could offer, in the midst of stalking terrors such as the world never saw before or since. That the mammals managed to maintain themselves is not surprising, for there is a teeming horde of small mammalian folk in the tiger-haunted jungles of India to-day; and that they did not dispute with the dinosaurs the realms of greater opportunity is but a logical assumption."

In conclusion it may be said of the Mesozoic mammals that they made less headway in the Mesozoic than did the reptiles in the Palæozoic; for they were all quite generalized in structure and of small size. There is evidence that the first mammals arose very soon after the reptiles became well established in the Permian. If this is so, we see that all of the vertebrate classes except the birds had their origin in Palæozoic times.

Cenozoic Mammals

The Cenozoic has been called the Age of Mammals, just as the Mesozoic is called the Age of Reptiles. The same great climatic or geological conditions that are assumed to have led to the extinction of the exuberant reptilian dynasties that flourished during the Mesozoic may be also given the credit for affording the mammals their first opportunity to "secure for themselves a place in the sun." After a lurking life in the shades and shadows they were able to emerge into the open and to invade the vast fields of opportunity vacated by the fallen races of reptiles. The small, warm-blooded mammalian races repeopled the wastes, gaining the upper hand over the few reptilian groups that remained, such as the lizards, snakes and turtles; these in turn took up the furtive life that the mammals left behind. The mammals had been under pressure during the entire Mesozoic, and when the pressure was removed they expanded marvelously.

The Archaic Mammals of the Cenozoic.—The mammals of the early periods of mammalian deployment during the Tertiary are usually called archaic mammals. They differ from modern mammals in the following particulars: 1, Their feet were conservative, showing little advance upon reptilian conditions; 2, their molar teeth were very little differentiated for the various feeding habits; 3, the brain, especially the part which is the main seat of intelligence (the cerebrum) was small in proportion to the size of the body and was reptile-like in many respects.

These archaic mammals went the ways of the Mesozoic reptiles to a considerable extent, in that they became large, robust, vegetative mechanisms with low intelligence and little adaptability. Like the dinosaurs they vanished completely from the face of the earth and left few descendants. Only those of small size and with comparatively unspecialized structures survived to become the ancestors of our modern mammalian faunas. Osborn sums up the situation as follows:

> "Nature deals in transitions rather than in sharp lines. We can not circumscribe the archaic mammals sharply, nor be sure as yet that some of them did not give direct descent to certain of the modernized mammals. Yet the animals of the basal Eocene of both Europe and North America are altogether of a very ancient type; they exhibit many primitive characters, such as extremely small brains, simple triangular teeth, five digits on the hands and feet, and prevailing plantigradism. They are to be collectively regarded as the first grand attempts of nature to establish insectivorous, carnivorous and herbivorous groups, or unguiculate (clawed forms) and ungulates (hoofed forms). The ancestors or centres of these adaptive radiations date back into the Age of Reptiles. At the beginning of the Eocene we find the lines all separated from each other, but not as yet very highly specialized. The specialization and divergence of these archaic mammals continue through the Eocene period and reach a climax near the top, although many branches of this archaic stock become extinct in the Lower Eocene. The orders which may be provisionally placed in this archaic group are the following:
>
> "Marsupialia.
>
> Multituberculata, Plagiaulacidæ.
>
> Placentalia.
>
> Insectivora. Insectivores not yet positively identified in the basal Eocene.
>
> Tæniodonta. Edentates with enamel teeth.
>
> Creodonta. Archaic families of carnivores.
>
> Condylarthra. Primitive light limbed cursorial ungulates.
>
> Amblypoda. Archaic, typically heavy limbed, slow-moving ungulates.

"This group is full of analogies, but is without ancestral affinities to the higher placentals and marsupials. There are forms imitating in one or more features the modern Tasmanian 'wolf' (*Thylacynus*), the bears, cats, hyænas, civets, and rodents of to-day, but no true members of the orders Primates, Rodents, Carnivora, Perissidactyla, Artiodactyla have been discovered."

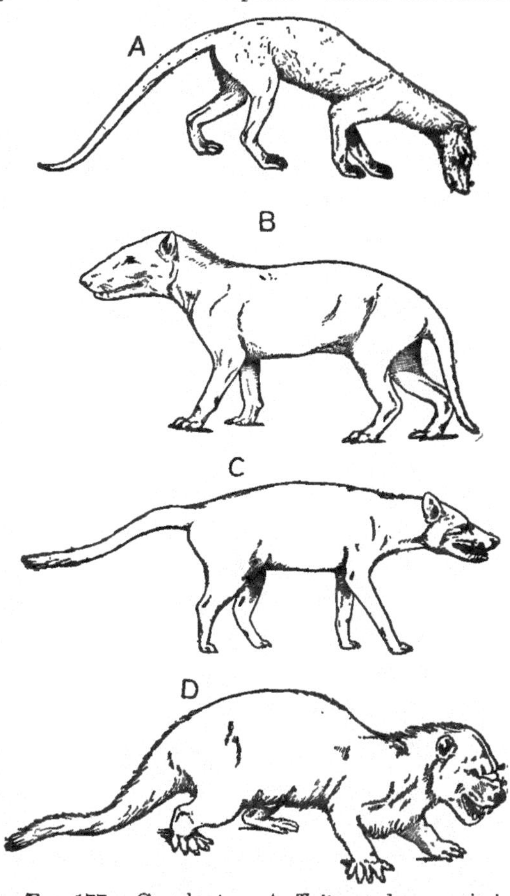

Fig. 177.—Creodonts. A, *Tritemnodon*, a primitive hyænodont, Middle Eocene, North America. (After Scott). B, *Hyænodon*, the last survivor of the archaic carnivores, Lower Oligocene, North America and Old World. (After Osborn). C, the dog-like *Dromocyon*, Middle Eocene, North America. (After Osborn). D, *Patriofelis*, Middle Eocene, North America. (All from Lull, after Osborn.)

The outstanding groups of archaic mammals are the *Creodonta*, the *Condylarthra*, and the *Amblypoda*. These three claim our attention.

The **Creodonta** (flesh-toothed) differ from the Condylarthra in having the skull and tooth characters of carnivores, and in having claw-like rather than hoof-like terminal phalanges. The most evident difference between the creodonts and modern carnivores is in the capacity of the brain-case; for, like all archaic mammals, they had small reptile-like brains. The teeth of the creodonts are also less specialized for rending of flesh than are those of the true carnivores. Of the six families of creodonts recognized by palæontologists, all but one became extinct before the dawn of the era of modernized mammals. Perhaps the best-known of the creodonts is the genus *Dromocyon,* which is shown in the illustration (Fig. 177, C). It is interesting to note that there were bear-like, dog-like, otter-like, cat-like, and hyæna-like creodonts.

The **Condylarthra** (knuckle-jointed) were archaic ungulates and differed from the creodonta mainly in adaptations for herbivorous diet. In form they closely resembled the creodonts, for both were rather generalized in structure. The most interesting of the condylarthrans is *Phenacodus* (Fig. 178), a form that for a long time was believed to be the five-toed ancestor of the horse, but is now known to be both too specialized in some respects and too late in its appearance to have the honor of playing this rôle. Only a few species of condylarthrans are known and these are grouped into two families. They range from the size of a fox to that of a large sheep. They had rather tusk-like, but small, canines and low-crowned grinding teeth of archaic pattern. The skull was long and low with a small brain case. The feet were five-fingered and of a primitive plantigrade form, with small hoofs. The genera that have been studied could not have given rise to modern ungulates, but the real ancestors of the ungulates must have been relatives of the condylarths.

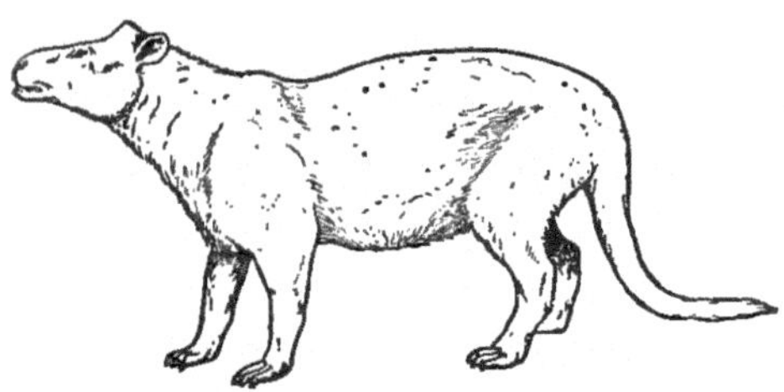

FIG. 178.—Cursorial archaic mammal, condylarth, *Phenacodus primævus.* Lower Eocene, North America. (From Lull, after Osborn.)

The **Amblypoda** (blunt-footed) were short-footed ungulate-like mammals, some of which attained a huge size, almost comparable with that of the elephants, but reminding one more of the rhinoceros or hippopotamus types. There were four families of amblypods that differed considerably among themselves.

The genus *Coryphodon* (Fig. 179) is one of the best known amblypods. It was probably a swamp-dweller, nearly as large as an ox, but much more thick-set and massive. The limbs were short and powerful, and the feet had spreading toes well adapted for swamp navigation. The skull was long and flat, and without horns. The canine teeth were tusk-like and were probably used for tearing out the succulent roots of water plants. Everything seems to indicate that it was a very sluggish and stupid creature.

Fig. 179.—A swamp-dwelling amblypod, *Coryphodon*, Lower Eocene, North America. (From Lull, after Osborn.)

Dinoceras (Fig. 180) represents another family of amblypods and appears to have been an end-product of a long line of specializations. It stood about seven feet in height, had very heavy, elephantine limbs and a massive body. The head was armed with two heavy horns and great tusks, which were doubtless used as a defense against the creodonts, the only contemporary animals that could have attacked a creature of such proportions.

The archaic mammals nearly all became extinct before the end of the Eocene. The causes of their extinction can only be conjectured.

Fig. 180.—Four-horned amblypod, *Dinoceras*, the culmination of its race, Upper Eocene, Wyoming. (From Lull, after Osborn.)

In some cases it seems probable that over-specialization combined with racial old age brought about extinction; in other cases it is more likely that the competition against the on-coming modernized mammals, which were more alert and intelligent, brought about their gradual elimination. Only a very few of the archaic mammals survived beyond Eocene times and these were rather generalized types, suitable to be the ancestors of the modernized mammals. There was probably considerable emigration on the part of certain of these surviving archaic types. It may well be, for example, that the marsupials of Australasia were the descendants of a very early group of multituberculate mammals that succeeded in reaching the Australasian peninsula before it was cut off from the Asiatic continent. The cutting off of the Australasian land bodies must have occurred before the modernized mammals reached that part of the world, for there are in those regions no true modernized mammals.

Origin of the Modernized Mammals

The modernized mammals include practically all existing placental mammals and their immediate ancestors, including: true carnivores, rodents, odd- and even-toed ungulates, elephants, sirenians and whales. To this list some authors would add edentates, bats, and insectivores.

> "In contrast with the archaic mammals," says Lull, "the modernized types are all creatures of high potentiality, and, where they became extinct, were rather the victims of circumstance than creatures that died because of lack of adaptability, although certain groups seem to have run a natural course and their extinction was heralded by evidences of racial senility.
>
> "As the archaic forms were characterized by lack of progressive brain and feet and teeth, so the modernized races were distinguished by the possession sometimes of one (primates), sometimes of two (elephants), again by all three (horses) of these destiny controlling organs, but in general the modernized animals were progressive, highly adaptable forms."

Place of Origin.—It is believed that the modernized mammals originated in the great Arctic Continent. The reasons for this belief are: first, there is a striking resemblance between the first European and the first North American modernized mammals; second, palæogeographers tell us that a fluctuating land bridge between the eastern and western continents existed from time to time, and between times was submerged; third, the climate of the Arctic regions was at one time warm, as is evidenced by the discovery of fossils of sub-tropical plants on the coast of Greenland.

Time of Origin.—All available evidence seems to point to the latter half of the Eocene period as the time when the modernized mammals arose. Some types evidently arose considerably later.

Migrations.—The spread of the modernized mammals must have been southward. This must have been so for two reasons: first, that was the only possible direction in which a group originating in the north could migrate; and second, because the increasing cold which culminated in the first glacial epoch, must have driven the majority of the mammals out of the northern regions. A few of the hardiest types still find these regions habitable. The migration occurred in several great waves, probably due to the alternating periods of cold and warm climate in the north. The groups least tolerant of cold probably migrated southward first and went farthest south; among these first migrants were probably the insectivores and primates; these were probably followed by the perissodactyls (horses and tapirs) and, somewhat later, by the true carnivores, especially the cat-like forms. The bears and rodents remained longer than the rest and still live well toward the north. To-day the modernized mammals have a world-wide distribution except in the oceanic islands, which they have no means of reaching.

Mammals of the Present

Brief Classification

Class Mammalia. "Beasts," "quadrupeds," "animals." Warm-blooded, hair-clad vertebrates with mammary glands.

Sub-Class I. Prototheria.—Egg-laying mammals.

Order 1. Monotremata.

Family 1. Ornithorhynchidæ.

" 2. Echidnidæ.

Sub-Class II. Eutheria.—Viviparous mammals.

Division I. Didelphia (Metatheria).—Marsupials.

Order 1. Marsupialia. Mammals that usually carry the young in a pouch; usually no placenta.

Sub-Order 1. Polyprotodontia.

Sub-Order 2. Diprotodontia.

Division II. Monodelphia (Placental mammals). Young never carried in a pouch; a true placenta, which nourishes the unborn fetus.

Section A. Unguiculata

Order 1. Insectivora.
" 2. Dermoptera.
" 3. Chiroptera.
" 4. Carnivora.
" 5. Rodentia.
" 6. Edentata (Xenarthra).
" 7. Pholidota.
" 8. Tubulidentata.

Section B. Primates

Order 9. Primates.
Sub-Order 1. Lemuroidea.
" 2. Anthropoidea.

Section C. Ungulata

Order 10. Artiodactyla.
" 11. Perissodactyla.
" 12. Proboscidia.
" 13. Sirenia.
" 14. Hyracoidea.

Section D. Cetacea

Order 15. Odontoceti.
" 16. Mystacoceti.

This classification, which follows closely that given by Osborn in "The Age of Mammals," departs widely from traditional lines. The grouping of five orders into the section Ungulata is decidedly novel; the separation of the Proboscidia from the Perissidactyla, and the inclusion of the Sirenia among the ungulates, are well founded; the distribution of the old group of Edentata among several distinct orders will doubtless meet with general approval, for it has long been felt that the old assemblage was artificial. But perhaps the most striking feature of the classification is the position assigned to the Primates—below the ungulates and the cetaceans instead of at the apex of the phyletic series where we have been accustomed to place them. This somewhat lowly position of the Primates is justified by the fact that, generally speaking, they are much less specialized than are either the ungulates or the whales. Even Man, apart from his remarkable brain and his upright position, is a comparatively unspecialized mammal. If this disrespectful treatment of lordly Man shocks the gentle reader, let him remember that several authorities have already assigned to the birds the distinction of being the most highly specialized vertebrate class; so the edge is taken off the contest for first honors in the second division.

SUB-CLASS I. PROTOTHERIA (MONOTREMATA, EGG-LAYING) MAMMALS

The modern representatives of this sub-class are few, consisting of but three genera of strange beasts native to Australasia. Some fragmentary remains of Multituberculata, already discussed in the section dealing with the Mesozoic mammals, have also been assigned by some authorities to this division.

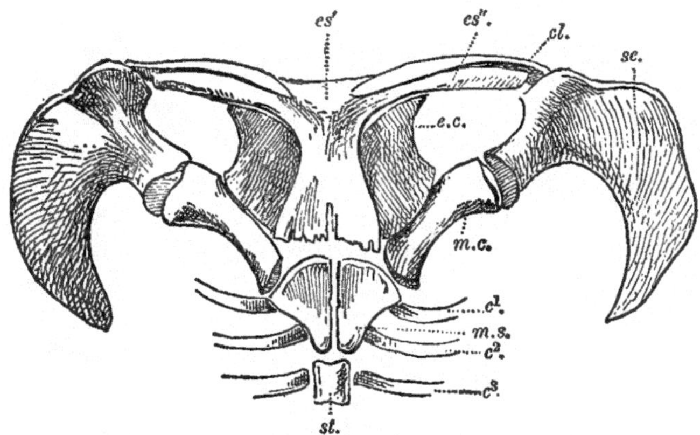

Fig. 181.—Pectoral arch and sternum of *Ornithorhynchus anatinus*. c'^1 c^2, c^3, first, second and third ribs; *cl*, clavicle; *ec*, epicoracoid; es^1 and es^2, prosternum (episternum) *m. c*, metacoracoid (coracoid); *m. s*, manubrium sterni; *sc*, scapula; *st*, sternebra. (From Wiedersheim.)

The **Monotremata** are mammals that lay large eggs with a shell, abundant yolk, and albumen (eggs practically reptilian in character); they have diffuse mammary glands without teats; the brain lacks the corpus callosum; the shoulder girdle has a large coracoid (Fig. 181) reaching to the sternum; an interclavicle is present; paired marsupial or epipubic bones extend forward from the pelvis; the vertebræ are for the most part without epiphyses; the ribs are one-headed, the tuberculum being absent; the mammary glands are modified sweat-glands and are not sebaceous; there is a shallow cloaca; one group (the Echidnidæ) has a temporary pouch for incubating the eggs. The oviducts are entirely separate throughout and open by two separate genital pores into the cloaca.

The majority of these characters hark back to a reptilian ancestry and are therefore to be considered as primitive. It is not believed, however, that the monotremes, as we know them, are at all close to an ideal prototypic mammalian condition; but rather that they are the end-products of a rather highly specialized side line of mammalian evolution, that came off from some early reptilio-mammal stock and that has retained some of the primitive characters of these ancestors. It is not thought, therefore, that the monotremes are in any sense ancestral to the Eutheria.

Family 1. Echidnidæ.—This family contains two genera, *Echidna* and *Proechidna*. *Echidna aculeata* (Fig. 182, D), the "Australian Anteater," is the best known species. It is found in New Guinea, Tasmania and Australia, and several local sub-species are distinguished. Its characters may be dealt with under two categories: those that are cænogenetic, adaptations for the anteating habit; and those that are palingenetic or primitive.

Echidna is a typical anteater in all of its adaptations. It has a heavy protective covering of quill-like spines, with an underlying layer of coarse hair. The snout is long and tapering, reminding one rather strongly of a bird's bill. The tongue is extremely long and extensible and is covered with a sticky salivary secretion, which holds the ants when the tongue is thrust into ant holes. The claws are very long and powerful and are used for tearing down ant-hills and for making burrows. As in anteaters of other orders, teeth are lacking. Two other characters seem in no way to relate *Echidna* to the anteating habit; these are first, a rudimentary tail, much like that of a bird, and second, a small spur connected with a peculiar gland on the heel, a structure whose function is not well understood. Of somewhat more fundamental importance are the following characters: the cerebral hemispheres are fairly large and well convoluted; there is a temporary marsupial pouch (Fig. 182, C), which seems to have no relation to the marsupium of the marsupials, but is more nearly

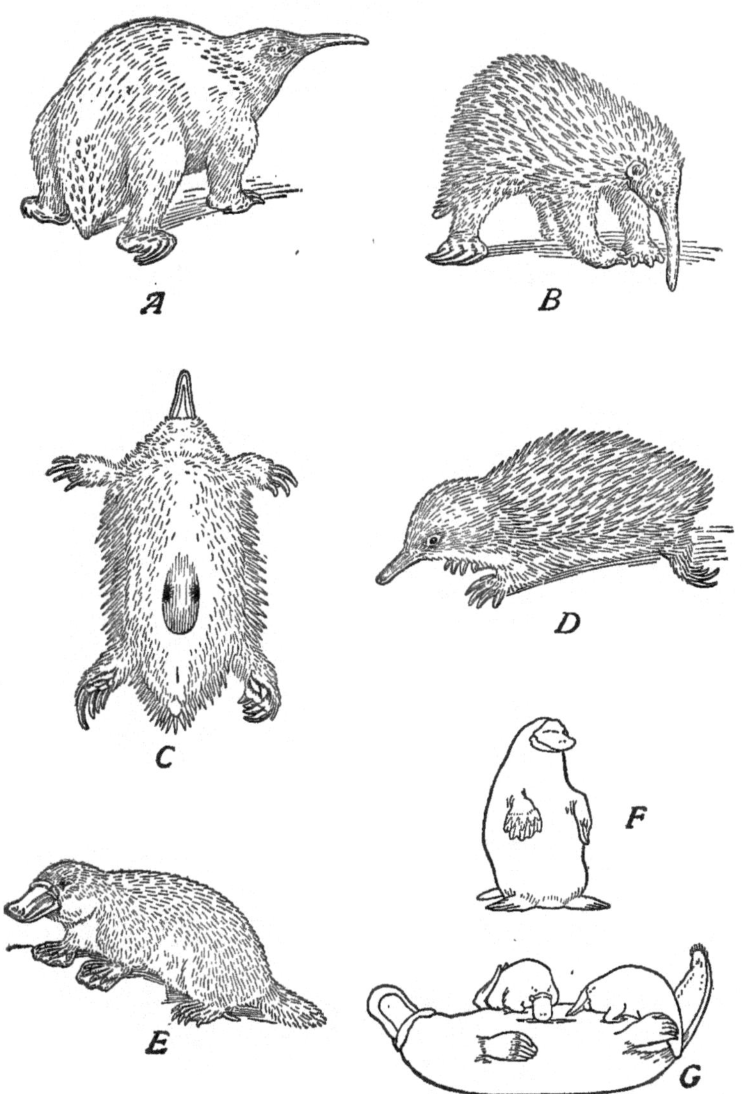

FIG. 182.—Group of Monotremata. A, *Proechidna bruijnii;* B, *Proechidna nigroaculeata;* C, *Echidna aculeata;* ventral aspect to show brood pouch; D, *Echdina aculeata;* E, *Ornithorynchus anatinus;* F, *Ornithorhynchus* standing up like a penguin; G, *Ornithorhynchus* female allowing young to obtain milky secretion from the diffuse abdominal mammary glands. (All redrawn, A, B, F, G, after Brehm; C, after Haacke; D and E, after Vogt and Specht.)

homologous with a teat; the temperature of the body is lower than in the higher mammals, and has a variation in health of at least 15° Centigrade, a character which seems to be intermediate between the poikilothermous and the homothermous conditions.

Proechidna, a New Guinea species, differs from *Echidna* in the following particulars: the toes on both fore and hind feet are reduced to three large and two rudimentary elements; the beak is longer and is curved downward; the back is more arched; the external lobe of the ear protrudes freely from the hair of the head. The combination of characters gives to the *Proechidna* a ridiculous resemblance to a miniature elephant. Two species, *P. bruijnii* (Fig. 182, A) and *P. nigroaculeata* (Fig. 182, B), are distinguished.

The **breeding habits** of the Echinidae are of especial interest. The egg is about half an inch long and has a leathery shell much like that of a tortoise. Only one egg is laid at a time and it is immediately transfêrred by the mouth of the mother to the brood pouch (see Fig. 182, C), where it undergoes a short incubation. When ready to hatch, the shell is broken, as in the bird, by means of a shell-breaking tubercle on the end of the snout; the mother then removes the broken fragments of shell. The just-hatched young is in a very immature and helpless condition and lies quietly in the pouch for some time, merely able to lap up the milky secretion that exudes from the walls of the pouch. After the young has reached a considerable size it is removed by the mother from time to time in order to give it exercise, but it is put back into the pouch to be suckled. There is among Echidnidæ really no need of a nest, for the egg is kept safely in a pouch. After a time, however, the mother leaves the young in the burrow while she pursues her nocturnal occupation of ant-hunting. This burrow with its enlarged terminal chamber is a safe retreat for the youngster when later he ventures forth to learn the anteating game for himself.

Family 2. Ornithorhynchidæ.—This family consists of but the single species *Ornithorhynchus anatinus* (Fig. 182, E), the Duck-bill Platypus, a native of Southern Australia and Tasmania. When the first specimen of this strange beast was exhibited in England it was believed to be a fake, on a par with the composite mermaids then in vogue. It was described as a furry quadruped with the bill and feet of a duck; a very apt characterization. The animal is about a foot and a half long, with a heavy coat of soft brown fur. The feet are five-toed and webbed, the webbing on the fore feet extending well beyond the tips of the toes, but that of the hind feet being about as it is in a water bird. Both feet are armed with sharp claws. The beak is very wide and flat and is covered with soft, naked skin that flares out at the base into sensitive flaps; this beak covering is highly sensitive owing to the abundance of sense organs that are scattered over its surface. There are no teeth in the adult, but instead, broad, horny plates line the inside of the bill; these are used for crushing the shells of bivalves and water snails, which constitute its chief food. The young platypus has a set of milk teeth, all molars and eight or ten in number; these are gradually worn off and then replaced by plates. The eyes are small and beady; there is no external ear lobe; the male has a spur on the heel like that of the Echidnidæ, but larger in size. The tail is large and dorso-ventrally flattened; it is used as a rudder in swimming.

The brain of *Ornithorhynchus* (Fig. 183) is the most primitive brain known for a living mammal. It is comparatively quite small, and the cerebral hemispheres are smooth and, like a reptile brain, entirely lacking in convolutions. The habits of this creature are purely aquatic, not unlike those of a muskrat. It lives in stagnant, weedy ponds or streams, feeding chiefly on mollusks, crustaceans, and worms that are secured by scooping up the muddy bottom with the spoon-like snout. Provender is stored in capacious cheek-pockets and is carried in this way to the burrow, where it is eaten at leisure. The burrow is dug deep into the bank of the stream, beginning below the water-line and sloping upward until at a distance of twenty-five to fifty feet it terminates in a large, dry chamber with top ventilation. The chamber is comfortably lined with reeds and rushes.

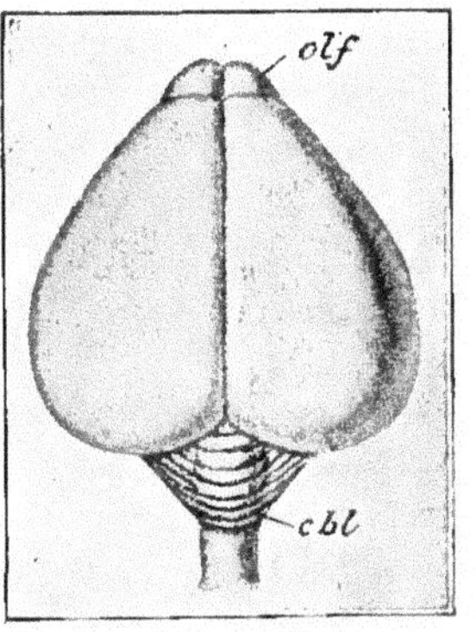

FIG. 183. — Brain of *Ornithorynchus*, dorsal view, natural size; *cbl*, cerebellum; *olf*, olfactory bulbs. Note lack of cerebral convolutions. (From Parker and Haswell.)

Breeding Habits.—The eggs to the number of two or three are laid in a nest of grasses, quite like a simple bird's nest. They are somewhat smaller than those of *Echidna* and have a rather hard, flexible shell, yellowish-white in color. They are incubated while still in the nest by means of the body heat of the mother; hence there is no brood pouch. When the young hatch they are fed by a milky secretion which exudes from the primitive abdominal milk-glands that are buried deep in the hair. The young simply licks off the drops of milk as they drip from the wet hairs. When the youngsters are older the mother lies on her back (Fig. 182, G) and the ludicrous little fellows climb on top of her in order to feed to better advantage.

SUB-CLASS II. EUTHERIA (VIVIPAROUS MAMMALS)

Definition.—Mammary glands are of the sebaceous type and are provided with teats; brain has a corpus callosum between the cerebral hemispheres; coracoid is vestigial and does not reach the sternum; there is no interclavicle; ribs are double headed; vertebræ have epiphyses; ovum is small; young are born alive.

The group includes both marsupials and the placental mammals. There is a much closer resemblance between the marsupials and the placentals than between the former and the monotremes; hence it has seemed justifiable to group the marsupials and the placentals in one sub-class.

DIVISION I. DIDELPHIA (METATHERIA)—MARSUPIALS

Definition.—Mammals with small eggs that are usually provided with a thin shell and a thin layer of albumen. The oviducts are enlarged into a pair of uterine pouches which are sometimes fused for a short distance. The distal parts of the oviducts remain entirely separate, giving a double vagina, a character responsible for the name Didelphia. The egg develops in the uterus, absorbing nutriment through its membranes. In rare cases a primitive allantoic placenta is present. The young is born in a very immature condition and is placed by the mother in the marsupium (Fig. 185, B) or brood pouch (not present in all marsupials), and is fed from the milk glands by means of a long tubular teat (Fig. 185, C and D) that is thrust down the throat, and to which the young is attached semi-permanently by means of a special larval mouth sucker. The marsupials have epipubic bones; have rudimentary corpus callosum; a shallow cloaca is present in at least some species. The skull has the following peculiarities that are useful in identifying fossil species: incompletely ossified palate, jugal bone reaching as far as the glenoid cavity; teeth more numerous than is typical for placentals; molars generally four on each side; usually but one tooth of the milk set, the fourth premolar, is functional.

In general it may be said that the marsupials occupy a position intermediate between the monotremes and the placental mammals. They have undergone an elaborate adaptive radiation, occupying in their native countries most of the life zones that are in other parts of the world occupied by the placentals. Their favorite life zone is the arboreal and they seldom invade the aquatic zones. There are some highly specialized cursorial types; some sub-terrestrial, fossorial types; and some semi-volant types. They have produced several giant forms, now extinct, but recent forms are small or of moderate size.

While they were at one time numerous and fairly well distributed over Europe and North America they are now almost confined to Australasia. A number of species of opossums and the rat-like *Cœnolestes* belong to the American continent, mostly to South America. None are found in Europe, Asia or Africa to-day.

It is believed by some authorities that the marsupials spread to Australia and South America over the hypothetical Antarctic land bridge and were subsequently cut off before the placental mammals were evolved. They have persisted in Australia largely because they have escaped competition with the larger and more capable modernized mammals that ruled the other continental bodies. A more carefully considered theory, however, would derive the marsupials from northern forms that migrated southward to escape the rigors of the early glacial epochs, and reached Australia and South America before the onset of the dominant placentals. The cutting off of Australasia gave them their best opportunity for adaptive expansion.

It is not now believed that the marsupials represent a stage in the evolution of the placental mammals; rather it is thought that they represent the adaptive radiation of a primitive mammalian stock that arose far back in the Mesozoic (probably during the Jurassic) and has had an evolution of its own, somewhat less successful and slower than that of the modernized groups. They show many evidences of racial senility and some of their supposedly primitive features may well be the products of regressive processes. The fact that there are only traces of the milk dentition, and the occurrence in some species of a transitory allantoic placenta, have been interpreted as retrograde conditions and as evidences in favor of the idea that at one time the marsupials were more fully diphyodont and had a true placental gestation.

Regarding the marsupials as a single order, we may divide them into two sub-orders: *Polyprotodontia* (many incisors), and *Diprotodontia* (two incisors). The first group is now believed to be the more primitive and the second more highly specialized and somewhat senescent.

Sub-Order I. Polyprotodontia

This group, which consists mainly of insectivorous and carnivorous types, is more primitive than are the herbivorous diprotodonts. The polyprotodonts are characterized by the possession of four or five incisors on each side of the upper jaw and one or two fewer in the lower jaw; both canines and molars have the typical carnivorous shape. They are confined to Australasia, with the exception of the American opossums.

Family 1. Didelphidæ (the Opossums).—Of all living marsupials the opossums appear to be the most generalized in both structure and habits. The *Virginia opossum* (Fig. 184, A), *Didelphys virginiana,* is the only North American member of the family and deserves special mention. It is distinctly arboreal, with a prehensile tail adapted for clinging to branches and for use as a hold-fast by the young, who wind their tails about the arched tail of the mother. The opossum is omnivorous, eating fruit, insects, birds, reptiles, and their eggs. There is a distinct pouch in which the young are suckled and carried. The animal is nocturnal in habit, sleeping in hollow trees during the day. The death-feigning instinct has received the proverbial description "playing 'possum." Important genera of the family are: *Didelphys, Marmosa, Chironectes, Peramys,* and *Philander;* there are about twenty-five species, all American. *Marmosa murina* is a tiny opossum about the size of a small rat; *Chironectes* is an aquatic type with webbed feet and about the size of a muskrat. It is the only aquatic marsupial (Fig. 184, D).

Family 2. Myrmecobiidæ (Banded Anteaters).—This small family is represented by a single species, *Myrmecobius fasciatus* (Fig. 184, B), an animal about the size of a cat, with only slight specializations for the anteating habit. Its snout is moderately prolonged; its tongue is very long and extensible and is covered with the customary sticky secretion; the tail is covered with long, coarse hair; the claws are only moderately heavy. Instead of being toothless like the anteaters of other orders they have an unusually large number of small teeth,

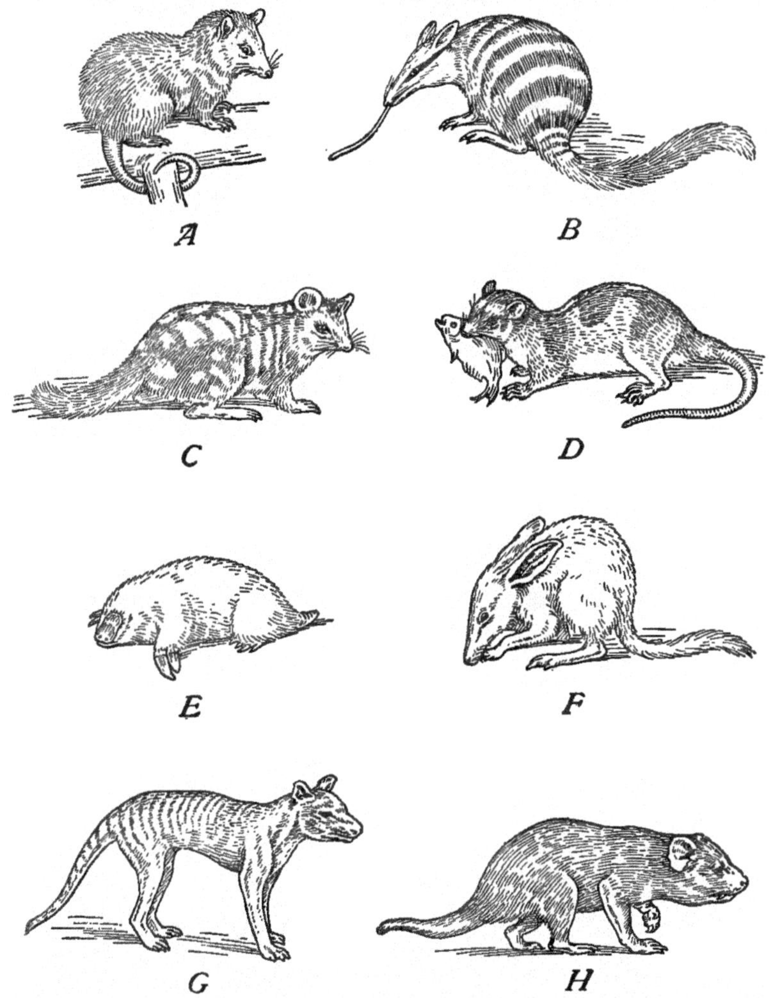

Fig. 184.—Group of Marsupials (Polyprotodonts). A, Virginia Opossum, *Didelphys virginiana;* B, Banded Ant-Eater, *Myrmecobius fasciatus;* C, Native Cat, *Dasyurus viverrinus;* D, Water Opossum, *Chironectes minima;* E, Marsupial Mole, *Notoryctes typhlops;* F, Rabbit Bandicoot, *Peragole lagotis;* G, Thylacine or Tasmanian Wolf, *Thylacinus cynocephalus;* H, Tasmanian Devil, *Sarcophilus ursinus.* (All redrawn, A after Vogt and Specht, D, after Lydekker; B, after Flower and Lydekker; E, after Beddard; others, after Brehm.)

ranging from 50 to 54. In this respect and in several others they resemble the Mesozoic marsupials. *Myrmecobius* has no pouch.

Family 3. Dasyuridæ (Carnivorous Masuspials).—This is a somewhat heterogeneous family of marsupials, ranging from mouse-like to badger-like types. They may or may not have a pouch. *Dasyurus viverrinus*, the "native cat" (Fig. 184, C) is less cat-like in appearance than marten-like. It feeds largely on birds and their eggs. *Sarcophilus ursinus*, the "Tasmanian devil" (Fig. 184, H), is an animal about the size and shape of a badger. It has the reputation of being one of the most ferocious of animals, with a devilish "yelling growl." Native Australians say, however, that it is rather a slinking than an openly pugnacious creature. *Phascologale* is a genus of small animals not unlike some of the smaller American opossums in appearance and habits. *Sminthopsis* is a genus of pouched mice. *Antechinomys* is a genus of jumping mice, with long ears and legs.

Family 4. Thylacynidæ (Thylacynes).—This family is represented by the single species *Thylacinus cynocephalus* (Fig. 184, G), which receives the name of the "*Tasmanian wolf.*" The creature is less like a wolf than like some of the smaller members of the Cat family, but the Australians must have some sort of "wolf," and this is the nearest approach that the marsupials can afford. It is a predaceous animal, almost as large as a small wolf, with a dog-like head and a series of tiger-like bands across the back and tail.

Family 5. Peramelidæ (Bandicoots).—There are three genera in this family. *Perameles* is a genus of twelve species of medium sized forms, with the pouch opening backwards. *Peragale* (Fig. 184, F) is a genus of two species of "rabbit bandicoots," which have the habit of burrowing in the soil for grubs and other soil insects. *Chæropus castanotis* is the "pig-footed bandicoot," also a burrowing form, with only two toes on the fore feet.

Family 6. Notoryctidæ (Marsupial or Pouched Moles).—This family is represented by a single species, *Notoryctes typhlops* (Fig. 184, E), a South Australian mole-like animal, with silky reddish-gold fur, which harmonizes with the color of the arid soil in which it burrows. It has a complete set of mole-like adaptations and leads a thoroughly mole-like life. The eyes are rudimentary; there are no external ear lobes; the fore feet are armed with extremely heavy burrowing claws, the third and fourth being much more conspicuous than the rest; the tail is very short and stumpy.

Sub-Order 2. Diprotodontia

The members of this division are mainly herbivorous. Their dentition is not unlike that of the rodents, the incisors being of the gnawing type, usually two pairs above and one pair below. The canines are either small or absent; the molars have either tubercles or transverse ridges. This group contains the largest and most highly specialized of the marsupials.

Family 7. Epanorthidæ.—This family consists of various extinct forms and the single living genus *Cænolestes* (marsupial shrews), the only American diprotodonts. It is native to Andean foot-hills of South America. The affinities of this genus are still somewhat in doubt, but Osgood, in an unpublished monograph on the genus, claims that it is in a sense intermediate between the polyprotodonts and the diprotodonts. It has a primitive diprotodont dentition, but a foot structure more like that of the polyprotodonts. Its resemblances to *Perameles* are rather striking, but these may be homoplastic in character. Osgood considers that the ancestor of *Cænolestes* was a North American form, which also may have given rise to the early diprotodont stock that migrated to Australasian territory. In general appearance *Cænolestes* is one of the most generalized of marsupials, reminding one more of the shrews than anything else. Many of its anatomical features are also very generalized, a fact that is in harmony with its close resemblance to a long extinct group, that lived in Miocene times. The name *Cænolestes* means "a modern representative of an ancient group."

Family 8. Phalangeridæ (Phalangers).—This is one of the largest marsupial families and consists mostly of arboreal forms. They are characterized by having five fingers and toes, with the second and third phalanges bound together by an integumentary bond; the hallux is usually opposable. The pouch is well developed; the tail is usually long. The following are some of the more important genera: *Tarsipes*, the long-snouted phalanger; *Acrobates*, the pigmy flying phalanger; *Distæchurus*, the pen-tailed phalanger; *Dromicia*, the dormouse phalanger; *Petaurus*, the true flying phalangers; *Tricosurus*, the true phalangers; *Phascolarctus*, the koala or marsupial bear.

The true phalangers (Fig. 185, F) are fairly large forms, more or less fox-like in form and sometimes known as "brush-tailed opossums." The flying phalangers are much like our flying squirrels in

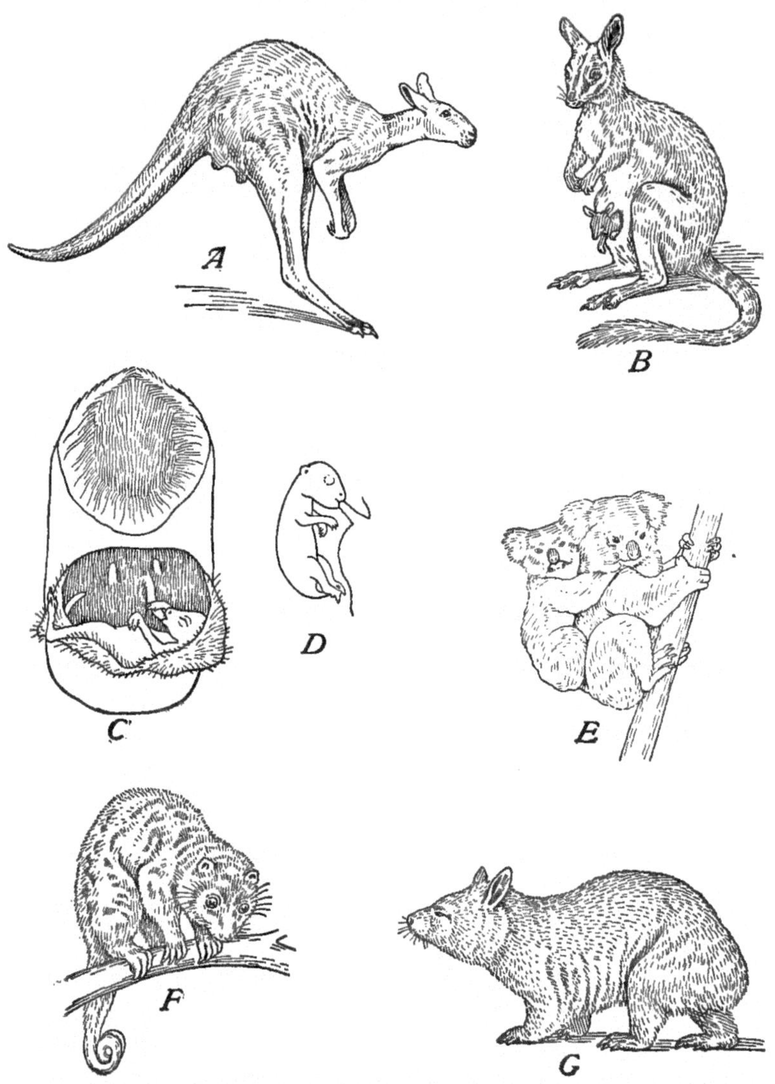

Fig. 185.—Group of Marsupials (Diprotodonts). A, Red Kangaroo, *Macropus rufus* (after Lydekker); B, Rock Wallabi, *Petrogale xanthopus* (after Vogt and Specht; C, Young Kangaroo attached to nipple in pouch of mother; pouch laid back to show interior (after Brehm); D, lateral view of same removed from pouch (after Parker and Haswell); E, Koala, *Phascolarctos cinereus*, carrying young on back (after Brehm); F, *Phalanger maculatus;* G, Wombat, *Phascolomys ursinus* (after Lydekker).

structure and habits; they are not genuine flyers but merely soarers that parachute from tree to tree by means of folds of skin stretched between the fore and hind limbs. The **koala** is a curious slow-moving, nocturnal animal, that feeds almost exclusively on the leaves of the gum tree. It has been called "marsupial bear," but is really more like a large "Teddy Bear" than anything else, as the illustration (Fig. 185, E) plainly attests.

Family 9. Macropodidæ (Kangaroos, Wallabies, etc.).—The kangaroos are mostly terrestrial forms, but some of them appear to be secondarily arboreal. The hind legs are very large and powerful and usually the fourth and fifth toes are much enlarged into a sort of hoof. The tail is always long and heavy at the base. *Macropus rufus* (Fig. 185, A) is the largest of the marsupials, attaining a length of five and a half feet, exclusive of the tail. They are very fleet of foot, progressing by great leaps of the long hind legs covering twenty feet at a jump. The fore legs are of no use in running and appear to be merely for grasping food and for handling the young. The genus *Petrogale* (Fig. 185, B) includes kangaroos that live among the rocks, using the long tail as a balancing pole as they leap from rock to rock. *Dendrolagus* (the tree kangaroo) is very different in its habits from any of the other members of the family. The foot structure indicates that the arboreal habit has been superimposed upon an ancestral cursorial habit, for there is the same great enlargement of the fourth and fifth toes as in the other kangaroos.

Family 10. Phascolomyidæ (Wombats).—This family consists of but one genus, *Phascolomys*. It is in general appearance something like a small bear (Fig. 185, G) or a heavily built marmot. It lives entirely on the ground and moves about with a sort of shuffling plantigrade gait much after the manner of a bear. It is shy and gentle, though it can put up a vigorous defense with teeth and claws if forced to do so. In habits it is nocturnal, spending the daytime in burrows or holes among the rocks.

CHAPTER X

MAMMALIA—*Continued*

DIVISION II. MONODELPHIA (PLACENTAL MAMMALS)

Definition.—This is the great group of present-day mammals, including about 95 per cent. of all living mammalian species. They are characterized by the following features: no marsupium; no epipubic bones; the young always nourished for a considerable time in the uterus by means of a placenta; no cloaca; always a good-sized *corpus callosum.*

The most primitive placental mammals are now believed to be more nearly representative of the ancestral mammalian prototype than are the monotremes or marsupials. Certain members of the order Insectivora have been selected as the most generalized of living mammals. Osborn selects as his mammalian prototype the tree shrew *Tupaia* (Fig. 186, B), while Lull selects as his, *Gymnura* (Fig. 186, A), a large rat-like animal related to the hedgehogs. The most specialized mammals are undoubtedly the whales, if structural modification be taken as the criterion; but Man outranks all other mammals in brain and nervous specialization, and therefore in intelligence.

SECTION A. UNGUICULATA (CLAWED MAMMALS)

ORDER I. INSECTIVORA (HEDGEHOGS, MOLES AND SHREWS)

These are primitive, rather small, furry animals, that feed almost exclusively on insects. They are for the most part nocturnal and terrestrial in habit, as the first mammals are believed to have been. Some of them have been specialized slightly for an arboreal habit; others have been rather profoundly modified for a fossorial life. In bodily proportions they are as a rule quite generalized, fitting well the rôle usually assigned to them of persistently primitive mammals.

The members of the **Shrew** family (Fig. 186, A and B) are rather rat-like in form and more or less plantigrade in attitude. There is nothing especially striking or noteworthy about these animals except their lack of specialized characters. It has already been pointed

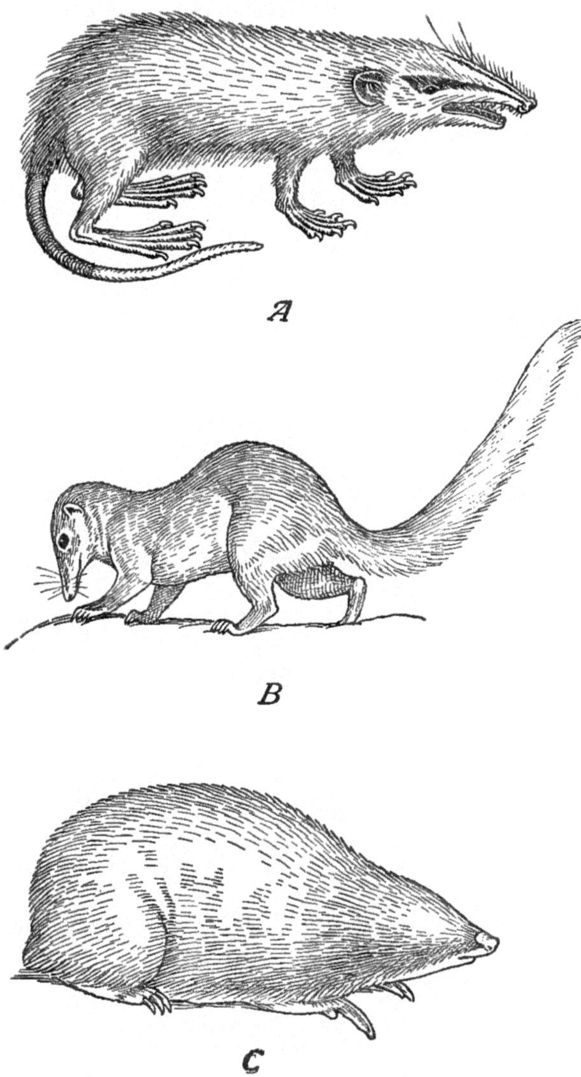

FIG. 186.—Group of Insectivora. A, *Gymnura rafflesii*, believed by Lull to be the most primitive insectivore (after Horsfield and Vigors); B, *Tupaia*, the Tree Shrew, considered by Osborn as near the prototype form of all higher placental mammals (after Osborn); C, Golden Mole, *Chrysochloris trevelyani* (after Günther). (All redrawn.)

out that various authorities on mammalian morphology have selected the shrews as the most generalized of living mammals.

The *Erinaceidæ*—*Erinaceus*, *Hylomys*, and *Gymnura* (Fig. 186, A)—are a little more specialized than are the shrews, though Lull considers the latter the most primitive living placental mammal. The true hedgehog is characterized by its armor of quills, which are much like those of the porcupine in structure.

Fig. 187.—*Galeopithecus*.—(From Parker and Haswell, after Vogt and Specht.)

The **True Moles** (Fig. 186, C) are profoundly specialized for a sub-terrestrial burrowing habit and resemble in their adaptations the marsupial mole. They have rudimentary eyes, no ear-lobes, short tail, and heavy digging claws. The golden mole (*Chrysochloris*) of South Africa is a beautiful creature with iridescent golden fur. Moles feed chiefly on earthworms and dig long tunnels just beneath the turf, and on this account are the bane of lawn-keepers and gardeners. No less than nine families of Insectivora have been distinguished, but lack of space forbids a detailed description of them.

ORDER 2. DERMOPTERA.—This is an order containing but a single species, *Galeopithecus volans* (Fig. 187), the so-called "flying lemur." It is a bat-like creature, nearly as large as a cat, with membranes stretched between the fore and hind legs, also between the head and the hand and between the tail and the hind feet. In certain respects it seems to be intermediate between the insectivores and the bats.

ORDER 3. CHIROPTERA (BATS).—Bats may be defined as true flying mammals in which the fingers of the fore limb are greatly elongated to support, like the ribs of a fan, a membraneous airplane. They do not merely soar or parachute like the flying lemur or the flying squirrels, but actually propel themselves with rapid wing strokes as effectively as do many of the birds. Extra planing surface is acquired by a stretch of membrane running from the hind limbs to the tail. The knees of bats are turned backwards, a position that would require dislocation of the hip in any other mammal. Many of the bats have large delicate ears and extremely complicated folds of sensitive membrane surrounding the nostrils (Fig. 188, B, C, D); these are believed to be organs of a sixth sense (kinæsthetic sense) that gives warning of the nearness of solid objects in the dark. It is said that bats living in caves that have absolutely no light, fly about in swarms at a high speed and never collide with one another nor with the walls or roof of the cave. Bats are divided into two sub-orders: Microchiroptera and Megachiroptera.

Sub-Order 1. Megachiroptera (Fruit-eating Bats).—These are rather large animals and are sometimes called "flying foxes." They occur in India, Australasia, Ceylon, Africa and Madagascar. The best known is *Pteropus,* a large bat with a wing-spread of over five feet, though the body is only about a foot in length. Their main food consists of figs and guava. They are distinctly social in habit and move about in droves of considerable size. Another well-known species is the collared fox-bat (*Xantharpyia collaris*) which is shown in its customary resting position with its young clinging to its abdomen (Fig. 188, A).

Sub-Order 2. Microchiroptera (Insectivorous Bats).—These are small bats (Fig. 188, B) with practically cosmopolitan range on account of their great powers of flight. At least five hundred species are known. They are decidedly nocturnal in habit, taking up the rôle of birds while the latter are asleep. "Blind as a bat" is a familiar aphorism that has its basis in the fact that the bats' eyes are so sensitive to lights of low intensity that they are blinded by the broad daylight. At night they skim rapidly and dexterously through the air catching insects on the wing with remarkable expertness. In the daytime they spend their time sleeping in caves or other dark sheltered places, hanging up-side-down by means of the claws of their hind legs. They are decidedly gregarious, living in colonies of thousands within the narrow confines of certain small caves. A common American species is the *Brown Bat* (*Eptesicus fuscus*); another common species of the eastern parts of North America is the *Little Brown Bat* (*Myotis lucifugus*), which is less than three and a half inches in length. The *Vampire* (*Desmodus rotundus*) is a bat of rather large size, native to South America. True to its reputation it lives the life of a blood-sucker, attacking horses and cattle and occasionally men. Its mode of attack is to fasten its razor-edged front teeth (Fig. 188, E) in the throat and to sever a vein or an artery, after which it proceeds to gorge itself with blood. One curious family of bats, the *Molossidæ*, are of interest because they have become secondarily terrestrial, appearing to be more at home on their feet than one would expect of a bat; for they run about almost like mice. This is quite in contrast to the usual situation among bats, which move about on the land with extreme awkwardness. When the typical bat crawls it hooks the thumb-nail in front and pushes with its feet behind, a pitiably helpless mode of locomotion.

ORDER 4. CARNIVORA (FLESH-EATING MAMMALS).—This is an immense order, characterized by large average size, predatory habits, and dominant position in the economy of nature. The largest carnivores, lions, tigers, and bears rank as kings among beasts. The cheek teeth are generally provided with sharp cutting edges, and the canines are characteristically large and curved. The brain is relatively large and complex; a fact that accords well with their high grade of intelligence. The clavicle is vestigial or absent, giving them a narrow-chested appearance. Digits are never less than four and are armed with curved claws. It is difficult to enter into a more detailed account of the order as a whole, because the two sub-orders, Fissipedia and Pinnipedia, are unlike in so many particulars.

Sub-Order 1. Fissipedia (*Terrestrial Carnivores*).—The dentition (Fig. 169) is probably the best diagnostic character of this group;

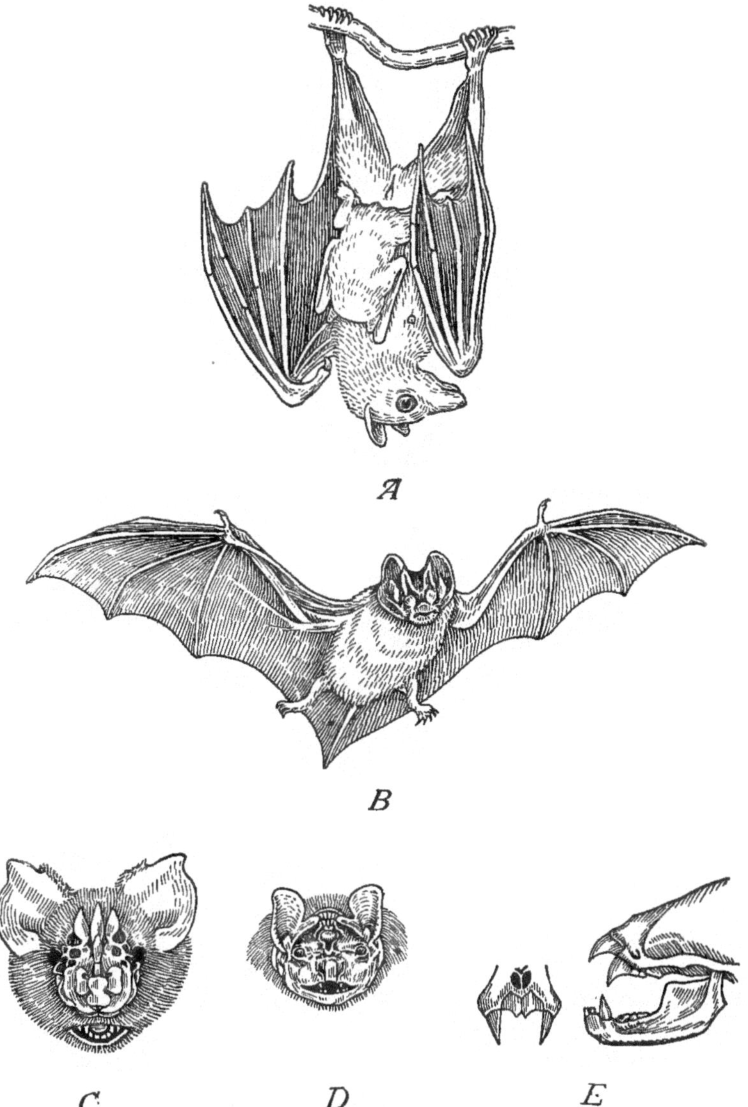

FIG. 188.—Chiroptera (Bats). A, Collared Fox-Bat, *Xantharpyia collaris*, and young. (After Sclater.) B, *Synotus barbastellus*. (After Vogt and Specht.) C, Face of *Triænops persicus*, showing nasal folds. (After Dobson.) D, Face of *Centurio senex*. (After Dobson.) E, Dentition of Vampire, *Desmodus rufus*, to show sharpness of teeth. (After Flower and Lydekker,.)

they have six incisors of small size in each jaw, canines are large and strong, the last premolar and the first molar are "carnassial" or cutting teeth, and the last two molars are crushing teeth. The fissiped carnivores have a world-wide distribution, being native to all of the large continental bodies except Australia. The principal family groups are: the cats, the civets, the hyænas, the dogs, the raccoons, the weasels, and the bears.

Family 1. Felidæ (Cats).—This is much the largest and most dominant of the carnivore families. The carnassial teeth are highly perfected shearing organs, canines especially long and curved, and molars are greatly reduced. The claws are retractile, an arrangement that gives the cats a quiet tread when stalking their prey. The typical genus *Felis* includes such cats as the lions, tigers, leopards, lynxes, jaguars, ocelots, pumas, and many smaller types. The domestic cat is believed to be a descendant of the eastern wild species, *Felis caffra*, first domesticated by the Egyptians and considered by them a sacred animal. The Canada lynx (Fig. 189, A) is a short-tailed, somewhat aberrant type of cat.

Family 2. Viverridæ (Civets).—The civets (Fig. 189, B) and their kin, which comprise this family are rather small, more or less cat-like carnivores that are native to Ethiopian and Oriental regions. The claws are incompletely retractile and they have more teeth than the true cats. The civets proper are decidedly feline in appearance and are usually marked with black and white spots or stripes. The *fossa* is a very cat-like carnivore; it is the largest carnivore native to Madagascar. The *mongoose* is a small, extremely active animal of oriental countries; it is noted for its ability to kill snakes, especially the deadly cobra.

Family 3. Hyænidæ (Hyænas).—These animals (Fig. 189, C) are in appearance and habits intermediate between the cats and the dogs. They are either spotted or striped. The voice is said to be almost human in sound and stories are told of human beings lured to their death by following their cries.

Family 4. Canidæ (Dogs).—The dog family includes the wolves (Fig. 189, D), foxes, coyotes, and the dingo of Australia, which is believed to be an imported species. The domestic dogs are believed to have been derived from several wild stocks, some of which may have become extinct. In many ways the dogs are the most primitive of the carnivores: the dentition is quite generalized, the claws are

FIG. 189.—Group of Fissiped Carnivora. A, Canada Lynx, *Felis canadensis* (after Fuertes). B, Civet Cat, *Viverra civetta* (after Beddard). C, Spotted Hyæna, *Crocuta maculata* (after Beddard). D, Gray or Timber Wolf, *Canis nubilus* (after Fuertes). E, Raccoon, *Procyon lotor* (after Fuertes). F, Badger, *Taxidea taxus* (after Fuertes). G, Otter, *Lutra canadensis* (after Fuertes). H, Largest of the bears, Alaska Brown Bear, *Ursus gyas* (after Fuertes.) (All figures redrawn, those after Fuertes in National Geographic Magazine, simplified and more or less modified.)

less specialized than in other groups and in several other ways they appear to resemble the ancestral carnivores. They have been associated with Man from a very early period, and are as cosmopolitan in their distribution as Man is, because wherever Man goes he takes his dogs.

Family 5. Procyonidæ (Raccoons).—This is an American family of carnivores that in some ways is intermediate between the dogs and the bears. They have plantigrade feet and grinding teeth like the bears, but in other respects are more like the dogs. The common *raccoon* (*Procyon*) is a familiar type (Fig. 185, E) around streams and lakes, where it catches crayfish, clams, and sometimes fish, without, however, going very far into the water.

Family 6. Mustelidæ.—This is a large family of bloodthirsty, predaceous creatures, including: weasels, pole-cats, badgers (Fig. 189, F), martens, wolverines, sables, minks, ermines, ferrets, stoats, skunks, otters (Fig. 189, G), and other less known types. For the most part they give off a nauseous musky odor, which is most marked in the skunks. They are among the most important of our fur-bearing animals. Representatives of the family are native to all the continental bodies except Australia and Madagascar.

Family 7. Ursidæ (Bears).—The bears (Fig. 189, H) are the largest of modern carnivores and are characterized most sharply by their plantigrade walk and the short tail. Most bears belong to the genus *Ursus*, but several other genera are distinguished, such as *Melurus*, the sloth bear of India, and *Æluropus*, a rare species native to Thibet. The bears are native to the Northern Hemisphere, few of them having crossed the equator.

Sub-Order 2. Pinnipedia (Seals and Walruses).—The animals of this sub-order are marine forms, in which there has been a secondary adaptation of the whole body for aquatic life. They are, however, much less radically modified than the Sirenia or the Cetacea. The Pinnipedia are characterized as follows: the greater part of the limbs are inclosed within the body skin; the claws are reduced and the digits are increased in number; the milk dentition is feeble and is shed early; the cranial cavity is large as compared with the face.

Family 1. Otariidæ (Sea-lions and Fur-seals).—These animals are gregarious and polygamous. The males (Fig. 190, B) are several times as large as the females (Fig. 190, C). As a rule they breed on rocky northern islands; and great numbers have in the past been slaughtered at this season. The governments of several nations have protected seals in their rookeries, and they are now multiplying satisfactorily.

Family 2. Trichechidæ (Walruses).—These are large, heavy-bodied forms (Fig. 190, A) with tusk-like canines in the upper jaws and a

Fig. 190.—Pinniped Carnivora. A, Pacific Walrus, *Odobenus obesus;* B, Male, and C, female, of Steller Sea-lion, *Eumetopias jubata;* D, Greenland Seal, *Phoca grœnlandica.* (All redrawn after Fuertes.)

mustache of heavy bristles on the upper lip. They are Arctic in habitat. On the whole they are more extensively modified for aquatic life than are the sea-lions.

Family 3. Phocidæ (The True Seals).—These animals have no external ears; the nostrils are dorsal in position; the hind limbs are intimately bound up with the short tail to make a sort of caudal fin, which is used as a very effective swimming organ. The fore limbs are rather small and fin-like, and the whole body is decidedly spindle-shaped. The seals are much more highly specialized for marine life than are either the sea-lions or the walruses. One of the commonest of the seals is *Phoca groenlandica* (Fig. 190, D), a small spotted animal about four or five feet long.

Extinct Carnivores

Representatives of the Mustelidæ have been found as far back as Eocene times; some Canidæ lived during Pliocene times. A whole

Fig. 191.—Extinct carnivore, *Smilodon*. (From Lull, after Knight and Osborn.)

family of cat-like creatures, the *Machærodontia*, lived from Eocene to Pleistocene times and are now extinct. The classic "saber-tooth" (*Smilodon*) is a characteristic example of this rather remarkable extinct family (Fig. 191), which was characterized mainly by the extreme modification of the teeth and skull in adaptation to the peculiar method of attacking with the saber-like upper canines. These huge teeth were thin and knife-like, with sharp edges. The method of using these teeth was evidently quite different from that employed by tigers; the prey was struck a downward, slashing blow, and was probably stabbed as though by a dagger. The upper jaw was especially modified to support these huge canine teeth, and the skull was radically altered to furnish attachment for the huge neck muscles that were used in driving in the daggers. The lower canines are much reduced in size.

ORDER 5. RODENTIA (GNAWING MAMMALS).—The rodents are for the most part rather small mammals, though a few of them have reached a considerable size. It has been claimed by some authorities that there are more species of rodents living to-day than of all other mammals combined. Unquestionably they are the most typical mammalian group to-day, as well as the most successful. Because they are so extremely prolific, because they are omnivorous, and because many of them lead a nocturnal burrowing life, they seem likely to be the main mammalian rivals of Man in the next geological period. The rodents are characterized by absence of canine teeth; and the incisors are long and strong, with persistently growing pulp and enamel confined chiefly to the anterior edge. This arrangement of the enamel makes the teeth wear down to a chisel edge, which is self-sharpening with use. The brain is smooth, with few furrows, and the intelligence is usually low. The testes are usually abdominal in position; the placenta is discoidal and deciduate. Two sub-orders are distinguished: Duplicidentata (Hares and Pikas) and Simplicidentata (Rodents Proper).

Sub-Order 1. Duplicidentata (Hares and Pikas).—These animals are characterized by two pairs of incisor teeth in the upper jaw, the inner being small and lying behind the outer. The tail is short. The group is regarded by some as a distinct order.

Family 1. *Leporidæ (the Hares)* are distinguished by long ears, long hind legs, and short though obvious tail.

Family 2. *Lagomyidæ (Pikas)* are distinguished by short ears, short hind legs, and no external evidences of a tail.

Sub-Order 2. Simplicidentata (True Rodents).—The members of this sub-order are divided into three sections: represented by squirrel-like, rat-like, and porcupine-like rodents.

Section 1. Sciuromorpha (Squirrel-like Rodents).—This large section includes the squirrels proper, the flying squirrels, the ground squirrels and chipmunks, the gophers, the prairie dogs, the marmots, the beavers, and others. The flying squirrels (Fig. 192, A) are parachuting animals, with a membrane stretched between the fore and hind limbs. The prairie-dogs are burrowing rodents of the western plains, that live in large colonies and share their burrows with ground owls and rattlesnakes, as well as other messmates. The habits of the beaver (Fig. 192, D) are too well known to require description here. They are nearing extinction on account of their highly desirable fur. No other rodent is so highly modified for aquatic life as is the beaver.

Fig. 192.—Group of Rodentia. A, Flying Squirrel, *Sciuropterus volucella* (after Lydekker); B, Long-tailed Marmot, *Arctomys caudatus* (after Beddard); C, Egyptian Jerboa, *Dipus jaculus* (after Lydekker); D, Beaver, *Castor fiber* (after Lydekker); E, Agouti, *Dasyprocta aguti* (after Beddard); F, European Porcupine, *Hystrix cristata* (after Beddard.)

Section 2. Myomorpha (Rat-like Rodents).—This is the largest modern group of mammals in point of numbers of species and of individuals. At least a hundred genera and nearly five hundred species have been distinguished. The group includes: dormice, field-mice, rats and mice proper, mole rats, jumping mice (Fig. 192, C), and the so-called African flying squirrels. They exhibit a very wide range of adaptive specializations, being terrestrial, sub-terrestrial, arboreal, cursorial and jumping, aquatic, and volant. They do serious damage to the world's food supply and are responsible for the spread of some of the worst plagues that Man has to contend with.

Section 3. Hystricomorpha (Porcupine-like Rodents).—This is a somewhat heterogeneous group and is not very well described as "porcupine-like," since many types appear quite unlike porcupines. There are eight families, including: "water-rats," cavies, guinea-pigs, agoutis (Fig. 192, E), chinchillas, ground porcupines, and tree porcupines. The **cavies** are South American and West Indian forms that reach a length of four or five feet. They are terrestrial in habit, with small ears and short tail. The **chinchilla** is a small squirrel-like animal native to the Andes; the fur is soft and gray and is highly prized. The **Canada porcupine** is a heavy-bodied terrestrial and arboreal form that gnaws off the bark of trees, eats water-lily leaves and roots. It is armed with short quills that are nearly hidden in the long fur. Its equipment is purely for passive defense, except that it lashes the tail and thus drives in its largest quills, when attacked. Dogs are often injured when they are unwise enough to attack the porcupine, for they get their mouths full of barbed quills that are extremely difficult to remove. The **European porcupine** (Fig. 192, F) is considerably larger than its American relative, having a body length of about three feet. It has quills nearly a foot in length, those on the tail being hollow so as to produce a rattling sound when the animal is disturbed. A great crest of coarse hair surmounts the head and hangs down like a mane. In spite of the prevalent reports to that effect, the porcupine never shoots its quills.

ORDER 6. EDENTATA (SLOTHS, ARMADILLOS, AND ANT-BEARS).—This group is believed to be a surviving remnant of an archaic group. They have become highly specialized in several ways and exhibit many evidences of racial senescence. The name of the order implies a total lack of teeth and is therefore not appropriate for either the armadillos or the sloths; the ant-bears alone are quite toothless.

The dentition of the toothed edentates is peculiar in that there are no incisors and the teeth in the definitive condition are without enamel. The testes are abdominal; the clavicle is always present; there is an additional pair of zygopophyses on the posterior dorsal and lumbar vertebræ. The edentates are strictly American in distribution and have been limited to this territory from the first. In adaptive characters the three main types differ widely from one another.

Sub-Order 1. Pilosa (Hairy Edentates).—The hairy edentates belong to two quite distinct families: The Myrmecophagidæ (ant-bears), and Bradipodidæ (sloths).

Family 1. Myrmecophagidæ (Ant-bears).—These are among the strangest animals now living. They are truly edentate, have a long slender snout, long sticky tongue, heavy front claws, and long, coarse hair, characters that we have already found to be adaptive features of the anteating type of mammal, no matter to what group it belongs. *Myrmecophaga tridactyla*, the *great ant-bear*, is a large animal with a total length from end of snout to tip of tail of at least seven feet. It is very powerful and quite formidable when attacked. One swipe of the great hooked claws has been known completely to eviscerate a large dog. *M, jubata* (Fig. 193 A) is somewhat smaller but quite similar. The *Tamandua* is a smaller ant-bear with arboreal habits and a long prehensile tail. *Cyclopes* is the smallest of the ant-bears.

Family 2. Bradipodidæ (Sloths).—The sloths (Fig. 193, B), in spite of their marked external differences, exhibit many fundamental resemblances to the ant-bears. They are highly specialized for arboreal life. Their strong hooked claws which are much like those of the ant-bears are used as hooks for suspending them from branches. They always progress up-side-down, hanging from the under side of a branch. In accord with this habitually inverted position the heavy hair slopes from the belly toward the back; similarly the hair on the limbs slopes from the feet towards the body. It seems likely that this peculiar position of the hair serves the purpose of effectually shedding the rain. An interesting fact has been discovered about the hair: it is green in color, due to the presence in the hollows of the individual hairs of numerous cells of a green alga. This greenish coloring doubtless serves as a protective adaptation. The face of the sloth is extremely flat, in very marked contrast with the elongated face of the ant-bears. There are only four or five teeth in each half jaw. The sloths are very peculiar in that they have an excessive number of dorsal vertebræ, as many as 23 being present in some species. The cervical vertebræ are also quite an exception for mam-

Fig. 193.—Edentata, Pholidota, and Tubulidentata. A, Great Anteater, *Myrmecophaga jubata;* B, Two-toed Sloth, *Cholœpus didactylus;* C, Texas Nine-banded Armadillo, *Dasypus novemcinctus texanus;* D, The Aard Vark, *Orycteropus capensis;* E, Short-tailed Pangolin, *Manis temmincki.* (All redrawn, A, D, E, after Lydekker; B, after Beddard; C, after Newman.)

mals, in that they depart consistently from the number seven, which is so characteristic for mammals, having six, eight, or nine. They are largely insectivorous in diet. *Bradypus*, the three-toed sloth, and *Chœlopus*, the two-toed sloth, are the best known members of the family.

Sub-Order 2. Loricata (Armored Edentates; Armadillos).—The living *armadillos* belong to the family *Dasypodidæ* and are much more numerous in species than are the Pilosa. At least seven genera and over twenty species have been distinguished. They are characterized by having a well-developed dermal skeleton, composed of numerous bony plates, in which hairs are imbedded, and which are covered with horny scales. They have numerous teeth, which in the adult are without enamel; but in the embryonic stages a well-defined enamel layer has been discovered, which subsequently wears off. Incisors are not found in the adult, but embryonic rudiments of these teeth have been described. The armadillos range from moderately large animals of three feet or more in length to small forms about the size of a rat. Only a few of the species can receive mention here. The little *Chlamydophorus* has a solid unjointed armature and is considered primitive in this respect. *Euphractus sexcinctus* (the Peludo) is a decidedly hairy type. *Tolypeutes* has three movable bands and rolls up into a ball. *Priodontes* is the giant among armadillos, being three feet long to the base of the tail and having thirteen movable bands in the armor.

Dasypus novemcinctus (the nine-banded armadillo) is the only North American armadillo and therefore deserves especial attention. It is really a South American species that has migrated northward through Central America and now inhabits Mexico and Southern Texas. It is a medium sized animal that lives in burrows in the daytime and forages for insects at night. Its ears are long and close together and remind one of a donkey's ears. It is a source of satisfaction to be able to contribute an adequate illustration (Fig. 193, C) of this interesting species to take the place of the atrocious figure of Flower and Lydekker, which was evidently drawn from a badly stuffed specimen. Perhaps this armadillo deserves especial mention on account of its unique embryological features. It produces regularly, with rare exceptions, four young at a birth, that are always all four of the same sex. A study of the early developmental history of the egg has revealed the fact that this is a case of *specific polyembryony*, in which the four individuals are produced from a single fertilized egg, that divides at a very early period into four embryos. There is a single chorion, but four separate amnia. This case is taken as evidence that in mammals sex is determined at the time of fertilization, since the four division products of a single egg are invariably of the same sex.

Extinct Edentata.—The best known extinct edentates are the giant ground sloths, of which *Mylodon* is a type, and the giant armadillos, of which *Glyptodon* is the classic example. *Mylodon* was as large and as heavy as a rhinoceros, and *Glyptodon* was sixteen feet long.

ORDER 7. PHOLIDOTA (SCALY ANTEATERS).—This is a small order formerly included within the order Edentata, but now given ordinal value on account of the discovery of morphological differences more fundamental than the resemblances that formerly led to their classification as edentates. The order consists of the *pangolins*, which are placed in the family *Manidæ* and the genus *Manis*. *Manis gigantea* is a fairly large and massive animal, about six feet in length, tail included. It is native to the islands of the Malayan Archipelago. The most striking feature of these animals is the scaly covering, or what appears to be an armor composed of large pointed, overlapping scales, which are really groups of fused hairs. Scattered hairs occur between these "scales." The species shown in the illustration is *Manis temminckii* (Fig. 193, E).

The pangolins are anteaters, and possess all of the characteristic adaptations already mentioned for several other anteaters: the long snout, sticky tongue, integumentary protection from ants, and heavy claws for digging into ant galleries. Their method of capturing ants is highly individual. After stirring up a colony of ants they are said to erect the scales so as to allow ants to crawl under the scales. The scales are then clamped down so as to hold the ants, and then the animal goes in for a swim. When submerged in the water the scales are lifted and the ants washed out so that they float about on the surface, where they are easily picked up by means of the long tongue.

ORDER 8. TUBULIDENTATA.—This order contains only the curious aard-vark, *Orycteropus* (Fig. 193, D) of South Africa. These curious animals were formerly classed as edentates, but are now known to be unique in a number of characters and have therefore been accorded ordinal value. They are anteaters and have the slender snout, long tongue, and strong claws characteristic of this habitus. The skin is very thick and is covered with sparse hair.

Section B. Primates (Mammals with "Nails")

Order 9. Primates (Lemurs, Monkeys, Apes and Man).—The traditional position allotted to the primates is the last and highest order of mammals, but it has come to be realized that the group is on the whole more generalized than several other orders, and in point of specialization belongs to a division just above that rather primitive section, Unguiculata. The primates may be defined as primarily arboreal animals with prehensile limbs; with thumb and great toe shorter than the other digits and more or less opposable to the latter; with plantigrade walking position of the feet; with terminal, flattened "nails" instead of claws; with hair covering the entire body except the palms and soles and parts of the face; with a single pair of pectoral mammæ; with the eyes directed anteriorly instead of laterally; the eye orbit completely surrounded with bone; a clavicle always present; the stomach simple; and the brain unusually large and well convoluted.

Probably the best among many classifications of the primates is that of W. K. Gregory:

Sub-Order 1. Lemuroidea (lemurs or "half-apes").
Sub-Order 2. Anthropoidea.
  Series 1. Platyrrhini (New World Apes).
    Family 1. Hapalidæ (Marmosets).
    " 2. Cebidæ (capuchins, howler monkeys, spider monkeys, etc.).
  Series 2. Catarrhini (Old World Apes and Monkeys).
    Family 3. Cercopithecidæ (monkeys, baboons, macaques, etc.).
    " 4. Simiidæ (man-like or anthropoid apes).
    " 5. Hominidæ (men).

Sub-Order 1. Lemuroidea (Lemurs).—The **lemurs** (Fig. 194, A) are much the most ancient and the most generalized of the primates, and therefore show less wide departures from the unguiculate mammals than do the anthropoids. They are exclusively arboreal, mostly nocturnal, and extremely timid and retiring. In appearance they strike one as intermediate between a squirrel and a monkey. The brain is comparatively unspecialized, the cerebral hemispheres being so small as not to cover the hind-brain. The second finger retains

FIG. 194.—Group of Primates. A, Smith's Dwarf Lemur, *Microcebus smithii;* B. Spider Monkey, *Ateles ater;* C, Drill or Mandrill, *Papio leucophæus;* D, Gibbon, *Hylobates lar;* E and F, Chimpanzee, *Pan pygmæus.* (Redrawn, A and B, after Beddard; rest after Lydekker.)

the ancestral claw, but the rest of the fingers have "nails." The lemurs have their headquarters in Madagascar, but are also found in small numbers in the tropical forests of Africa and in Malaysia.

During the Eocene period they lived both in North America and in Europe, a fact indicative of the antiquity of the group. Two persistent relics of that Eocene lemuroid fauna are the living genera *Tarsius* and *Chiromys*.

Chiromys madagascariensis, the "aye-aye," is a rather squirrel-like animal with long incisor teeth; a bushy tail; the thumb only has a "nail," the other digits being provided with claws; the mammæ are abdominal, a primitive position; it has but one young at a birth. The "aye-aye" has a plaintive voice resembling the name; it leads a prowling, furtive life, always in pairs. A nest of twigs is made in the tops of trees.

Tarsius spectrum, a native of the Malay Islands, is a remarkably strange little creature, with enormous eyes that give it the appearance of wearing spectacles, a character from which it derives its specific name. The digits are armed with adhesive pads and have small flat nails. The tail is long and tufted at the end. They live in pairs in holes in hollow trees, and are mainly insectivorous and decidedly nocturnal. The mother carries the young about by taking hold of the neck skin with the teeth, after the manner of a mother cat. *Tarsius* has an almost smooth cerebrum and a low order of intelligence.

The more modernized lemurs may be exemplified by the ruffed lemur, the mouse lemur, and the slow loris. Of all the lemurs the *ruffed lemur* (*Lemur varius*) is probably the most monkey-like. It has a rather long, bushy tail, a fox-like face and the full primate dentition. The voice is loud; they are diurnal as well as nocturnal in habit. The *mouse lemur* (*Chirogale coquereli*) is a native of Madagascar; it is very small in size, with soft, fluffy fur and of generalized proportions. The *slow loris* (*Nycticebus tardigradus*) is an aberrant lemur, native of East Indian and Malayan territories. It is extremely deliberate in its movements, moving about among the trees chattering and whistling as though without a care in the world. Like other lemurs it is looked upon with superstitious dread by the natives, who regard it as a beast of ill omen.

Sub-Order 2. Anthropoidea (*Monkeys, Apes, Man*).—The anthropoids are decidedly more highly organized than are the lemurs. They are characterized by the possession of: 32 to 36 teeth; completely closed orbit; pectoral mammæ; prehensile hands and feet (except in Man); cerebral hemispheres richly convoluted and covering the cerebellum.

Series 1. Platyrrhini (New World Apes).—These primates are distinguished by the broad nasal septum; the thumb is not opposable, but usually reduced to a small vestige; the tail is long and prehensile; there are no cheek pockets or pouches; there are no callosities on the ischium.

Family 1. Hapalidæ.—These are the marmosets, animals about the size of squirrels, quite extensively used as pets. They have a very generalized diet, eating fruit, eggs, and insects, and have claws instead of nails on the digits.

Family 2. Cebidæ.—Most of the common South American monkeys (Fig. 194, B) belong to this family. Several species of them are familiar to everyone as the accessory of the Italian organ-grinder. They are all rather slender and have exceptionally long, more or less prehensible tails. The howler monkeys are noted for their prodigious voice, which is produced by means of a specially modified sounding apparatus. The commonest of the Cebidæ are the capuchins, companions of the hand-organ.

Series 2. Catarrhini (Old World Apes and Man).—This series of primates is characterized by: narrow nasal septum, with nostrils directed downward; all have 32 teeth, as in man; non-prehensile or rudimentary tail; the great toe fully opposable, except in man; the thumb is always opposable.

Family 3. Cercopithecidæ (baboons, mandrills and macaques).—The baboons and macaques (Fig. 194, C) are characterized by: quadrupedal habit of locomotion; more or less dog-like heads; ischial or rump callosities; no vermiform appendix; narrow chests, a character associated with the quadrupedal habit; very large canine teeth; cheek pockets. They are omnivorous in diet, as are the other Catarrhini. One of the most striking characters of members of this family is the brightness of their coloring, especially that of nose, cheeks, and rump. Bright blue, scarlet and lilac colors are the commonest tints. In habits they combine those of the arboreal with those of the terrestrial types. They are good fighters and are able to cope with many of the predaceous terrestrial animals that inhabit the Asiatic and African forests.

Family 4. Simiidæ (Anthropoid Apes).—The members of this family have long been objects of especial interest on account of their close relationship to Man. In no sense are they to be thought of as ancestral to Man; rather it would appear that they are "cousins," derived from a common ancestral stock. Doubtless, were we to discover this common ancestor, we should be inclined to call it an ape, but it certainly was not very much like any of the present-day apes.

The family may be defined as follows: tail rudimentary as in Man; no cheek pouches; no ischial callosities except in the gibbon; arms longer than legs; the great toe fully opposable; the shoulders broad; bipedal habits; always a vermiform appendix; hair mainly on the ventral side of the body and on the limbs. The number of species is not great and there is so general an interest in them that we may spare the space to give a brief description of the principal ones.

The **gibbon** (Fig. 194, D), *Hylobates*, is a rather small ape with exceedingly long arms, that touch the ground when it stands erect. It has small rump callosities similar to those of the baboons. Its dentition is adapted for a fruit-eating habit, though the canines are large and saber-like for self-defense. The skull is rounded and without the sagittal crest characteristic of the gorilla. It has a very erect posture both in walking and in sitting, the head being set upon the neck much as in Man. The gibbon has a tremendous voice, much more voluminous than that of Caruso, though it weighs not more than about sixty pounds. It lives in heavily wooded mountain slopes, remaining largely in the trees, through which it is capable of making wonderful speed. With its long arms it swings along with a hand-stride of twenty to forty feet, and never misses a hold, though it must calculate the distances with great nicety or fall from great heights to the ground. Any animal that can use its arms and hands in this way must have a finely developed brain back of it; indeed the gibbon's brain development is exceptional, especially in the visual and tactual centers. When walking on the ground the gibbon walks erectly but very awkwardly, balancing itself by touching the knuckles of the hands to the ground. It is evidently about nine-tenths an arboreal creature, using the ground only when trees are not available.

The **orang** (Fig. 195), *Simia satyrus*, is a large ape native to Sumatra and Borneo. It is short and stocky, and has reddish hair. Though it is only about four feet in height it has an arm-spread of over seven feet. The head is short and broad and the eyes very close together. The skull has a sagittal crest for the attachment of the powerful neck muscles; the jaw is deep and massive and is used both for tearing open fruits and in fighting. The hands are the chief weapons, and are relied upon rather than the teeth. The heavy weight of the orang makes it a less efficient climber than is the gibbon; and its mode of climbing is much more deliberate and man-like. It builds its nest in trees by breaking off branches and arranging

FIG. 195.—The Orang-utang, *Simia satyrus*, sitting in its nest. (From Weysse, after Shipley and McBride.)

them platform-fashion in the crotch where two large limbs meet. The orang appears to be the only purely herbivorous member of the apes; its diet consists exclusively of fruits. On the ground it runs on all fours in an awkward and ineffective way. Its intelligence has been experimentally shown to be greater than that of any other creature except Man.

The chimpanzee (Fig. 194, E and F), *Pan pygmæus,* is an African ape with black hair and a height of about five feet; it is less bulky than the orang. These characters make the chimpanzee a better climber than the orang, though not so good as the gibbon. The head is larger than that of the orang, and the brow ridges are very prominent. There is a pronounced sagittal crest on the skull for the attachment of the neck musculature. The jaws are prognathous and resemble those of prehistoric Man. It builds nests much like those of the orang. Some authorities distinguish several species of chimpanzees. They are largely but not exclusively fruit-eaters. Their range is rather limited, being confined to central equatorial Africa.

The **gorilla** (Fig. 196), *Gorilla gorilla,* is much the largest and fiercest of the anthropoid apes. It is native to the tropical African forests and is confined to a very restricted territory. It stands about five feet in height, but is so massive in build that it frequently reaches a weight of between four and five hundred pounds. If it had legs in proportion to its arms and trunk it would be a giant of at least seven feet in height. The gorilla has become as highly specialized as a muscular brute as has Man as a creature of intelligence and finesse. The skull has a much heavier sagittal ridge than that of the other apes, and this is accompanied by a neck musculature of tremendous strength. The jaws are prognathous and very powerful, with large canine teeth, and the brow ridges are very prominent. All of these characters are much more pronounced in the old males than in the young males or in the females; a condition that suggests strongly their highly specialized character. The gorilla is a "negro" ape in the sense that the skin is black and the hair black and coarse. In habits the gorilla appears to be transitional between the arboreal and the terrestrial types. Both hands and feet approach the human type, especially in young specimens, though the great toe remains completely opposable. Gorillas are gregarious, living in bands of considerable size, with an old male at the head of each band. They will not run from Man or from any other creature, but stand their ground and put up a ferocious fight with both hands and teeth. The statement has often been made that the gorilla uses sticks or clubs in fighting, but this has never been confirmed by a reliable authority. From the purely brute physical standpoint the anthropoids have attained a higher degree of specialization than any other primate, but they fall far short of Man in nervous specialization.

Family 5. Hominidæ (Men).—The human family is, structurally speaking, very closely related to the Simiidæ; in fact the Simiidæ and the Hominidæ are more closely related than are the Simiidæ

FIG. 196.—The Gorilla, *Gorilla gorilla*. (From Lull, after mounted specimen in Philadelphia Academy of Sciences.)

and the Cercopithecidæ. The chief differences between Man and the anthropoid apes are viewed as the direct result of the acquisition by Man of terrestrial habits, erect posture, and larger brain, all of which acquisitions are undoubtedly closely correlated. These primary human adaptations are accompanied by secondary changes. Erect posture, for example, involves a series of adjustments, such as alterations in the curvatures of the spine, changes in the structure of the legs, loss of grasping power of the great toe, and increased length of legs. The following comparison between Man and the anthropoid apes is made by Gregory:

> "The anthropoids are chiefly frugivorous and typically arboreal; when upon the ground they run poorly and (except in the case of the gibbons) use the fore limbs in progressing. Thus they are confined to forested regions. Man, on the other hand, is omnivorous, entirely terrestrial, erect, bipedal and cursorial, an inhabitant primarily of open country. The anthropoids use their powerful canine tusks and more or less procumbent incisors for tearing open the rough rinds of large fruits and for fighting. Primitive man, on the contrary, uses his small canines and more erect incisors partly for tearing off the flesh of animals, which he has killed in the chase with weapons made and thrown or wielded by human hands. These implements and weapons also usually make it unnecessary for man to use his teeth in fighting and functionally they compensate for the reduced and more or less defective development of his dentition."

Although some authors recognize four species of Man, the best authorities now admit of but a single species, *Homo sapiens*. Possibly the minor divisions are the equivalent of sub-species, races, or varieties. Four races are distinguished by Lull:

Australian race: skull long; eyebrows very prominent; teeth large, especially the canines; tall and long-limbed; skin brown; hair black, long and wooly. Habitat: Australia, Dekkan, Hindustan.

Negroid race: skull long; forehead round; nasal bones flattened; teeth sloping; skin, eyes, and hair black; hair short and wooly. Habitat: Madagascar and Africa from the Sahara desert to Cape of Good Hope.

Mongolian race: skull broad and short; nose flat; eyes small and oblique; stature short and thick-set; skin golden brown; hair coarse, straight and black; beard scanty. Habitat: east of a line drawn from Lapland to Siam; Chinese, Tartars, Japanese, Malays, Esquimos, North and South American Indians.

Caucasian race is usually subdivided into three varieties:

A. **Mediterranean**: short; slender; long-headed; with hair and eyes dark brown to black.

B. **Alpine**: medium height; stocky build; round-headed; hair and eyes dark brown or black, but in the North often hazel or gray, probably due to admixture with the northern varieties.

C. **Nordic**: tall; long-headed; hair flaxen, red, or light brown; eyes blue, gray, or green.

Habitat of Caucasian race: mainly Europe and North America: includes Moors, Berbers, Egyptians, Kurds, Persians, Afghans, Hindus, Turks, Jews and Armenians.

The Immediate Ancestors of Man

According to Gregory, Man arose from an early, large-brained anthropoid stock, not far from the chimpanzee-gorilla group. Evidences point toward central Asia as the place of origin and early development of the pre-human Hominidæ. The time of origin is believed not to have been later than early Pliocene and not earlier than Miocene times; thus dating back some hundreds of thousands of years. The earliest fossil remains of the Hominidæ consist of the relics of the Java "ape-man," *Pithecanthropus erectus* (Fig. 197). Fragmentary remains of this creature, consisting of a skull-cap, a thigh bone, and two upper molar teeth, indicate that it was intermediate between the most primitive type of present-day man and the highest of the living apes. McGregor has reconstructed busts of *Pithecanthropus*, of the most primitive of extinct human species (*Homo neanderthalensis*), and of Homo sapiens, a series which strikingly shows the gradual evolution away from apish and toward human features (Fig. 198).

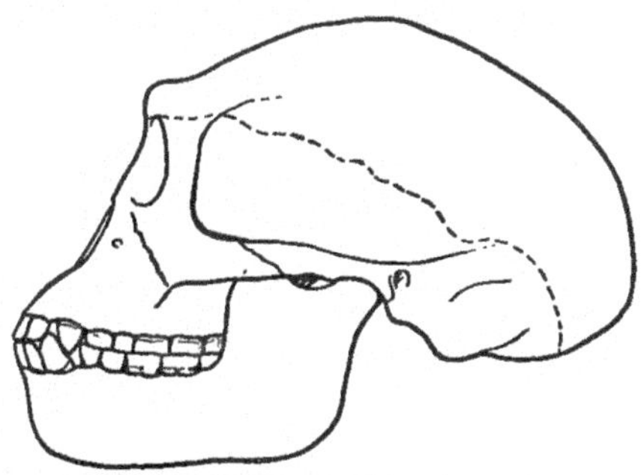

Fig. 197.—Skull of the Java ape-man, *Pithecanthropus erectus*. (From Lull, after Dubois.)

The science of anthropology concerns itself with the study of races of man, past and present, a field that cannot be more than touched

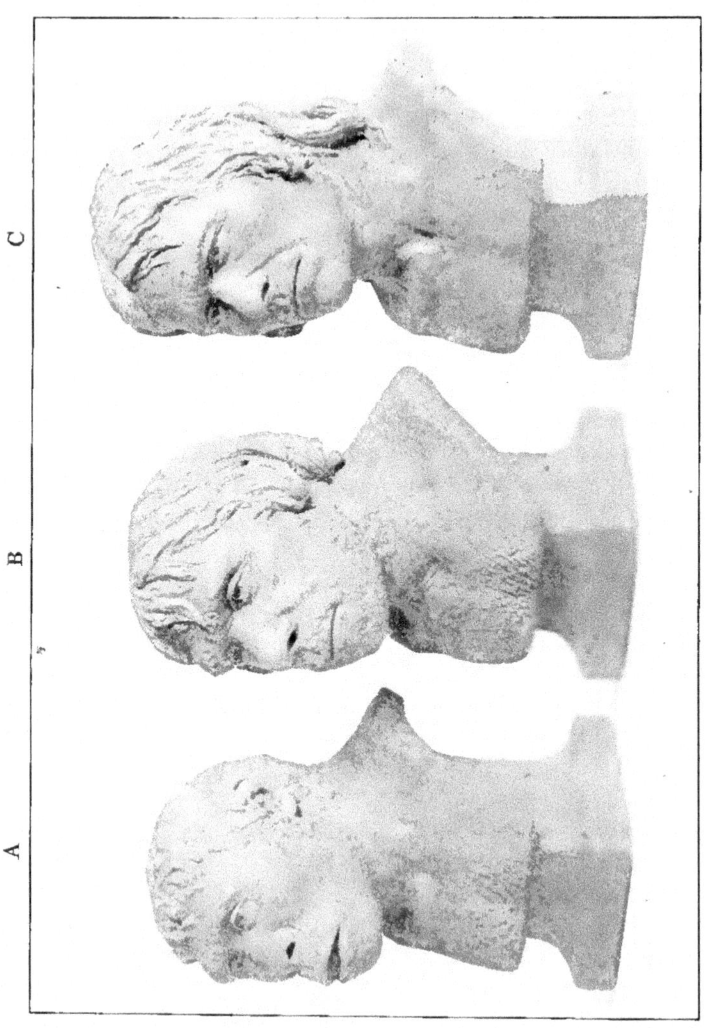

Fig. 198.—Series of busts representing prehistoric men, after J. H. McGregor. A, *Pithecanthropus erectus*, the Java ape-man; B, *Homo neanderthalensis*, the Neanderthal man of Europe; C, *Homo sapiens*, the Crô-Magnon man of Europe. From Lull.

upon in a volume dealing with vertebrate zoölogy. Our main purpose has been to place man in his appropriate setting among his fellow mammals.

SECTION C. UNGULATA (HOOFED MAMMALS)

This immense assemblage of specialized forms includes: two orders of extinct mammals already dealt with, Condylarthra and Amblypoda; four orders of present-day terrestrial mammals, Artiodactyla, Perissodactyla, Proboscidia, and Hyracoidea; and one order of marine mammals, Sirenia. The Ungulates are on the whole the most highly specialized of terrestrial mammals, just as the Cetacea are the most highly specialized of aquatic mammals.

ORDER 10. ARTIODACTYLA (EVEN-TOED UNGULATES).—The mammals of this group are: swine, hippopotami, peccaries, camels, deer, moose, elk, giraffes, pronghorns, cattle, buffalos, gnus, antelopes, gazelles, yaks, sheep, ibex, goats, and many other less well known types. It is a major assemblage of animals, whose size on the average is large. They are purely terrestrial, though some of them are mud-loving; for the most part they are cursorial, though some are heavy-bodied and not very fleet of foot. They have hoofs on two or four toes. The stomach usually has several chambers in adaptation to a purely herbivorous diet.

Group 1. Suina (Swine-like Ungulates).—This group consists of three families, represented respectively by the hippopotamus, swine proper, and peccaries. The *hippopotamus* (Fig. 199, A) is a large heavy-bodied aquatic "hog," with four hoofs on each foot. It is native of Africa, as is also the *pigmy hippopotamus*, a dwarf species found in Liberia. The swine proper include the European wild hog (Fig. 199, B), the wart hog, and several other types. The domestic varieties of hog have been derived from several wild species. The peccaries are swift, cursorial, hog-like creatures, that run in large packs, and on account of their sheer numbers, are said to be very dangerous to meet.

Group 2. Ruminantia (Ruminants).—These ungulates "chew their cud," by which is meant that they swallow their food rapidly and afterwards regurgitate it into the mouth for further mastication. Three assemblages of these forms are distinguished: A, *Tragulina* (Mouse Deer); B, *Tylopoda* (Camels, Lamas); C, *Pecora* (Deer, Antelopes, Oxen, Giraffes, Goats and Sheep).

The **chevrotanis** or mouse deer are intermediate between the swine and the ruminants, and are the most primitive of the ruminants. The **camels** (Fig. 199, C) are a small group of well-known types, confined to arid regions of the Old World. Camels are not known in the wild state; all are domestic or feral. The ancestral history of the camel family is now almost as well worked out as that of the horse. Proverbial for the camels are two characters: that of living for long periods without water, and the use of the fatty humps for food when compelled to fast. Both of these characters may be considered as adaptations for desert life and have made it a highly valuable beast of burden and transport across the arid trails of the Asiatic and African deserts; on this account they have earned the cognomen "ships of the desert." The camel is very valuable for its hair, which is used in making fabrics highly prized for their richness, softness, and wool-like characters. The llamas are creatures with camel-like characters, but more generalized in several respects; they might be called the camels of the New World, for they are native to South America. They are of value chiefly for their rather thin hair, which is coarser than that of the camel and is the material out of which are made vicuna or alpaca fabrics. The llama has the disgusting habit when irritated of forcibly spitting the contents of its stomach at the object of its annoyance.

The **deer family** is a very large one and includes such well-known types as elk, moose, reindeer, etc. They are characterized by the possession of antlers in the male sex, and in the reindeers in both sexes. The antlers vary in degree of elaborateness in the different genera, ranging from the small, unbranched horns, as in *Cervulus*, to the complex branching antlers of the elk (Fig. 199, D). In all cases they are solid bony structures, as opposed to the hollow horns of the Bovidæ. About sixty species of deer are known, the majority of which are Old World forms. The moose is the king of the deer family,.on account of its great size and its fighting qualities. The reindeer is the most northerly of the deer, occupying circumpolar territory. The musk-deer is an exceptional type in that it has no horns, but instead is possessed of long, sharp tusks, probably used in digging roots for food.

The **giraffe family** (Fig. 199, E) is a small family of highly specialized ruminants distinguished by their great height, long neck, and slender legs. The horns differ from all others in that they are merely prominences of the frontal bones of the skull covered with skin and hair. Africa is the home of the giraffe, as well as that of the okapi, a small less specialized member of the giraffe family, which is more like an antelope in general appearance.

The **cattle family** (Bovidæ) is much the largest family of rumi-

FIG. 199.—Group of Artiodactyla (Even-toed Ungulates). A, Hippopotamus, *Hippopotamus amphibia;* B, Wart Hog, *Phocochærus æthiopicus;* C, the Bactrian Camel, *Camelus bactrianus;* D, Wapiti, or American Elk, *Cervus canadensis;* E, *Giraffa camelopardalis;* F, Ibex, *Capra sinaitica.* (Redrawn and somewhat modified; A, C, after Lydekker, B, E, F, after Beddard, D, after Fuertes.)

nants. It includes oxen, sheep, goats (Fig. 199, F) and antelopes. The most prominent distinguishing character of the group is the horns, which are hollow and composed of chitin, and are usually present in both sexes. A large number of the Bovidæ have been domesticated, and from the human standpoint are the most important of all animals. The members of the group are so familiar that no description of the different species is necessary.

ORDER 11. PERISSODACTYLA (ODD-TOED UNGULATES).—In this group the middle digit of both fore and hind feet is preëminent and carries most of the weight. The axis of the limb passes through the third digit. The teeth of the odd-toed ungulates are usually lophodont, a type characterized by the presence of enamel ridges running back and forth across the grinding surface. The present-day perissodactyls are grouped into three families: Equidæ, Tapiridæ, and Rhinocerotidæ.

Family 1. Equidæ (horses, asses, and zebras).—The members of the **horse family** (Fig. 200, A) are characterized by the possession of but a single functional toe, the third toe, on each foot. The second and fourth toes are represented by vestigial remnants, called "splint bones." The molar teeth are highly complex in structure and wear down through most of the life of the individual, so that the age of any specimen may be arrived at by the amount of wear upon the teeth. All of the modern Equidæ are placed in the single genus *Equus*. Perhaps the most convincing record of the ancestry of any vertebrate group is that of the horse. With respect to toes, teeth, and general form, the gradual perfection of the present highly specialized cursorial type may be traced back through an unbroken line of ancestors to a very generalized ungulate type with four functional toes, generalized teeth, and comparatively small size. The horse has played and is still playing an extremely important rôle in the progress of human civilization. Next to cattle and sheep the horse has been the most important domesticated animal; but if present tendencies furnish a reliable criterion of the future, the horse is likely to be displaced by the motor-driven vehicle.

Family 2. Tapiridæ (Tapirs).—The **tapirs** (Fig. 200, B) are the most generalized of modern odd-toed ungulates. They are characterized by moderate size and by a short proboscis produced by elongation of nose and upper lip. The dentition is more generalized than that of the horses, there being forty-two teeth, a number very close to that of the most primitive eutherian mammals. There are four toes on the fore feet and three on the hind feet. The tapirs are confined to South and Central America and to the Malay Peninsula.

Family 3. Rhinocerotidæ (*Rhinoceroses*).—This family consists of a few species of large, massive animals, whose general appearance is familiar to all (Fig. 200, C). They are distinguished by the presence

Fig. 200.—Group of Perissidactyla (Odd-toed Ungulates). A, Burchell's Zebra, *Equus burchelli;* B, American Tapir, *Tapirus terrestris;* C, African Rhinoceros, *Rhinoceros bicornis.* (Redrawn and modified: A, B, after Beddard; C, after Lydekker.)

of median "horns" on the nose; but the structures are not true horns, being composed of masses of agglutinated hair fastened to a roughened patch of the nasal bones. There are usually three, sometimes four, toes on the fore feet, but in either case the third toe is the most important; the hind feet always have three toes. The upper lip is long and more or less prehensile, but not elongated into a proboscis as in the tapirs. The skin is extremely thick and the hair very sparse. They are fierce and intractable, charging at an enemy with great fury and stopping at nothing. Only guns of large caliber and hard-hitting qualities will stop their mad rush. They have a fairly wide distribution, being native to both India and Africa. The fossil record of the ancestry of the rhinoceros is almost as complete as that of the horse, and the two groups appear to converge upon a common ancestral group. The early rhinoceroses must have looked more like horses than the present forms, which have grown heavy of limb and body and are no longer typically cursorial.

ORDER 12. PROBOSCIDIA (ELEPHANTS).—This group comprises the largest and in many respects the most highly specialized of terrestrial mammals. They are characterized by the elongation of the nose and upper lip into a very long trunk; by the possession of five functional digits on both fore and hind feet; by the specialization of the incisor teeth of the upper jaw into great tusks; and by the extreme type of lophodont molar teeth. The skull is immensely thick and the bones contain large air cavities; there is no clavicle; the cerebral hemispheres are much convoluted, but they do not cover the cerebellum; the testes are abdominal in position.

Elephants walk with the legs stiff, almost as if they were jointless, an adaptation for bearing the great weight; for it would require great muscular effort to support the huge bulk of these animals upon a bent type of limb. Two families of Proboscidia are distinguished: *Elephantidæ* and *Dinotheridæ.* The latter were Miocene forms characterized by great downwardly directed tusks of the lower jaw.

There are but two living species of elephant, the Indian elephant (Fig. 201), *Elephas indicus,* and the African elephant (Fig. 202), *E. africanus.* The African species is the larger, and has much larger ears. The largest specimen on record is probably the notorious "Jumbo," which was about eleven feet high at the shoulder. African elephants are wild and intractable as compared with their Indian cousins; and therefore are seldom seen in circus parades. The Indian elephant is the common circus elephant, a smaller and more manageable type. In its native country it is used extensively as an equipage and as a beast of burden. As a species, however, they are not dependable, some being vicious and others perfectly docile in disposition. In nature they are creatures of the jungle and are purely herbivorous. They are capable of defending themselves against all enemies except Man. An encounter between an elephant and a tiger is one of the finest gladiatorial contests that the world affords.

Elephants have been credited with extraordinary intelligence, but they are much less wise and sagacious than they have been painted. Undoubtedly they have a good brain, but their capacity to reason is

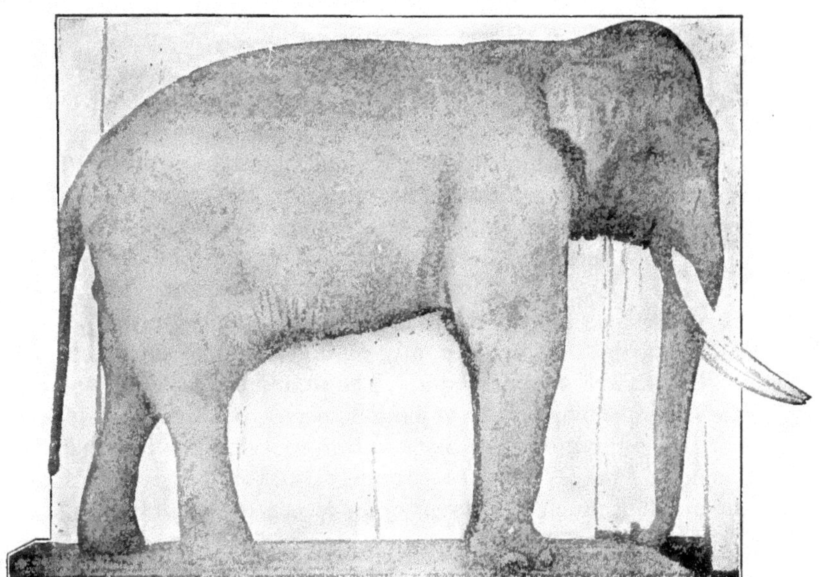

FIG. 201.—Indian Elephant, *Elephans indicus.* (From Weysse.)

FIG. 202.—African Elephant, *Elephans africanus.* (Redrawn after Beddard.)

quite rudimentary. The tenacity of their memory is well authenticated, for they have been known to cherish an injury for years. In all probability an extraordinarily keen sense of smell plays a prominent part in their memory, an enemy being associated with a special odor. Even in human beings, whose sense of smell is at best rudimentary, memories of all sorts are inextricably bound up with odors.

Elephants live to a great age, probably in the neighborhood of two hundred years. In this connection the peculiar arrangement of the molar teeth is of interest; for as the molar teeth that first emerge are worn off by long years of use other molars gradually replace them. The grinding teeth are arranged as though in the arc of a circle, so that only two or at most three on each jaw are in contact at one time. When the front ones wear out the rest move up and take their places, until in very old animals only the last teeth are present. This dentition is by far the most specialized found among vertebrates.

Among the best known **recently extinct types of elephants** are the mammoth and the mastodon. The **mammoth** was more nearly like the Indian elephant than any other species, but was much larger. Its tusks were enormous, one being known to weigh two hundred and fifty pounds. These tusks are extremely durable as is demonstrated by the fact that much of the ivory now in use in the form of billiard balls, etc., has been made from them, though their original owners have been dead for thousands of years. The **mastodon** was about as high as the Indian elephant, seven to nine feet, but was much more stockily built and longer bodied. The tusks were sometimes as much as nine feet or more in length.

The evolution of the peculiar characters of modern elephants is well shown in a series of extinct forms, as represented by Lull (Fig. 203). The earliest proboscidian appears to have been a form like *Mœritherium* (Fig. 203, F′), which, though rather generalized in most respects, shows the beginnings of elephantine characters in the air cells in the back of the skull, in the enlarged second incisors or incipient tusks, and the primitive lophodont molars. It was, however, only about three and a half feet high. Transitional stages are shown in *Palæomastodon* (Fig. 203, E′), in *Trilophodon* (Fig. 203, D′), and in *Stegodon* (Fig. 203, C′), in which all of these characters have approached several steps nearer the present condition, as shown in upper figure (Fig. 203, A′).

Order 12. Sirenia (Dugongs and Manatees).—The sirenians are now looked upon as an aquatic offshoot of an early ungulate stock distantly related to the proboscidians. The traditional position of these aquatic mammals has always been alongside the Cetacea (whales), but the resemblances between these two aquatic groups

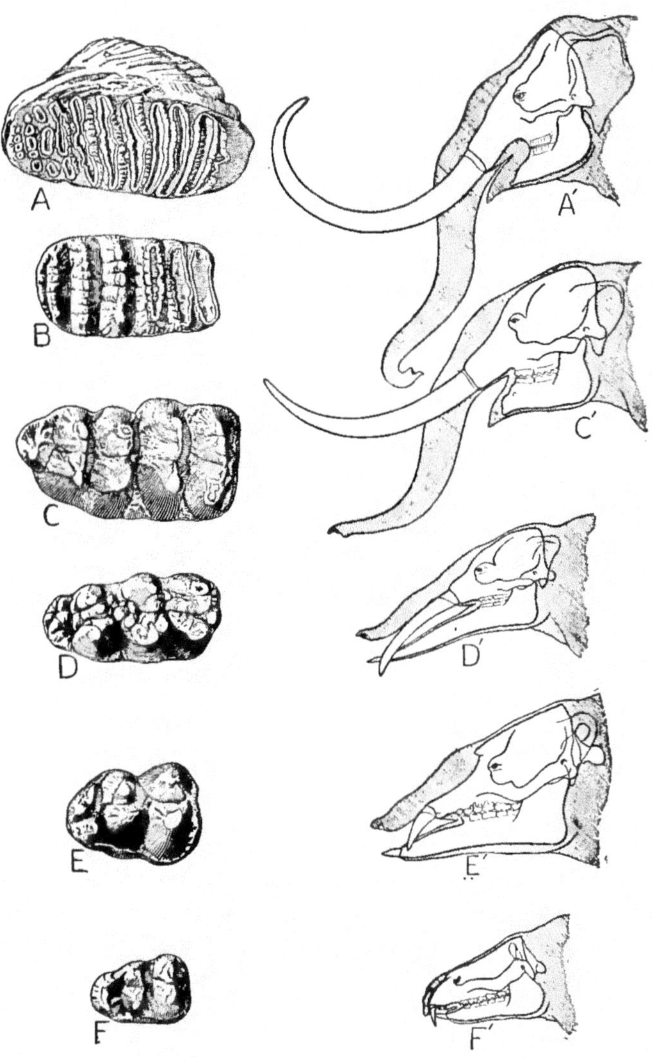

Fig. 203.—Evolution of head and molar teeth of mastodons and elephants. A, A′, *Elephas*, Pleistocene; B, *Stegodon*, Pliocene; C, C′, *Mastodon*, Pleistocene; D, D′, *Trilophodon*, Miocene; E, E′, *Palæomastodon*, Oligocene; F, F′, *Mœritherium*, Eocene. (From Lull.)

are evidently largely homoplastic, or parallel adaptations to a similar habitat. Both *dugongs* and *manatees* are large, almost hairless mammals, with hind limbs absent, and with the tail flattened into the semblance of a caudal fin or a fluke. The nostrils are on the upper surface of the snout; there are no clavicles; the stomach is complex and resembles that of the ungulates; the testes are abdominal in position; the mammæ are pectoral as in elephants.

The **manatees** (Fig. 204) are fairly abundant in fresh waters along the Atlantic coasts of North America and Africa. They are said to be especially numerous among the lagoons of the Florida Everglades. The use of their flesh as meat has been strongly urged; for they feed upon nothing but sea-weeds, of which there is an inexhaustible supply. The flesh is said to compare favorably with beef. The manatees have but six cervical vertebræ; there are as many as twenty molar teeth, which seem to continue to increase during life. In these two respects they are unique among mammals.

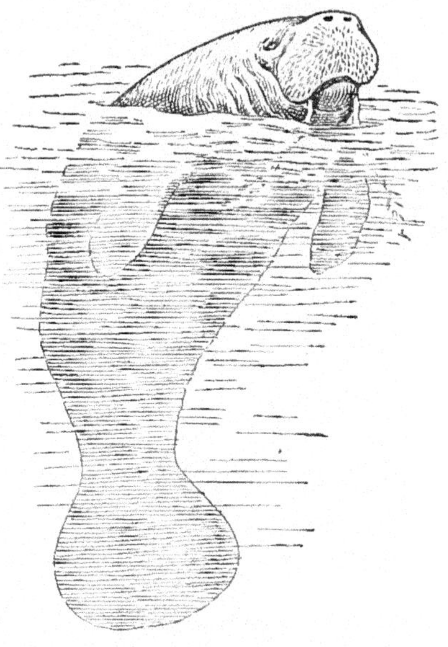

Fig. 204.—Florida Manatee, *Trichechus latirostris*. (Redrawn after Fuertes.)

The **dugong** (Fig. 205), *Halicore*, is an oriental and Australian species, with whale-like tail-flukes instead of the rhomboidal type of tail paddle seen in the manatee. It is more extensively specialized for aquatic life than the manatee, for the nostrils are more dorsal, the tail is more fish-like and the digits have no claws. It is said that the dugong is responsible for most of the mermaid legends, for when the female holds her young to her pectoral breast by means of one flipper while swimming with the other, she presents a slightly human resemblance.

Order 13. Hyracoidea (Coneys).—This small order consists of but one living genus of primitive ungulates. The **coney** (Fig. 206)

(*Hyrax* or *Procavia*) bears a strong resemblance to certain rodents, the short ears and reduced tail being especially like those of the cavies.

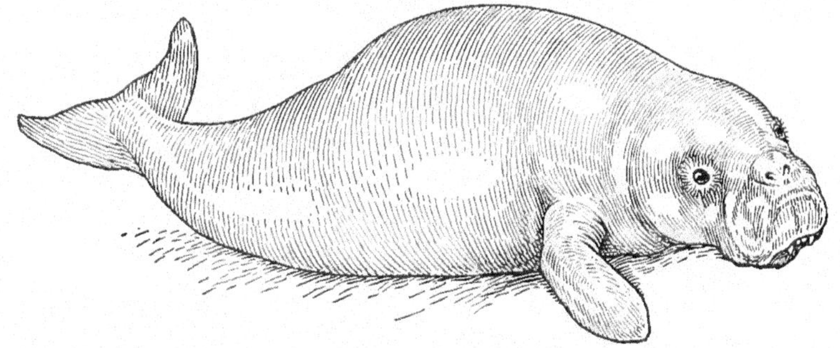

Fig. 205.—Dugong, *Halicore dugong.* (Redrawn after Lydekker.)

They are unlike the ungulates and like the rodents in that the incisor teeth grow from persistent pulps. In certain other respects they resemble primitive ungulates.

Fig. 206.—Coneys or hyraces, *Hyrax abyssinicus.* (From Lull, after Brehm.)

Some of the coneys live among rocks, while others are partly arboreal. The Scriptures describe them as "exceeding wise" and as "feeble folk," but the observation that he "cheweth the cud but divided not the hoof" is without foundation on either count; for they are not ruminants, and there are four hoofs in front and three behind.

Section D. Cetacea (Whales and Dolphins)

This assemblage of large aquatic mammals is profoundly modified for marine life. They are unquestionably the most highly specialized structurally of all mammals, although certain of their characters are persistently primitive. In older classifications they have usually been placed among the earlier orders, because they are least like Man, who was looked upon as the ultimate goal of organic evolution. Is it too serious a blow to human complacency to have to cede the honor of being placed at the top of the systemic ladder to the whales? The statement that the whales are the most highly specialized mammals is backed up by the following criteria of specialization: 1, the whales are farthest removed from the generalized types of mammals in all of their adaptive characters; 2, they have undergone losses of such typical mammalian structures as hair, teeth (in some groups), claws, and hind limbs; 3, the skeleton of the fore limbs is progressively specialized by the addition of several digits; 4, they have reached a size unrivaled in the world's history, far surpassing that of the giant reptiles of Mesozoic times; 5, the stomach is one of the most complex among mammals; 6, the skull of some of the whales is the most asymmetrical and otherwise specialized among mammals.

Three orders of Cetacea are distinguished: *Zeuglodontia* (extinct generalized whales), *Odontoceti*, and *Mystacoceti*.

Order 14. Odontoceti (Toothed Whales).—This order includes the sperm whales, narwhals, beaked whales, porpoises and dolphins. They are characterized by the presence of teeth and absence of whalebone; by the possession of a single nostril or blow hole; by asymmetry of the skull; and by having some of the ribs two headed.

The **sperm whale** or **cachalot** (Fig. 207, C), *Physeter*, is probably the largest animal that ever lived, and the writer was fortunate enough to have been able to examine and to record the measurements of what is now believed to have been the largest specimen ever authentically described. This was the well-known Port Arthur whale, that came ashore on the north coast of the Gulf of Mexico in March, 1910. This animal measured on a straight line from snout to end of flukes (not following curvatures as is usually done) sixty-three and a half

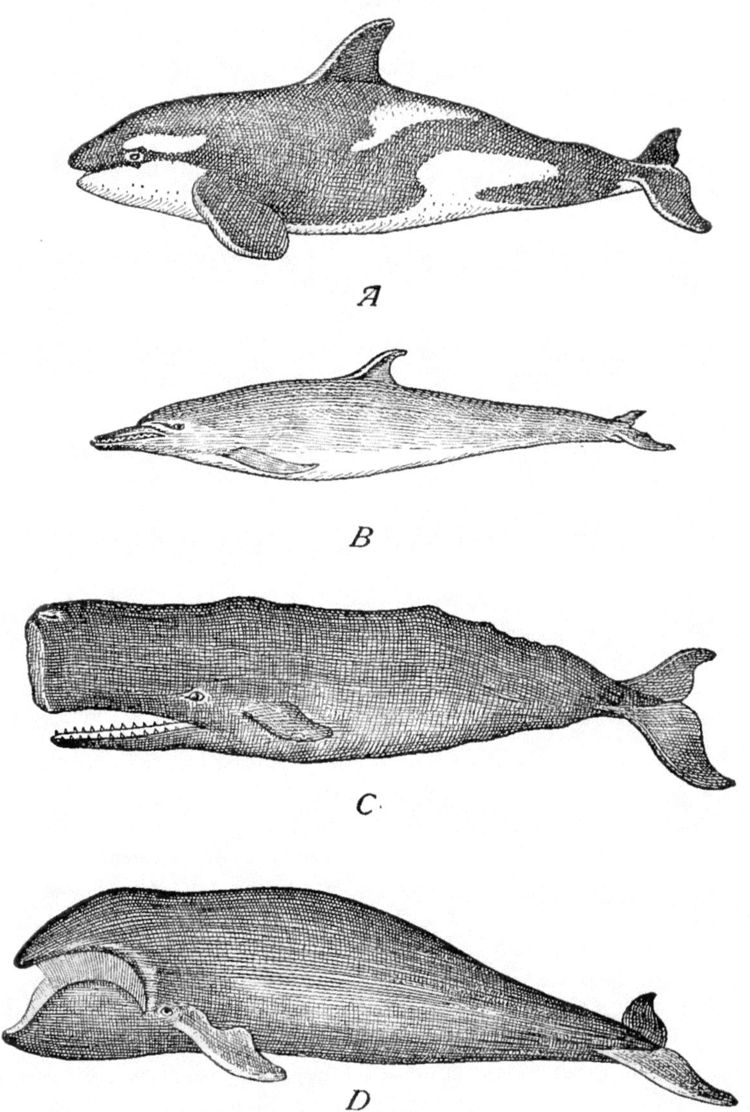

FIG. 207.—Group of Cetacea. A, Killer Whale, *Orca gladiator* (after True); B, Common Dolphin, *Delphinus delphis* (after Reinhardt); C, Sperm Whale, *Physeter macrocephalus*; D, Southern Right Whale, *Balæna australis*. (C and D, after Beddard.)

feet. Its circumference in front of the flippers was thirty-seven feet; it was twelve feet in height at the shoulders. This enormous animal did not impress one as a long slender type, but as distinctly stocky, retaining its great diameter from the end of the snout to within about fifteen feet from the tail. The lower jaw of the sperm whale is long and narrow and is armed with from forty to forty-eight conical teeth that fit into the toothless groove of the upper jaw. A large cavity in the skull is filled with a liquid oil, *spermaceti,* which is a valuable product. This reservoir of light oil is believed to be largely of hydrostatic value, in that it must be quite buoyant. The huge skull is the most highly modified skull known for a mammal. The right maxillary and left nasal bones are much larger than their fellows, the right nasal being vestigial. The top of the skull has a great bony crest running diagonally instead of mesially as in other skulls. The cervical vertebræ are largely fused into a short immovable neck. The sperm whale is valuable for spermaceti, for oil made from blubber, and for ambergris; the latter is a very valuable product said to be worth its weight in gold, and is a cumulative byproduct of intestinal digestion, having a composition somewhat like cholesterin. Ambergris is used in imparting long-lasting quality to fine perfumes and even minute quantities add value to considerable volumes of perfume. The food of the sperm whale consists largely of giant squids, as may be judged by the remains of the latter found in the whale's stomach.

One of the most fish-like of the toothed whales is the killer (Fig. 207, A), *Orca,* a small species that has the reputation of killing larger whales.

Beaked whales are animals of moderate size, seldom more than thirty feet in length; they have a prolonged muzzle armed with numerous teeth. They are quite slender and doubtless have done duty as "sea serpents." **Dolphins** and **porpoises** (Fig. 207, B) are small whales of rather generalized structure. They have teeth in both jaws, and the head is more mammal-like than that of other whales. According to Flower, there are nineteen genera of these small whales, and they comprize a considerable majority of all existing cetaceans. They are distinctly gregarious, running in schools of considerable size. Their habit of leaping out of the water at intervals makes them an interesting sight for ocean travelers. Closely allied to the porpoises is the **narwhal,** a form in which the teeth are reduced to a single tusk in the upper jaw, which protrudes out in front like a spear. This tusk is twisted in structure like a rawhide ox-whip and is limited to the males, who use it in fencing contests among themselves.

ORDER 15. MYSTACOCETI (WHALEBONE WHALES).—The **whalebone** or **baleen whales** (Fig. 207, D) are the last word in adaptive specialization among mammals. The teeth are rudimentary and functionless, present in the young but replaced in the adult by baleen. The nostrils are paired; the skull is symmetrical; the sternum is single; the ribs are one headed, articulating only with the transverse processes of the vertebræ. The group is composed exclusively of large forms, the only one that is less than a giant being the pigmy right whale, which is only about fifteen feet in length. The baleen or whalebone is a horny material developed from the epithelial lining of the mouth cavity. It is disposed in curtain-like plates (Fig. 208), frayed out into fringes at the bottom. The plates reach a length of twelve or more feet and are triangular, with the greatest width at the top or attached end. As many as three hundred and seventy blades or curtains, placed with their edges an inch or so apart, have been counted in a single mouth. The function of the baleen is that of a strainer. The great beast rushes through the water with the mouth wide open, gathering in fishes or whatever else happens to be in the way. Then the mouth closes and the water is forced out between the sheets of baleen, while fishes, etc., are retained and swallowed. Such huge creatures require vast quantities of food and cannot become very numerous. Formerly whalebone was a commercial product of some importance, used chiefly as stays in women's garments. Many substitutes, however, have been discovered and, moreover, stays have gone out of fashion; so that the market value of the

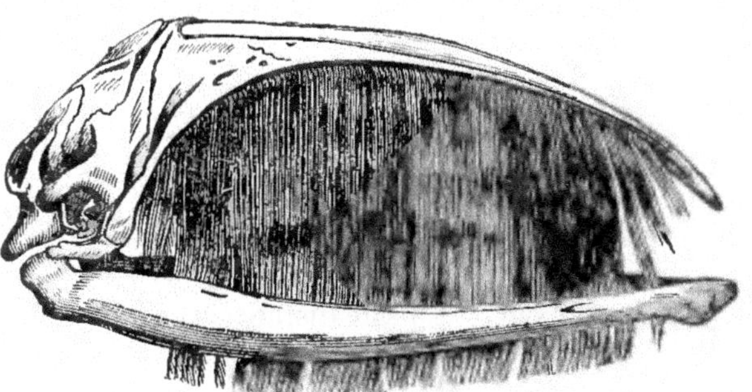

FIG. 208.—Skull of Baleen Whale, *Balæna mysticetus*. (From Weysse, after Claus and Sedgwick.)

commodity has been greatly depressed. A single large whale produces several tons of whalebone, and, since a ton used to be worth about ten thousand dollars, the capture of a single baleen whale meant a small fortune to the whaler.

Rorquals are a type of whalebone whale with comparatively small heads, a distinct dorsal fin, and with a throat deeply corrugated into longitudinal furrows. The flipper has only four fingers, but each finger is very long, having many extra joints. They range in length from forty to nearly seventy feet; one species has a record of eighty-five feet in length. The cervical vertebræ are all separate.

Right Whales are the more typical baleen whales. They have no dorsal fin; the head is very large, being about one-fourth of the entire length; the baleen is very long; the throat is not corrugated; the cervical vertebræ are fused into a solid mass. The Greenland right whale is, perhaps, the best known of all whales. It has a very limited distribution, being confined to the Arctic Ocean. It grows to be about seventy feet in length. The pursuit of whaling used to be one of the most romantic and dangerous of human occupations; but with the advent of whaling guns, with which the great creatures may be harpooned at a safe distance, the danger is largely eliminated, though much of the romance remains. The southern right whale, a close relative of the Greenland species, has a wide range, avoiding only the Arctic regions. The two species never occur in the same territory. It is less prized by the whaler on account of the relatively short and coarse whalebone.

Whales as a whole are much less numerous than they were a century ago and it seems probable that, unless some protection is given them, they are likely to become extinct before another century rolls by. Man seems to have no compunctions in his lust for commercial profit, and even these noble creatures of the deep may soon go the ways of the giants of ages past.

THE DEVELOPMENT OF MAMMALS

It is much more difficult to give a concise account of development of mammals than of any other of the vertebrate classes, because there is such a wide range of diversity of conditions. In the first place it will be recalled that some of the mammals lay large eggs essentially as in reptiles, that others have a sort of uterine gestation without establishing any definite structural connection between the fetal and the uterine membranes, and that still others have a well-defined type of placental gestation. We may quickly dispose of the situation involved in the egg-laying mammals by saying that their mode of development is essentially sauropsidan, and need not be repeated here.

The **marsupials** present a wide variety of conditions. Their eggs though minute are somewhat larger than those of the monodelphian mammals. The embryo has a brief period of uterine gestation, though no fixed nor definite uterine attachment is established. In

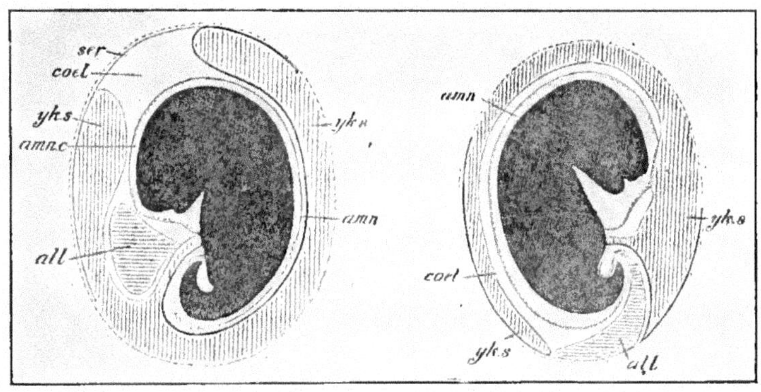

FIG. 209.—Diagram of the embryo and fetal membranes of the marsupial *Hypsiprymnus rufescens* (on the left). *all*, allantoic cavity; *amn*, amnion; *amn. c*, amniotic cavity; *coel*, extra-embryonic cœlom; *ser*, chorion or serous membrane. (From Parker and Haswell, after Semon.)

FIG. 210.—Diagram of embryo and fetal membranes of *Phascolarctus cinereus* (on the right). Letters as in fig. 209. (From Parker and Haswell, after Semon.)

most marsupials a large part of the surface of the egg is covered over by the compressed and expanded yolk-sac, as in *Hypsiprymnus* (Fig. 209). In *Phascolarctus* (Fig. 210) the allantois is in contact with part of the surface. In *Perameles* (Fig. 211) a primitive type of allantoic placenta is formed and sends out vascular outgrowths into the maternal tissues, much as does the Träger or primary placenta of the rodents and armadillos, among true placental mammals. Just here it may not be out of place to recall that it is believed by several leading authorities that the conditions found in the marsupials of to-day are not primitive but largely degenerate, and that

Perameles with its primitive placenta represents a more nearly primitive condition than any other living marsupial so far studied. Such a view would involve the corollary that both modern marsupials and modern placental mammals have been derived from a primitive placental ancestry, possibly akin to the insectivores.

Conditions in Placental Mammals.—Some of the simpler types, such as that of the pig and the horse, are not unlike those seen in the marsupial, *Perameles*, but in others, as for example the primates, the armadillos, etc., the conditions are very much modified. In earlier stages, however, the differences are slight.

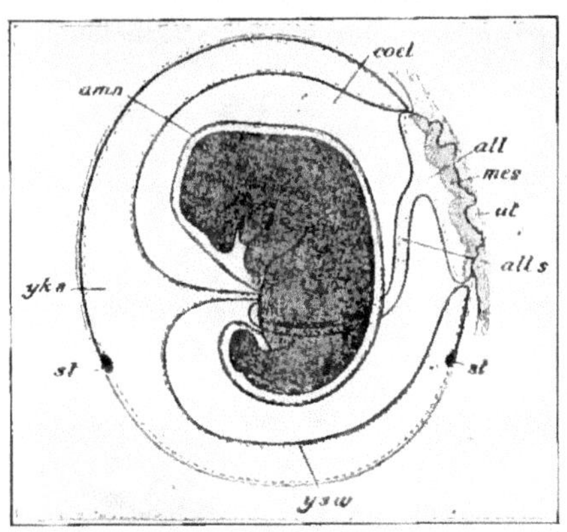

FIG. 211.—Diagram of the embryo and placenta of the marsupial *Perameles obesula*. Letters as in fig. 209. In addition —*all. s*, allantoic stock; *mes*, mesenchyme of outer surface of allantois fused with mesenchyme of serous or chorionic membrane; *st*, sinus terminalis; *ut*, uterine wall. (From Parker and Haswell, after Hill.)

The egg of the placental mammal is extremely small and essentially yolkless, yet many changes take place that seem to occur with reference to a large yolk supply. The embryo is developed from a small region of the blastula, and is cut off from the extra-embryonic area, with which it remains connected by a slender yolk-stalk. There is a fairly large yolk-sac, without any yolk content, upon which a vitelline circulation develops up to the point of blood formation and then goes no further. Amnion, chorion, and allantois form much as in birds, though secondary modifications of all of these membranes are found in various groups. All of these conditions seem to admit of but one interpretation: that the small yolkless mammalian ovum is the lineal descendant of a large-yolked egg similar to that of the monotremes or the reptiles, and that the yolk has been lost in connection with the habit of uterine gestation. With all the conservativeness of the typical germ cell, the mammal egg persists in behaving much as though it had a large supply of yolk upon which it had to depend for nourishing the embryo.

Cleavage and Early Development in a Placental Mammal.—It is not easy to compare the cleavage (Fig. 212) of the mammalian ovum with that of any other form. It appears deceptively simple, but we know that this apparent simplicity is a camouflage, for subse-

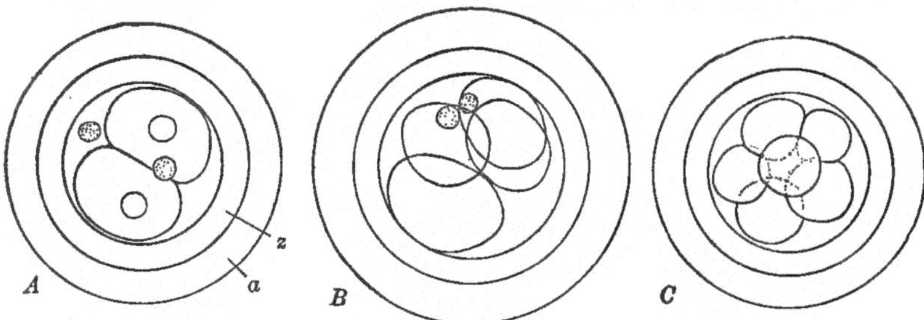

Fig. 212.—Cleavage of the ovum of the rabbit. A, Two-cell stage, 24 hours after coitus, showing the two polar bodies separated. B, Four-cell stage, 25½ hours after coitus. C, Eight-cell stage. *a*, albuminous layer derived from the wall of the oviduct; *z*, zona radiata. (From Kellicott, after Assheton.)

quent events reveal that the apparent holoblastic cleavage gives results that are similar to those resulting from a sauropsidan type of meroblastic cleavage. It appears that the first two cleavages are total and equal, just as in Amphioxus. After that the cleavages

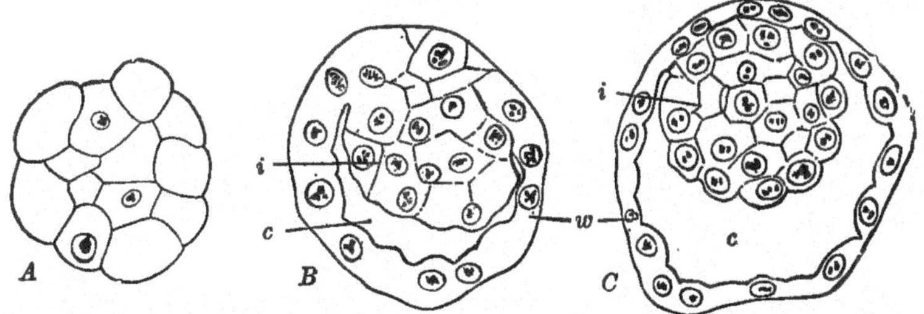

Fig. 213.—Morula and early blastodermic vesicles of the rabbit. The zona radiata and albuminous layer are not shown. A, Section through a morula stage, 47 hours after coitus. B, Section through very young vesicle, 80 hours after coitus. C, Section through more advanced vesicle, 83 hours after coitus; taken from uterus. *c*, cavity of blastodermic vesicle; *i*, inner cell mass; *w*, wall of the blastodermic vesicle (trophoblast). (From Kellicott, after Assheton.)

are not easy to follow, since the cells seem to shift about and not to retain their original positions.

The blastula stage takes the form of a solid mass of cells, the morula (Fig. 213, A), in which a peripheral layer of cells, the trophoblast, is distinguished from the inner-cell-mass. Subsequently (Fig. 213, B and C) the trophoblast separates from the inner-cell-mass except at the animal pole and a large cavity filled with fluid appears between the two layers. The trophoblast layer is a temporary structure serving as a sort of primitive placenta for the young embryo and helping the latter to gain its first connection with the uterine membrane. A specialized region of the trophoblast, called the Träger, sends short papillæ into the uterine mucosa, opening the way for the true placental villi that come later. The inner-cell-mass forms the entire embryo, together with the embryonic membranes, amnion, chorion, allantois, and yolk-sac. At first a round ball of cells, the inner-cell-mass flattens out to form a thin lens-shaped mass in contact with the attached part of the trophoblast, or Träger. Later two layers form, ectoderm and endoderm, by a sorting out of two types of cells, or a migration inwards of the endoderm cells. This process is the equivalent of the first step in gastrulation, but cannot readily be compared with the equivalent process in any other type of embryo. Once the two-layered germinal disk, early gastrula, is formed, the remainder of the process of embryogenesis is much like that of the Sauropsida and need not be further described.

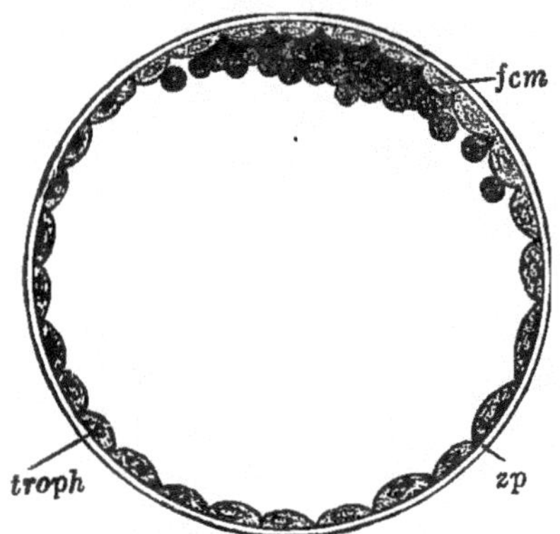

Fig. 214.—Section through the fully formed blastodermic vesicle of the rabbit. *fcm*, granular cells of inner cell mass; *troph*, trophoblast; *z. p*, zona pellucida (From Kellicott, after Quain.)

The development of the embryonic membranes, however, differs in many ways from that seen in the bird. The layer of endoderm, at first confined to the upper part of the vesicle, spreads until it forms a complete inner lining for the trophoblast. The gut of the embryo is pinched off from the upper part, leaving an empty yolk-sac below, connected with the gut-endoderm by a slender yolk-stalk. The amnion sometimes forms as in the chick (Fig. 216), by a fold of the somatopleure, which also produces the outer layer or chorion; but sometimes the amnion forms by means of a cavity opening up in the midst of the ectodermic mass, a short-cut method used by the insectivores (Fig. 215), rodents, armadillos and man. The allantois forms as in birds, but frequently remains rudimentary as in man (Fig. 217); but in some cases, as in the rabbit (Fig. 216), it forms a primitive type of allantoic placenta much like that seen in the marsupial, *Perameles*.

The formation of the true chorionic placenta is a complicated process. The mesodermic layer of the chorion, which becomes highly vascular, and becomes connected with the embryonic circulation, sends out branching processes, chorionic villi, into the uterine tissues, which penetrate the uterine lymph cavities and absorb liquid nutri-

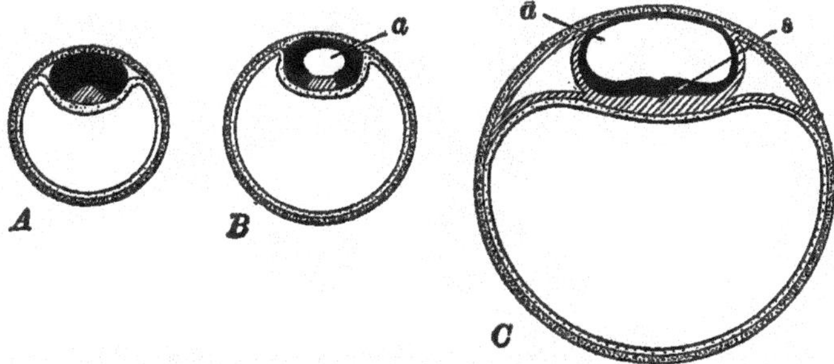

Fig. 215.—Diagram of the formation of the amnion in the Insectivores. Black; embryonic ectoderm; heavy stipples, trophoblast; light stipples, endoderm, oblique ruling, mesoderm. A, before the appearance of the amniotic cavity; inner cell mass differentiated into ectoderm and mesoderm; endoderm extending completely around the wall of the vesicle. B, The amniotic cavity (*a*) appearing in the ectoderm. C, Enlargement of the amniotic cavity. Mesoderm expanded and split into somatic and splanchnic layers, separated by the cœlom. *s*, primitive streak. (From Kellicott, after Keibel.)

ment directly from the maternal supply. The maternal tissues become thick and congested in these regions, and the fetal and maternal tissues together constitute the definitive placenta. The entire chorion is at first provided with simple villi, but later only certain regions retain the villi and act as placental areas. Frequently the placental area is discoidal in shape, as in the primates, in some of the edentates, and in many of the rodents; sometimes the placental area is band-like or zonary, as in the carnivores; and in the case of the ungulates it is cotyledenous, in which case thick knots of villi are scattered over almost the entire chorion, separated by extensive non-villous areas.

Parturition or birth takes place at widely different stages of maturity in the different mammalian groups. In some species, as in cattle and horses, the young at birth are well advanced and, within a few

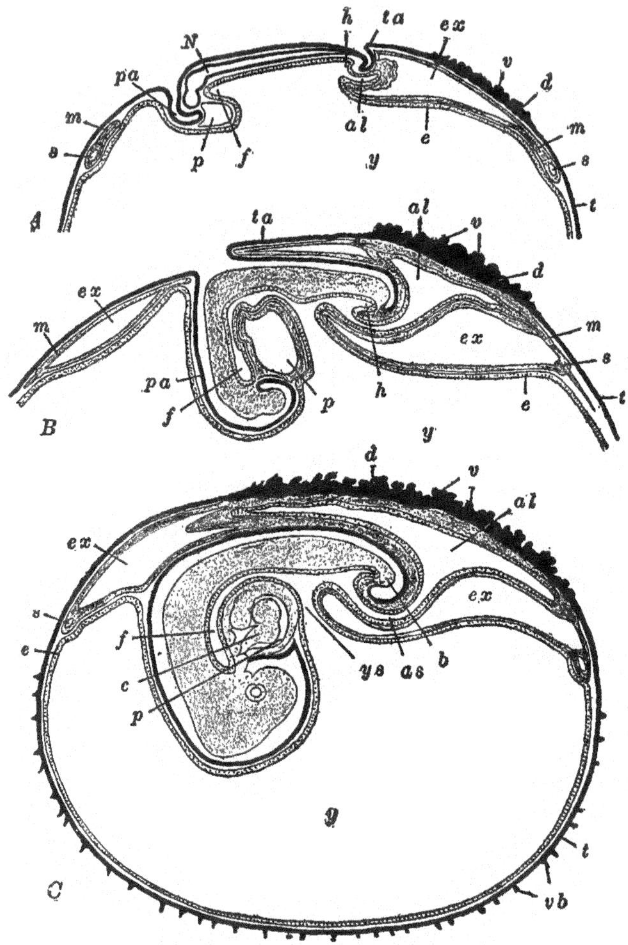

FIG. 216.—Diagram of the formation of the embryonic membranes and appendages of the rabbit. A, at the end of the ninth day; B, early the tenth day; C, at end of tenth day. Ectoderm, black; endoderm, dotted; mesoderm, gray. *al*, allantois; *as*, allantoic stalk; *b*, tail bud; *c*, heart; *d*, trophoderm; *e*, endoderm; *ex*, exocœlom; *f*, foregut; *h*, hind-gut; *m*, mesoderm; *N*, central nervous system; *p*, pericardial cavity; *pa*, proamnion; *s*, marginal sinus (sinus terminalis); *t*, trophoblast; *ta*, tail-fold of amnion; *v*, trophodermal villi; *vb*, trophoblastic villi; *y*, cavity of yolk-sac; *y. s*, yolk-stalk. (From Kellicott, after Van Beneden and Julin.)

hours after birth, are able to walk or even to run, and require little parental care except in connection with mammary feeding. In other species, as in the carnivores and rodents, the young are born naked,

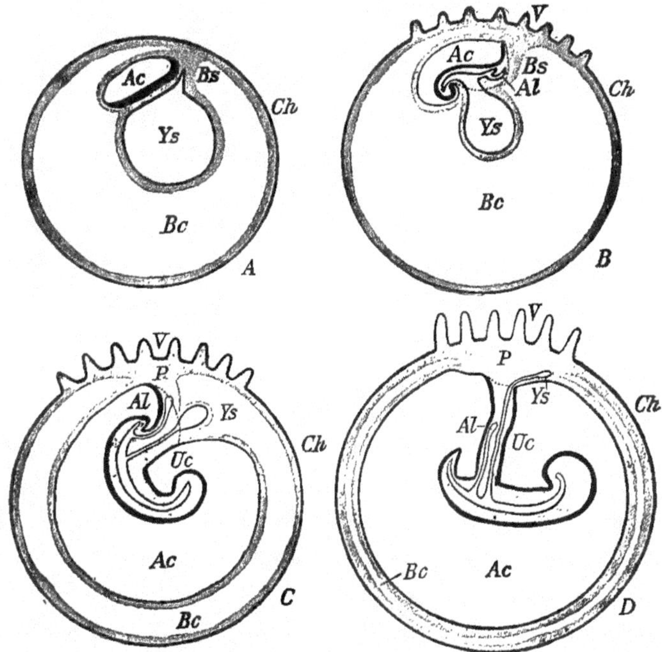

FIG. 217.—Diagram illustrating the formation of the umbilical cord and the relations of the allantois and yolk-sac in human embryo. The heavy black line represents the embryonic ectoderm; the dotted line marks the line of transition of the body (embryonic) ectoderm and that of the amnion. Stippled areas, mesoderm. *Ac*, Amniotic cavity; *Al*, allantoic cavity; *Al*, allantois; *Bc*, exocœlom; *Bs*, body stalk; *Ch*, chorion; *P*, placenta; *Uc*, umbilical cord; *V*, chorionic (trophodermic) villi; *Ys*, yolk-sac. (From Kellicott, after McMurrich.)

blind, and helpless and need much care for a considerable period. The human infant, while not as immature as some of those just mentioned, is decidedly helpless and needs care longer than any other creature.

PARTIAL LIST OF LITERATURE

Andrews, E. A.: An Undescribed Acraniate: Asymmetron lucanum, 1893.
Ayers, H.: Concerning Vertebrate Cephalogenesis, 1890.
Beddard, F. E.: Mammalia,—Cambridge Natural History, Vol. X, 1909.
Beebe, C. W.: The Tetrapteryx Stage in the Ancestry of Birds, 1915.
Bensley, B. A.: On the Evolution of the Australian Marsupialia, with Remarks on the Relationships of Marsupials in General, 1903.
Boulenger, G. A.: The Tailless Amphibia of Europe, 1897.
Boulenger, G. A.: Teleostei; Cambridge Natural History, Vol. VII, 1904.
Bourne, G. C.: An Introduction to the Study of the Comparative Anatomy of Animals, 2 vols., 1900.
Brehm, A. E.: Tierleben; allgemeine Kunde des Tierreichs. 4th edition, revised. Vol. I, Säugetiere, 1912.
Bridge, T. W.: Fishes, Cambridge Nat. Hist., Vol. VII, 1904.
Bronn, H. G.: Klassen und Ordnungen des Thierreichs.
Caldwell, H.: Embryology of Monotremata and Marsupialia, 1887.
Case, E. C.: The Permo-Carboniferous Red Beds of North America and their Vertebrate Fauna, 1915.
Cerfontaine, P.: Recherches sur la developpement de l'Amphioxus, 1906.
Chamberlin, T. C.: On the Habitat of the Early Vertebrates, 1900.
Child, C. M.: Individuality in Organisms, 1915.
Child, C. M.: Senescence and Rejuvenescence, 1915.
Cope, E. D.: Batrachia of North America.
Cope, E. D.: Article on Amphibia in Riverside Natural History.
Cope, E. D.: Crocodiles, Lizards, and Snakes of North America, 1900.
Coues, E.: Key to North American Birds, 1887.
Dean, B.: Fishes Living and Fossil, 1895.
Dean, B.: Fin-fold Origin of Paired Fins, 1896.
Dean, B.: Development of Garpike and Sturgeon, 1895.
Dean, B.: Chimeroid Fishes and Their Development, 1906.
Dendy, A.: Outlines of Evolutionary Biology, 1912.
Dickerson, M. C.: The Frog Book, 1907.
Ditmars, R.: The Reptile Book, 1907.
Duckworth, W. L. H.: Prehistoric Man, 1912.
Evans, A. H.: Birds; Cambridge Natural History, Vol. IX, 1909.
Flower, W. H.: Osteology of Mammals, 1885.
Flower, W. H., and Lydekker, R. Mammals, 1891.
Gadow, H.: Amphibia and Reptiles; Cambridge Natural History, Vol. VIII, 1901.

Gadow, H.: Volume on Birds in Bronn's Tierreich.

Gegenbaur, C.: Vergleichende Anatomie der Wirbeltiere, 2 vols. 1898 and 1901.

Goode and Bean.: Oceanic Ichthyology, 1895.

Goodrich, E. S.: Cyclostomes and Fishes; Lankester's Zoölogy, Vol. IX, 1909.

Gregory, W. K.: Present Status of the Origin of the Tetrapoda, 1915.

Gregory, W. K.: Studies on the Evolution of the Primates, 1916.

Gregory, W. K.: Theories of the Origin of Birds, 1916.

Günther, A.: Introduction to the Study of Fishes, 1880.

Harmer, S. F.: Hemichordata; Cambridge Natural History, Vol. VII, 1904.

Hatschek, B.: The Amphioxus and its Development; English Translation, 1893.

Hay, O. P.: The Batrachians and Reptiles of the State of Indiana, 1892.

Hegner, R. W.: College Zoölogy, 1912.

Herdman, W. A.: Ascidians and Amphioxus; Cambridge Natural History, Vol. VII, 1910.

Hertwig, O.: Manual of Zoölogy, 1910.

Holmes, S. J.: The Biology of the Frog, 1914.

Huxley, T. H.: Manual of the Anatomy of Vertebrated Animals, 1872.

Jordan, D. S.: Guide to the Study of Fishes, 1905.

Jordan, D. S.: Fishes, 1907.

Jordan D. S., and Evermann, B. W.: Fishes of North America, 4 vols. 1900.

Jordan, D. S., and Evermann, B. W.: The Aquatic Resources of the Hawaiian Islands. Part. I, The Shore Fishes; Bul. U. S. Fish Com., Vol. XXIII, 1905.

Jordan, D. S., and Kellogg, V. L.: Evolution and Animal Life, 1908.

Keibel, F., and Mall, F. P.: Manual of Human Embryology, 1910.

Kellicott, W. E.: Outlines of Chordate Development, 1913.

Kingsley, J. S.: Textbook of Vertebrate Zoölogy, 1899.

Kingsley, J. S.: Systematic Position of the Cæcilians, 1902.

Kingsley, J. S.: Comparative Anatomy of Vertebrates, 2nd ed., 1917.

Knowlton, F. H.: Birds of the World, 1909.

Lankester, E. R.: Treatise on Zoölogy, Vol. IX; Cyclostomes and Fishes (by Goodrich), 1909.

Lydekker, R.: The New Natural History.

Lillie, F. R.: The Development of the Chick, 1908.

Locy, W. A.: Contribution to the Structure and Development of the Vertebrate Head, 1895.

Lucas, F. A.: The Beginnings of Flight, 1916.

Lull, R. S.: Volant Adaptations in Vertebrates, 1906.

Lull, R. S.: Organic Evolution, 1917.

Marshall and Hurst: Practical Zoölogy, 1905.

Marshall, A. M.: The Frog: an Introduction to Anatomy, 1896.

Matthew, W. D.: The Arboreal Ancestry of the Mammals, 1904.
Matthew, W. D.: Climate and Evolution, 1915.
Minot; C. S.: Human Embryology, 1892.
Moody, R. L.: The Coal Measures Amphibia of North America, 1916.
Morgan, T. H.: The Development of the Frog's Egg, 1897.
Neal, H. V.: Segmentation of the Nervous System in Acanthias, 1898.
Nelson, E. W.: The Larger North American Mammals; Nat. Geog. Mag., 1916.
Newman, H. H.: The Habits of Certain Tortoises, 1906.
Newman, H. H.: Spawning Behavior and Sexual Dimorphism in Fundulus heteroclitus and Allied Fishes, 1907.
Newman, H. H., and Patterson, J. T.: Development of the Nine-banded Armadillo, 1910.
Nopcsa, F.: Ideas on the Origin of Flight, 1907.
Osborn, H. F.: The Age of Mammals, 1910.
Osborn, H. F.: Men of the Old Stone Age, 1915.
Osborn, H. F.: The Origin and Evolution of Life, 1918.
Parker and Haswell: A Text-book of Zoölogy, 1910.
Patten, W.: The Evolution of the Vertebrates and Their Kin, 1912.
Pirsson, L. W., and Schuchert, C.: A Text-book of Geology, 1915.
Reighard and Jennings: Anatomy of the Cat, 1901.
Reynolds, S. H.: The Vertebrate Skeleton, second edition, 1913.
Sarasin, P. and F.: Entwicklungsgeschichte und Anatomie der ceylonischen Blindwühle Ichthyophis, 1890.
Scott, W. A.: A History of the Land Mammals of the Western Hemisphere, 1913.
Shipley and McBride: Zoölogy, 1904.
Summer, F. B., Osborn, R. C., and Cole, L. J.: A Biological Survey of the Waters of Woods Hole and Vicinity; Bul. Bur. of Fisheries, Vol. XXXI.
Walter, H. E.: The Human Skeleton, 1918.
Weber: Die Säugetiere, 1904.
Weysse, A. W.: Synoptic Text-book of Zoölogy, 1909.
Wiedersheim, R.: Comparative Anatomy of Vertebrates (English translation), 1907.
Wilder, H. H.: History of the Human Body, 1909.
Willey, A.: Amphioxus and the Ancestry of the Vertebrates, 1894.
Williston, S. W.: American Permian Vertebrates, 1911.
Williston, S. W.: Water Reptiles of the Past and Present, 1914.
Williston, S. W.: The Phylogeny and Classification of Reptiles (short paper), 1917.
Wilson, J. T., and Hill, J. P.: Development of Ornithorhynchus, 1907.
Woodward, A. S.: Outlines of Vertebrate Palæontology.
Zittel, K. von: Textbook of Palæontology, 2 vols., 1902.

INDEX

(Asterisks indicate pages on which occur illustrations of the subject mentioned)

Aard Vark, 375 *, 377, 378
Acanthiuridæ, 146
Acanthodei, 110
Acanthodidæ, 110
Acanthopterygii, 98, 110, 111, 146–150
Accessory respiratory organs of fishes, 108, 109
Acipenser, 133; eggs of, 164; *A. ruthenus* 132 *
Acris gryllus, 201
Acrobates, 357
Adaptations, laws of, 16–20; aquatic, 16, 17 *; climbing (scansorial), 18 *; cursorial, 18; fossorial, 18
Adaptive changes incident to terrestrial life, 175, 176
Adaptive radiation, 19; of reptiles, 211 *; of snakes, 258, 259; of marsupials, 353
Adelochorda (see Cephalochordata)
Adobenus obesus, 369 *
Æluropus, 368
Æpyornithes, 287
Aglossa, 196–198
Agnathostomata, 69
Agouti, 372
Air-bladder, of Teleostomi, 128; of *Polypterus*, 131
Albatross, 293 *, 294, 295
Alcedo, foot of, 280 *
Alecteromorphæ, 289
Alimentary system, of Amphioxus, 38; of tunicate, 51; of *Balanoglossus*, 62; in *Squalus*, 114; of fishes, 101; of turtle, 230, 231; of bird, 267
Allantoida, 68, 69
Allantois, 68, 69; in reptiles, 212, 213 *; in birds, 320, 321 *; in placental mammals, 409, 410 *, 411 *; in Man, 411 *
Allen, B. W., 195
Alligator Gar, 133, 134
Alligator mississippiensis, 245 *
Allotheria, 338
Alopecias vulpes, 122 *
Alosa, eggs of, 164
Altrices, 323
Alytes obstetricans, 197 *, 198
Ambergris, 402
Amblypoda, 340, 342–344
Amblyrhynchus cristatus, 250 *, 252
Amblystoma tigrinum, 189, 190 *, 192
Amia, 109, 133, 167; *A. calva*, 133
Amiidæ, 133
Ammocœtes larva, 38
Amnion, 68, 69; in reptiles, 212, 213 *; in birds, 319, 320 *, 321; in placental mammals, 409 *, 410 *, 411 *
Amniota, 68, 69, 95 *, 96
Amphibamus, 180 *
Amphibia, 13, 33, 68, 69; present and past status, 173, 174; origin of, 175, 176; characters of, 183, 184;
Amphibian ancestry of mammals, theory of, 333
Amphignathodon, 199
Amphioxus, 31, 32, 33, 34 *, 35, 36 *, 37 *, 38, 39 *, 40 *, 41 *, 42, 43 *, 44 *, 45 *, 46 *, 47 *, 48, 71, 72, 75
Amphioxus theory of vertebrate origin, 72–75
Amphiuma means, 186, 187 *
Amphiumidæ, 186, 187
Anabas scandens, 108 *, 109, 150 *
Anableps, 166
Anacanthini, 111, 145, 146
Anallanloida, 68, 69
Anamnia (Anamniota), 68, 69
Anapsida, 215
Anastomus, beak of, 282 *
Ancistrodon piscivorus, 257
Angel-shark, 122 *, 123

Anglers, 111, 150, 151
Anguis fragilis, 250 *, 251
Annelid theory of vertebrate origin, 77 *, 78, 79
Anseres, 296, 297
Anseriformes, 289, 296, 297
Ant-bears, 374, 375 *
Ant-eaters, adaptations of, 19; toads, 202; marsupial, 354, 355 *, 356; Edentate, 374, 375 *; scaly, 375 *, 377
Antelopes, 392
Antiarchi, 82, 112, 171, 172
Anthropoidea, 346, 378, 380–388
Anura, 196–203
Apes, 378; New World, 381; Old World, 381–384; anthropoid, 381–384
Ape-man, 387
Apoda, 184–186
Apodes, 111, 140, 141
Appendicularia, 59 * (see Larvacea)
Apterygiformes, 286, 287
Apteryx australis, 285 *, 286, 287
Arachnid theory of vertebrate origin, 79–82
Arachnid and vertebrate compared, 81 *
Arboreal origin of flight, theories of, 273–275
Archæopteryx, 272, 273, 274 *, 275, 276–278; *A. lithographica*, 276, 277 *, 278 *
Archæornithes, 276–278
Archaic, birds, 289; mammals, 339–344
Archegosaurus, 181
Archenteron, 47
Arctomys candatus, 372 *
Arius, eggs of, 164
Armadillos, 376
Aromochelys odorata, 237
Arthrodira, 112, 172
Arthropod ancestry of vertebrates, theory of, 79–82
Artiodactyla, 346, 389–392
Ascidia mammillata, 52 *
Ascidiacea, 33, 50 *, 51 *, 52 *, 53, 54 *, 55 *, 56 *
Ascidiæ compositæ, 55 *
Ascidiæ luciæ, 55, 56 *
Ascidian, a typical, 50 *, 51 *; development of, 52 *, 53, 54 *; colonial, 55 *, 56 *; tadpole larva of, 52 *, 53
Aspidonectes spinifer, 240 *, 241
Asymmetron, 35, 48
Asymmetry, of *Amphioxus*, 48
Ateles ater, 379 *
Athecæ, 234–235
Atrial funnel (see Atriopore)
Atrial cavity, 39, 51
Atriopore, 39, 50, 51 *
Atrium, of *Amphioxus*, 38, 39, 40 *
Auditory capsules, in *Squalus*, 114
Auditory organs, of fishes, 101; of *Squalus*, 118
Auk, Great, 302
Auricularia larva, 64
Autostylic (skull), 125
Aves, 33, 69, 260–323; characterized, 260; compared to aëroplane, 260–265; compared with reptiles, 265, 266; list of characters of, 266–271; origin of, 271–276
Axial gradient theory, 9–11
Axial gradient in development, 9, 10
Axis, of polarity (primary axis), 1–3; antero-posterior, 1–3; apico-basal, 1–3; dorso-ventral (secondary axis), 4 *; bilateral (tertiary axis), 4, 5 *
Axolotl, 189, 190 *
Aye-aye, 380

Baboons, 381
Backward retreat of lower functions, 8
Badger, 367 *, 368
Balæna australis, 401 *; *B. mysticetus*. 403 *
Balæniceps, beak of, 282 *
Balanoglossus, 31, 60, 61 *, 62 *, 63, 65, 83; *B. clavigerus*, 62 *
Baleen, 403
Ballistapus rectangulus, 151 *
Bandicoots, 355 *, 356
Baptanodon, 218 *
Baptornis, 279, 281
Bartlett, 196, 198
Basiliscus americanus, 248 *, 252
Basilisk, helmeted, 248 *, 252
Bass, 111, 146
Bats, 363–365
Bat-Fish, 150
Bathymal Sea Devils, 111, 151
Batoidei, 111, 123–125
Bdellostoma, 91 *; eggs of, 164 *
Bdellostomidæ, 106
Beebe, C. W., 273

Belone, 111, 145
Bendire, Major, 298
Bernicla ruficollis, 293 *
Birds (see Aves); archaic, 276, 277; modern, 278–323; feet of, 280 *; beaks of, 282 *; present status of, 281; sex-dimorphism of, 281; of Paradise, 310, 311 *; future of, 312, 313; migration of, 313; geographic distribution of, 314
Birkenia, 169 *, 171
Bitterns, 295
Blastodermic vesicle, of rabbit, 408 *
Blastopore, 47
Blastula, of *Amphioxas*, 42, 43 *; of frog, 205, 206 *
Blennius, eggs of, 164
Blind-worms, 184
Boa-constrictors, 258
Boidæ, 258
Bombinator igneus, 197 *, 198
Bothriolepis, 80 *, 81, 82, 169 *, 171
Botryllus violaceus, 55 *
Boulenger, 137
Bovidæ, 390, 392
Bow-fin, 111, 134 *
Brachiosaurus, 218, 221
Bradipodidæ, 374
Bradypus, 376
Brain, Amphioxus, 38, 39 *; of Cyclostomata, 88; of Myxinoidea, 90; of Lamprey, 92, 93 *; of Pisces, 101; of Elasmobranchii, 118 *; of *Scyllium*, 118 *; of Teleostomi, 128; of turtle, 231; of alligator, 244 *; of pigeon, 270, 271; of rabbit, 330 *; of dog, 332 *; of mammals, 331; of *Ornithorhynchus*, 351 *
Branchial basket, of Cyclostomata, 88; of myxinoids, 90
Branchial clefts, of cyclostomes, 88
Branchiostoma (see Amphioxus)
Branchiosaurus, 180, 181 *; *B. amblystomas*, 182 *; *B. salamandroides*, 182 *
Breeding habits, of Amphioxus, 42; of lamprey, 94 *; of *Amia*, 134; of stickle-back, 144; of sea-horse, 144; of *Rhodeus*, 135 *, 137; of salmon, 138, 139; of killifishes, 142, 156; of cod, 146; of *Lepidosiren*, 158, 159; of bass and pickerel, 166; of *Ichthyophis*, 185 *, 186; of *Desmognathus*, 188 *; of *Pipa americana*, 196, 197 *, 198; of *Alytes*, 197 *, 198, 199; of *Hyla faber*, 201; of *Sceloporus*, 251; of *Echidna*, 350; of *Ornithorhynchus*, 351, 352
Bridge, T. W., 156, 157, 158
Brontosaurus *, 220
Brood-pouch, of sea-horse, 144 *
Bubo virginianus, 307 *
Bufo lentiginosus, 197 *, 199; *B. vulgaris*, 199
Bufonidæ, 199
Burbot, 146
Bustard, great, 301
Bustard quails, 299
Buzzard, Turkey, 297, 298

Cachalot, 400, 401 *, 402
Cacops, 180 *, 183
Cæcilia, 185 *
Cæcilians (see Apoda)
Cænogenetic characters, 24
Cænolestes, 357
Calamichthys, 111, 128, 129, 130 *, 159
Callechelyx luteus, 159
Callorhynchus antarcticus, 126 *
Camel-birds (see Ostriches)
Camels, 389, 390; Bactrian, 391 *
Camelus bactrianus, 391 *
Canidæ, 366, 367 *, 368
Canis nubilis, 367 *
Capitulum, 326
Capra sinaitica, 391
Caprimulgi, 306
Capuchins, 378, 381
Caracara, 298
Carapace, of turtle, 226, 227 *, 228
Carinatæ (see Neornithes Carinatæ); classification of, 289
Carnivora, 364–371; fissiped carnivores 364–368; pinniped carnivores, 368–371; extinct, 370
Carp, 111, 140
Cassowary, 284, 285 *, 286
Castor fiber, 372 *
Casuarias uniappendiculatus, 285 *
Catarrhini, 378, 381–384
Cat-fishes, 111, 140
Catosteomi, 111, 143, 144, 145
Cats, 366
Caudal fin, of fishes, 100; types of, 104, 105 *; of Holocephali, 127

Cave and deep-sea animals, 18
Caviar, 133
Cavies, 373
Cebidæ, 378, 381
Centurio senex, 365 *
Cephalaspis, 169 *
Cephalization in Vertebrates, 7, 8
Cephalochordata, 33
Cephalodiscus, 31, 32, 33, 60, 65 *, 66 *
Ceratobranchinæ, 202
Ceratobranchus guentheri, 202
Ceratodus (see *Neoceratodus*)
Ceratopsia, 223
Cercopithecidæ, 378, 381
Cervulus, 390
Cervus canadensis, 391 *
Cetacea, 346, 400–404
Chælopus didactylus, 375 *, 377
Chæropus castanotis, 356
Chamæleon vulgaris, 253, 254 *, *C. pumulis*, 254 *
Chamæleontes, 253–255
Chamberlin, T. C., theory of the stream origin of fishes, 73–75
Chameleons, 253–255
Characinidæ, 140
Charadriiformes, 289, 301–305
Chelodina, 241
Chelone imbricata, 235 *, 240; *C. mydas*, 240
Chelonia, 216, 233–242
Chelonidæ, 239, 240
Chelydosaurus, 181
Chelydra serpentina, 235 *, 236, 237
Chelydridæ, 236–237
Chelys fimbriata, 240 *
Chevrotani, 389
Child, C. M., vii, 8, 10, 161
Chimæras, 111, 125–127
Chimæra monstrosa, 126 *; *C. colliei*, 127
Chimpanzee, 378 *, 384
Chinchilla, 373
Chirogale coquereli, 380
Chiromys madagascariensis, 380
Chironectes, 354; *C. minima*, 355 *
Chiroptera, 346, 363–365
Chlamydophorus, 376
Chlameidoselachus, 107, 118; *C. anguinius*, 122 *
Chondrostei, 104, 106, 111
Chordata, the phylum characterized, 31; classification of, 33
Chorion, in birds, 319, 320 *; in mammals, 408, 409
Chrysemys picta, 238; *C. marginata*, 238
Chrysochloris trevelyani, 361 *, 362
Ciconia alba, 293 *
Ciconiiformes, 289, 295, 296
Cinosternidæ, 237
Cinosternum pennsylvanicum, 235 *
Circulatory system, of *Amphioxus*, 39, 41 *; of Pisces, 100; of *Squalus*, 116; of Amphibia, 177; of turtle, 231; of crocodile, 243; of bird, 267, 269 *, 270; of mammal, 333
Cistudo carolina, 235 *, 238, 239
Civets, 366
Cladista, 111, 129
Cladoselache, 71, 103, 105; *C. fyleri*, 119 *, 120
Cladoselachidæ, 110, 119
Clarias, 108,* 109
Claspers of sharks, 113; of Holocephali, 125, 126
Cleavage, of *Amia*, 165; of *Neoceratodus*, 166 *, 167, 168; of *Lepidosteus*, 165; of frog, 204, 205, 206 *; of chick, 316 *, 317; of placental mammal, 407 *; of rabbit, 407 *
Climbing Perch, 150 *
Clupeidæ, 138, 139
Cobra, 256 *, 259
Coccosteus, 171 *, 172
Cod, 145, 146 *
Cœlom, 1, 42; development of, 45 *, 46 *
Cœlomoducts, 41
Colii, 308
Coloration, of fishes, 110; of flounders, 148; of birds, 312
Columba, beak of, 282 *; *C. livia*, 304
Columbæ, 302, 304
Colubrinæ, 258
Colymbiformes, 289, 292, 293
Colymbimorphæ, 289
Colymbus gracilis, 293 *
Condylarthra, 340, 342 *
Coneys, 398, 399 *
Continuous fin-fold theory, 102; of Wiedersheim, 103 *; of Kingsley, 104 *
Convergence or parallelism, law of, 16
Cope, E. D., 137
Copperhead, 257

Copulation, in elasmobranchs, 113, 114; in Pœciliidæ, 142, 166; in snakes, 256
Coraciæ, 306 *
Coracias garrulus, 307 *
Coraciiformes, 289, 306–309
Coraciomorphæ, 289
Cormorants, 295
Corpus callosum, 325, 330 *, 331
Coryphodon, 342, 343 *
Corythosaurus, 218, 221
Costa, 33
Coues, 292, 293
Coyote, 366
Cranes, sandhill, 301
Craniata, classification of, 33
Cranium, of Cyclostomata, 88; of Lamprey, 93; of *Polypterus*, 130 *, 131; of stegocephalian, 182 *; of turtle, 229, 230 *; of *Sphenodon*, 233 *; of *Archæopteryx*, 278; of mammal, 328 *; of cynodont, 334 *; of *Ptilodus*, 338; of *Balæna*, 403
Creodonta, 340, 341 *, 342
Crex pratensis, 303 *
Cricotus, 180 *, 183
Crocodilus americanus, 245; *C. niloticus*, 246
Crocodilia, 242–246; habits of, 244
Crocodilidæ, 245, 246
Crocuta maculata, 367 *
Crossopterygii, 105, 106, 111, 127, 128–131; as ancestors of Amphibia, 174, 175
Crotalus durissus, 256 *, 258
Cryptobranchus alleghenienisis, 186, 187 *; *C. japonicus*, 186, 187
Cryptodira, 236–240
Cryptopsarus couesii, 136 *
Crypturiformes, 287, 288
Ctenoid scales, 109, 110
Cuckoos, 305
Cuculi, 305
Cuculiformes, 289, 306, 307
Cursorial origin of flight, theory of, 272–273
Cycloid scales, 109, 110
Cyclomyaria. 60
Cyclopes, 374
Cyclostomata, 33, 68, 69, 73, 87–96
Cynodontia, 216, 333, 334, 335, 336
Cynognathus, 215 *, 216
Cyprinidæ, 140
Cyprinodon, 166
Cypselus, foot of, 280 *; beak of, 282 *
Cystignathidæ, 201, 202

Dasypodidæ, 376
Dasyprocta aguti, 372 *
Dasypus novemcinctus, 375 *, 376
Dasyuridæ, 356
Dasyurus viverrinus, 355 *, 356
Deal-Fish, 150
Deep-sea Fishes, 136 *
Deer, Mouse, 389; family, 390; musk, 390
Degeneracy, a criterion of racial senescence, 21
Delphinus delphis, 401 *
Dendrobatinæ, 202, 203
Dendrobatinus, 203
Dendrochirus, 104; *D. hudsoni*, 149 *
Dendrolagus, 359
Dentition, of mammal, 328–331
Dermatemydæ, 237
Dermochelys coriacea, 234, 235 *
Dermoptera, 346, 362
Desert-dwellers (adaptations of), 18
Desmodactyli, 309, 310
Desmodus rotundus, 364; *D. rufus*, 365 *
Desmognathus fuscus, 188 *
Development, of Amphioxus, 43–48; of Tunicate, 52–54; of salpian, 57 *, 58; of myxinoids, 91; of lamprey, 94, 95 *, 96; of *Neoceratodus forsteri*, 166; of salmon, 167; of *Hylodes*, 202 *; of frog, 203–209; of birds, 314–323; of mammals, 404–411
Devil fish, 125
Diadectes, 216
Diaphragm, of mammals, 325
Diapsida, 214
Diemictylus viridescens, 191
Didelphia, 345, 352–359; definition, 352
Didelphidæ, 354–355
Didelphis virginiana, 354, 355 *
Dimetrodon, 215 *, 216
Dingo dog, 366
Dinoceras, 343 *
Dinornis, 287
Dinornithiformes, 287
Dinosaurs, Carnivorous, 219, 220; Herbivorous, 220–223; extinction of, 223, 224

Diomedea exultans, 293 *
Diphycercal caudal fin, 104, 105 *; in *Pleuracanthus*, 121
Diplacanthidæ, 110, 119
Diplocaulus, 180 *, 181
Diplosoma, 55 *
Dipneusti, 107, 112, 154–159
Dipnoi, 112
Diprotodontia, 345, 357, 358 *, 359
Dipterus, 156
Dipus jaculus, 372 *
Discoglossidæ, 198, 199
Dissorophus, 181
Distœchurus, 357
Divergence, law of (adaptive radiation), 19
Diving origin of flight, theory of, 275, 276
Dodo, 304
Dog-fishes, 113
Dog-sharks, 112
Dolichoglossus kowalevskii, 62 *
Doliolum, 56, 57 *, 58; life cycle of, 57 *, 58; *D. tritonis*, 57 *
Dollo, 235
Dominance and subordination, 11
Draco volans, 248 *, 249, 250
Drepanaspis, 169 *, 170
Dromœus novæ-hollandiæ, 285 *
Dromicia, 357
Dromocyon, 341 *, 342
Dromotherium sylvestre, jaw of, 337 *
Ducks, 297
Dugongs, 396, 397, 398 *
Duplicidentata, 371

Eagle, golden, 298
Eagle ray, 124 *, 125
Echidna aculeata, 348, 349 *
Echidnidæ, 345, 348–350
Edaphosaurus, 216
Edentata, 345, 373–377; extinct, 377
Eels, 111, 140, 141; electric, 140
Eel-like form, a criterion of racial senescence, 21
Eggs, of Amphioxus, 42; of tunicate, 52; of myxinoids, 91; of lamprey, 94; of fishes, 101; of Teleostomi, 128; of sturgeons, 133; of *Amia*, 134; of fishes, 163, 164 *, 165; of ling, 165; of *Lepidosteus*, 165; of frog, 204; of *Chelydra*, 236; of *Aromochelys*, 238; of *Aspidonectes*, 242; of Ostrich, 283; of bird, 315 *
Elaps, 257
Elasmobranchii, 112–127
Elasmosaurus, 218 *, 219
Electric organs, of rays, 123
Electric rays, 123, 124 *
Elephant birds, 287
Elephantidæ, 394, 395 *, 396; extinct, 396
Elephas indicus, 394, 395 *; *E. africanus*, 394, 395 *
Eleutherodactyli, 310
Elk, 390, 391 *
Elopidæ, 138
Emberiza, beak of, 282 *
Embryo, of chick (35 Somites), 318 *; five days old, 321 *; of *Hypsiprymnus*, 405 *; of *Phascolarctus*, 405 *; of *Perameles*, 406 *
Embryonic membranes, of reptile, 212, 213 *; of bird, 319, 320 *, 321; of marsupials, 405 *, 406; of rabbit, 410 *; of man, 411 *
Emeu, 284, 285 *, 286
Emys orbicularis, 235 *
Endocrine glands and racial senescence, 22
Endostyle, 37, 38, 40 *, 51, 96; of Ammocœtes, 95, 96
Engystoma carolinense, 202
Engystomatidæ, 202
Enteropneusta, 33, 84, 85
Environment in relation to evolution of characters, 28, 30
Epanorthidæ, 357
Eptesicus fuscus, 364
Equidæ, 392
Equus, 392; *E. burchelli*, 393 *
Erinaceidæ, 362
Erinaceus, 362
Eryops, 180 *, 183
Esocidæ, 142
Esox masquinongy, 142
Eudyptes chrysocoma, 293 *
Eulampis jugularis, 307 *
Eumetopias jubata, 369 *
Eunotosauria, 234
Euphractus sexcinctus, 376
Eusuchia, 242
Eutheria, 345, 352; definition, 352
Evans, A. H., 289

Evolution of foot from fin, 176 *
Evolutionary advances of reptiles, 211–212
Exonautes gilberti, 145 *
External gills, 106 *, 107, 108, 185 *, 207
Eyes, of Amphioxus, 38; of myxinoids, 90; of lampreys, 92; of fishes, 101; of bird, 271

Falcinellus, beak of, 282 *
Falco, foot of, 280 *; beak of, 282 *
Falconiformes, 289, 297, 298
Falcons, 298
Faserstrang, 77
Fierasfer, 111, 142; *F. acus*, 142 *
Feathers, of pigeon, 261; 266 *, 267
Feet, of Birds 280 *; of mammals, 337
Felidæ, 366
Felis canadensis, 367 *; *F. caffra*, 366
File-fishes, 112
Finfoot, 301
Fins, of fishes, 100, 102–105
Fishes, coloration of, 110; integument of, 109, 110; life zones of, 97; types of body form, 97, 98, 99 *; structural features of, 100–102; fins of, 102–105; respiratory organs of, 106–109; swim-bladder of, 108, 109; classification of, 110–112
Fissipedia, 364–368
Flamingoes, 295, 296
Flat-fishes, 147
Flounders, 111, 146, 147 *
Flower and Lydekker, 376
Flying dragon, 248 *, 249, 250
Flying-fishes, 145 *, 150
Flying frog, 203
Flying gurnards, 150
Food concentrating mechanism, of *Amphioxus*, 36, 37; of tunicates, 50, 51 *
Fool-fishes, 151
Fossa, 366
Four-wing theory of origin of flight, 273–275
Frigate birds, 295
Frilled shark, 122 *
Frog, bull, 200 *, 203; leopard, 200; * Javan flying, 200 *, 203; European brown, 203; cricket, 201
Fuertes, L. A., figures after, 367 *, 369 *, 391, 398
Fulica, foot of, 280 *
Fundulus heteroclitus, 142 *, 160, 161, 166

Gadow, H., 183, 184, 225, 226, 232, 253, 254, 256
Gadus morrhua, 146 *
Galeopithecus volans, 362 *, 363
Gall bladder, of Ammocœtes, 95, 96; of *Squalus*, 116
Galli, 299, 300
Galliformes, 289, 299, 300
Gallus gallus, 300
Gambusia, 166
Gannets, 295
Ganoid scales, 109, 110
Ganoin, 110
Gar-pike, 111, 133
Garrod and Forbes, 309
Gaskell, 80
Gasterosteus aculeatus, 144 *
Gastrostomus bairdii, 136 *
Gastrula, of Amphioxus, 42, 43 *; of teleost, 167 *, 168; of frog, 205, 206 *; in the bird, 317
Gavialidæ, 244–245
Gavialis gangeticus, 244, 245 *
Geckones, 247
Geckos, 247; wall gecko, 248 *
Geese, 297, 295 *
Generalized types of vertebrates, 12, 13; of fishes, 159–160; of birds, 312
Geologic time, main divisions of (chart), 29 *; scale, 26 *
Gephyrocercal caudal fin, 131
Giant land tortoises, 239, 240 *
Giant size, a criterion of racial senescence, 20, 21
Giantism, possible causes of, 22
Gibbon, 378 *, 382
Gila monster, 250 *, 252
Gill, T. N., 141
Giraffa camelopardus, 391 *
Giraffe, 390, 391 *
Glandiceps hacksi, 62 *
Glass snake, 250*, 252
Globe-fishes, 152
Glossobalanus, 64, 84
Glyptodon, 377
Gnathonemus, 139
Gnathostomata, 69
Goats, 392

Goat-suckers, 306
Gobiidæ, 149
Goby, 111, 146, 149
Goethe, 7
Golden Age of Reptiles, 217
Gonads, of Amphioxus, 42; of Cyclostomata, 88; of fishes, 102
Goodrich, 154
Gorilla gorilla, 384, 385 *
Graptemys, 227 *; *G. geographica*, 238
Grebes, 292, 294
Gregory, W. K., theory of origin of flight, 275; 378, 387
Gruiformes, 289, 301
Guinea-fowls, 299
Gulls, 302
Gulper-eels, 141
Gymnarchus, 139
Gymnophiona (see Apoda)
Gymnothrax waialuæ, 141 *
Gymnotidæ, 140
Gymnotus electricus, 140
Gymnura, 360, 361 *; *G. rafflesii*, 361 *, 362
Gypogeranus, 298
Gyrfalcons, 298

Haddock, 146
Hag-fishes, 87, 89 *, 90, 91 *
Hake, 146
Halicore, 398, 399 *
Hammer-head sharks, 121, 122 *
Hapalidæ, 378, 381
Haplomi, 111, 142
Harrimania, 84
Harriota raleighana, 126 *
Harrington, 129
Hatteria (see *Sphenodon*)
Hawksbill turtle (see *Chelone*)
Head-fishes, 98, 152, 153 *, 161
Heart, of crocodile, 243 *
Hedgehogs, 360, 362
Hellbender (see *Cryptobranchus*)
Heloderma horridum, 250 *, 252
Hemichordata, 33, 60, 61 *, 62 *, 63 *, 64, 65 *, 66 *, 67 *, 68
Hemimyaria, 60
Hemipodes, 299
Heptanchus, 106
Hermaphrodism, in tunicates, 52; in hag-fishes, 90
Herons, 295
Herrings, 111, 139
Hesperornis, 279 *, 280, 281
Heterocercal caudal fin, 104, 105 *
Heterodontus, eggs of, 164
Heteromi, 111, 142, 143
Heterosomata, 148
Heterostraci, 168, 169, 170
Hexanchus, 106
Hippocampus guttatus, 144 *
Hippopotamus amphibius, 391 *
Hirudo rustica, 311 *
Hoactzin, 300
Hogs, 389; European wild, 389; wart 389, 391 *
Holocephali, 104, 106, 107, 111, 125–127
Holostei, 104, 106, 133, 134
Hominidæ, 378, 385–388
Homo sapiens, 386; varieties of, 386, 387; *H. neanderthalensis*, 387, 388 *
Homocercal caudal fin, 105 *
Homologies as evidence of phylogeny, 23
Hornbill, 303 *, 306
Horses, 392
Hubrecht, 82
Humming birds, 307 *, 308
Hutton, 294, 295
Huxley, T. H., 333
Hyænas, 366, 367 *
Hyænidæ, 366
Hyænodon, 341 *
Hydromedusa maximiliani, 240 *, 241
Hyla versicola, 199, 200 *; *H. faber*, 201; *H. gœldii*, 201 *; *H. arborea*, 200 *
Hylidæ, 199–201
Hylobates lar, 379 *, 382
Hylodes martinicensis, 202 *
Hylomys, 362
Hypochordal rays, 104
Hypsiprymnus, embryo of, 405 *
Hyracoidea, 346, 398, 399
Hyrax abyssinicus, 399 *
Hystricomorpha, 373
Hystrix cristata, 372 *

Ibex, 391 *
Ibis, 295
Ichthyophis glutinosus, 185 *, 186
Ichthyopsida, 68, 69
Ichthyorniformes, 289–291

Ichthyornis victor, 290 *
Ichthyosauria, 17 *, 216, 217, 218
Ichthyotomi, 110
Icterus baltimore, 311 *
Idiacanthus ferox, 136 *
Iguana tuberculata, 248 *, 253
Iguanodon, 221, 222 *
Incus, 325
Insectivora, 340, 346, 360
Integument, of fishes, 109, 110; of Teleostomi, 128; of turtle, 226–228; of Crocodilia, 242, 243; of mammals, 325–328; of armadillos, 376
Internal gills, 107 *, 108; of frog larva, 207, 208 *
Isospondyli, 137

Jacana, 302
Jerboa, 372 *
Jordan, D. S., 137, 138, 144, 147
Jungle-fowls, 299, 300

Kangaroos, 358 *, 359
Keeled Birds (see Neornithes carinatæ)
Kellicott, W. E., viii, 43, 44, 45, 46, 47
Kestrels, 298
Killifishes, 111, 142 *
Kingsley, J. S., viii, 103
Kite, red, 293 *
Kiwi (see *Apteryx*)
Knowlton, F. H., 289, 305
Koala, 358 *, 359

Labidosaurus, 215 *, 216
Labyrinthodonta, 181
Lacerta viridis, 248 *, 249
Lacertæ, 247–253
Lacertilia, 246–255
Lagomyidæ, 371
Lampreys (see Petromyzontia)
Lanarkia, 169 *, 170
Lancelets (see *Amphioxus*)
Lappet-fins, of *Cladoselache*, 120
Lari, 302
Larva, of *Amphioxus*, 46 *, 47 *; of *Petromyzon*, 38; of *Polypterus*, 129, 130 *; of *Protopterus*, 155 *, 156; of *Lepidosiren*, 155 *, 156; of frog, 206 *, 207, 208
Larvacea, 33, 50
Lasanius, 169 *, 171
Lateral line organs; of Cyclostomata, 88; of fishes, 101; of *Squalus*, 118
Leatherback turtle, 234–235 *
Lemur varius, 380
Lemur, 378, 379 *, 380; Smith's dwarf, 379 *; ruffed, 380; mouse, 380
Lemuroidea, 346, 378, 379 *, 380
Lepidosiren paradoxica, 112, 155 *, 156, 158, 159
Lepidosteidæ, 133
Lepidosteus, 134 *; eggs of, 164
Leporidæ, 371
Life zones of fishes, 97
Lillie, F. R., 24
Limax lanceolatus (see *Amphioxus*)
Limicolæ, 302
Limulus, 77, 79, 170
Linophryne lucifer, 136 *
Lissamphibia, 180, 183
Liver diverticulum of *Amphioxus*, 34 *, 39
Liver, of Ammocœtes, 95; of *Squalus*, 116
Lizards, 246–255; wall, 248 *, 249; Galapagos sea, 250 *, 252
Lizard-tailed bird, 277, 278
Llamas, 390
Lobe-fin, 105
Lobe-finned ganoids (see Crossopterygii)
Loons, 292, 293 *, 294
Loricata, 376, 377
Loxomma, 180, 182
Lull, R. S., viii, 20, 21, 79–81, 223, 224, 338, 339
Lung-fishes, 112, 154–159
Lutra canadensis, 367 *
Lyre-birds, 310, 311 *, 312

Macaques, 381
Machærodontia, 370
Mackerel, 111, 147
Macrochelys temmincki, 236
Macropodidæ, 359
Macropus rufus, 358 *, 359
Mœritherium, 396, 397 *
Malacopterygii, 111, 135, 137–139
Malleus, 325
Mammalia, 13, 33, 68, 69, 324–411; characterization of, 324, 325; distinguishing characters of, 325, 326;

Mesozoic, 336–339; Cenozoic, 339–404; classification of modernized, 345, 346; placental mammals, 360–404
Mammary glands, 325
Mammoth, 396
Manatees, 396, 397, 398 *
Mandril, 379 *, 381
Manidæ, 377
Manis, 377; *M. gigantea*, 377; *M. temmincki*, 375 *, 377
Marmosa, 354
Marmosets, 377, 381
Marmot, 372 *
Marsipobranchii, 88
Marsupialia, 345
Mastodon, 396
Matamata, 240 *
Matthew, W. D., 219, 220
McGregor, J. H., 387, 388
Medullary Plate, definition of, 32, 45 *, 47
Megachiroptera, 363
Megalichthys, 182
Megalops atlanticus, 138 *
Melurus, 368
Menura superba, 311 *
Merganser, 297
Mergus, foot of, 280 *; beak of, 282 *
Merostomata, 79, 80 *, 81, 82
Mesænatides, 299
Mesite, Madagascar, 299
Metamerism in Vertebrates, 6, 7
Metamorphosis, of Amphioxus, 48; of tunicate, 53, 54 *; of lamprey, 96; of frog, 207, 208 *, 209
Metatheria, 345, 352–259; definition, 352
Miastor, 195
Microcebus smithii, 379 *
Microchiroptera, 363–365
Micromeres, 42
Micropodii, 308
Microstomum, 11
Milvus ictinius, 293 *
Moa, 287
Moles, pouched, 356; true, 360, 361 *, 362
Moloch horridus, 250 *, 252
Mollosidæ, 364
Mongoose, 366
Monitor, Cape, 250 *, 252, 253
Monkeys, howler, 378, 381; spider, 378
Monodelphia, 345, 360–404; definition, 360
Monotremata, 345, 347–352
Moose, 390
Moray, 141
Morula, of rabbit, 407 *
Mucous rope of *Amphioxus*, 37, 38
Mucous sacs, of myxinoids, 90
Mud-minnow, 142
Mud-puppy (see *Necturus*)
Müller, Johannes, 64
Multituberculata, 340
Muskalunge, 142
Mustelidæ, 368
Mustelus, 113
Mycteria, foot of, 280 *; beak of, 282 *
Myliobatidæ, 124, 125
Myliobatis aquila, 124 *
Mylodon, 377
Myocommata, 42
Myomorpha, 373
Myotis lucifugus, 364
Myotomes, 42
Myrmecobiidæ, 354–356
Myrmecobius fasciatus, 354, 355 *, 356
Myrmecophaga jubata, 374, 375 *
Myrmecophagidæ, 374
Mystacoceti, 346
Myxine, 89 *, 90; eggs of, 164 *
Myxinidæ, 106
Myxinoidea, 89 *, 90, 91 *, 92

Naja tripudians, 256 *, 259
Nannemys guttata, 238
Narwhal, 402, 403
Nasal apertures, in *Squalus*, 114
Nasal sacs, of Teleostomi, 128
Naso-pituitary sac, 90, 92
Necturus maculatus, 192, 193 *
Nematichthys, 141
Nemertean theory of vertebrate origin, 82, 83 *
Neoceratodus, 112, 155 *, 157, 166, 168
Neornithes, 278–323; N. Odontolcæ, 279, 280, 281; N. Ratitæ, 279, 281–288; N. Carinatæ, 279, 288–312
Neoteny (see Pædogenesis)
Nephridia, of Amphioxus, 41 *; of *Squalus*, 116; of fishes, 102
Nests, of lamprey, 95; of stickleback, 144; of *Amia*, 134; of *Hyla faber*, 201;

of *Chelydra*, 236; of *Aromochelys*, 238; of *Chrysemys*, 238; of *Aspidonectes*, 242; of alligator, 245, 246; of crocodile, 247; of lizard, 249; of flamingo, 296; of hornbill, 306; of birds, 323; of *Ornithorhynchus*, 352
Neural groove, 32, 45 *
Neural tube, 32, 45 *
Neurenteric canal, 47
Neurocœl, 32, 45 *
Nictitating membrane, of turtle, 226
Nidicolæ, 323
Nidifugæ, 323
Nopcsa, F., 272
Northern diver, 293 *
Nostril, median, of myxinoid, 90
Notochord, definition of, 31; of Amphioxus, 38, 45 *; development of, 45 *; of Cyclostomata, 88
Notodanidæ, 112
Notoryctes typhlops, 355 *, 356
Notoryctidæ, 356
Nototrema marsupium, 200 *, 201
Nuttall, 304
Nycticebus tardigradus, 380
Nythosaurus larvatus, 334 *

Odontocetæ, 346, 400, 402
Odontolcæ (see Neornithes Odontolcæ)
Ogilvie-Grant, 299
Oikopleura, 59 *
Olfactory, funnel of *Amphioxus*, 38
Olfactory lobes, of *Squalus*, 118
Olfactory nerves, of Amphioxus, 38
Olfactory organs, of fishes, 101; of *Squalus*, 118
Olms (see *Proteus*)
Operculum, 107; of Holocephali, 126; of Teleostomi, 127; of frog larva, 207
Ophidia, 255–259
Opisthocomi, 300
Opisthoglyphs, 257
Opisthomi, 111, 150
Oral, funnel of Amphioxus, 38; of tunicate, 50, 51 *; of lamprey, 92
Oral hood (see oral funnel), 48
Orang-utang, 382, 383 *
Orca gladiator, 401 *
Orientation of embryo in bird's egg, 317 *
Origin, of fishes, 72–75; of Amphibia, 174, 175; of reptiles, 213, 214; of birds, 271–276; of flight, 272–276; of mammals, 332–336; of modernized mammals, 344; of Man, 387
Oriole, Baltimore, 311 *
Ornithischia, 219, 221
Ornithopoda, 221
Ornithorhynchidæ, 345, 350–352
Ornithorhynchus anatinus, 157; pectoral girdle of, 347 *; 349 *, 350, 351, 352
Orycteropus capensis, 375 *, 377
Osborn, H. F., viii, 16, 17, 19, 20, 25, 26, 27, 29, 71, 86, 97, 98, 99, 176, 210, 211, 212, 274, 340, 341, 345, 346
Ostariophysi, 111, 140
Osteolepida, 111, 129
Osteolepis, 156
Osteostraci, 170, 171
Ostrachion schlemmeri, 152 *
Ostrachodermi, 80 *, 82, 112, 168–171
Ostrich Dinosaur (see Struthiomimus)
Ostriches, 283, 284
Otariidæ, 368, 369
Otter, 367 *, 368
Owen, R., 7
Owls, 306; great horned, 307 *; screech, 306
Oxen, 392

Paddle-fish, 111
Pædogenesis, 14, 22; in Larvacea, 59; in *Ammocœtes*, 96; in fishes, 163; in urodeles, 189, 194, 195
Pair wing theory of origin of flight, 273
Palæogeography in relation to evolution, 28
Palæohatteria, 232
Palæomastodon, 396, 397 *
Palæospondylidæ, 112
Palingenetic character, 24
Pallas, 33
Pan pygmæus, 379 *, 384
Pangolins, 377
Papio leucophæus, 379
Paradisea minor, 311 *
Parapsida, 215
Parasuchia, 216
Parent stream theory, 138
Parrots, 303 *, 305, 306
Partridges, 299
Passer domesticus, 311 *

Passeriformes, 289, 309–312
Passerine birds, 309–312
Patriofelis, 341
Patten, W., 80–82, 171
Pea-fowls, 299
Pecora, 388, 391
Pecten, of bird, 271
Pectoral fins, 100; of *Squalus*, 113
Pectoral girdle, of *Squalus*, 114; of Teleostomi, 127; of turtle, 228, 229 *; of *Ornithorhynchus*, 347 *
Pediculati, 111, 150, 151
Pelargomorphæ, 289
Pelicanus, beak of, 282 *; 295
Pelobates cultripes, 197 *, 199
Pelobatidæ, 199
Pelvic fins, 100
Pelvic girdle, of *Squalus*, 114; of Teleostomi, 127; of turtle, 228, 229 *
Peragale lagotis, 355 *, 356
Peramys, 354
Perameles, 356, embryo of, 405, 406 *
Peramelidæ, 356
Perch, 111, 146
Percosoces, 111, 145
Perennibranchiata, 186
Perissidactyla, 346, 392–394
Permian reptiles, 214–217
Petaurus, 357
Petrels, 294
Petrogale xanthopus, 358 *, 359
Petromyzon, 93 *; *P. wilderi*, 95 *; egg of *P. marinus*, 164
Petromyzontia, 91–96
Phœnicopterus, beak of, 282 *
Phäeton, foot of, 280 *
Phalanger maculatus, 358 *
Phalangeridæ, 357
Phenacodus primævus, 342 *
Phaneroglossa, 196, 198–203
Pharomacrus moccino, 307 *
Pharyngeal clefts, definition of, 31; in Amphioxus, 38; in tunicates, 50; in *Myxine*, 90; in *Squalus*, 114
Pharynx, of Amphioxus, 38; of tunicates 50, 51 *
Phascolarctus, 357, 358 *; *P. cinercus*, 358 *
Phascologale, 356
Phascolomydidæ, 359
Phascolomys, 359
Phasianus, foot of, 280 *; *P. colchicus*, 303 *
Pheasants, 299, 303 *
Philander, 354
Phoca grœnlandica, 369 *, 370, *P. vitulina*, 370
Phociæ, 370
Phocochœrus æthiopicus, 391 *
Pholidogaster, 180
Pholidota, 346, 377
Phoronidia, 33, 60, 67, 68
Phoronis, 33, 60, 67, 68
Phosphorescent organs, in fishes, 18, 137
Photostomias guernei, 136 *
Phrynosoma cornutum, 250 *, 251
Phyllopteryx eques, 145 *
Physeter macrocephalus, 400, 401 *, 402
Pica, 372
Pici, 308
Pickerel, 111
Picus, foot of, 280 *
Pigeon, rock, 304; passenger, 304, 305
Pike, 142
Pilosa, 374, 375 *, 376
Pinnipedia, 368–370
Pipa americana, 196, 197 *, 198
Pipe-fish, 111, 143, 144
Pisces, 33, 68, 69, 97
Pithecanthropus erectus, 387 *, 388 *
Placenta, allantoic, 409; chorionic, 409, 410
Placoid scales, 109, 110; in *Squalus*, 114
Plagiaulacidæ, 340
Plagiostomi, 111, 121–125
Plastron, of turtle, 226, 227 *, 228
Platalea, beak of, 282 *
Platurus laticaudatus, 256 *, 259
Platypus (see *Ornithorhynchus*)
Platyrrhini, 378, 381
Platysternidæ, 238
Platysternum, 238
Plectognathi, 112, 146
Plesiosauria, 216, 219
Pleuracanthidæ, 110, 119
Pleuracanthus ducheni, 120 *, 121
Pleurodira, 241
Pleuronectes cynoglossus, 147 *
Pleuropterygii, 110
Podiceps, foot of, 280 *
Pœciliidæ, 142
Pollock, 146

Polyembryony, specific in armadillos, 376, 377
Polyodon folium, 132 *
Polyprotodontia, 345, 354, 355 *, 356
Polypterus, 108, 111, 128, 129, 130 *, 131, 159; *P. bichir*, 129; *P. senegalus*, 129, 130 *
Porcupines, 372 *, 373
Porcupine-fish, 112, 151, 152
Porpoise, 17 *, 402
Præcoces, 323
Prairie-hen, 299
Pre-oral body cavity, of *Amphioxus*, 48
Pre-oral pit, of *Amphioxus*, 48
Primates, 346, 378–388; classification of, 378
Priodontes, 376
Pristidæ, 123, 124 *, 125
Pristis antiquorum, 124 *
Pro-avis hypothetical cursorial ancestor of birds, 272 *
Proboscidia, 346, 395 *, 396
Procellariiformes, 289, 294, 295
Procyon lotor, 367 *, 368
Procyonidæ, 368
Proechidna bruinjnii, 349 *, 350; *P. nigroaculeata*, 349 *, 350
Pro-mammals, time, place, etc., of origin, 335, 336
Pronephros, in Myxinoidea, 90; of *Ammocœtes*, 95
Proteidæ, 192–194
Proteus anguineus, 192, 193 *
Protopoda, 214
Protopterus œthiopicus, 112, 155, 158 *P. annectans*, 155 *; *P. dolloi*, 158
Prototheria, 345, 347–352
Prosauria, 231, 232
Psephurus, 132
Pseudobranchus striatus, 194
Pseudosuchia, 242
Psittacus crithacus, 303 *
Psittaci, 305
Ptarmigan, 299
Pteraspis, 170
Pterobranchia, 33, 60
Pterocles, 302
Pterodactyl, 218, 224
Pterodon, 224
Pteroglossus, beak of, 282 *
Pteropus, 363
Ptilodus gracilis, skull of, 338 *
Ptychodera, 64, 84
Puffer, 112, 151, 152
Puffin, 302
Pyrosoma, 56 *
Python, 256 *, *P. seba*, 256 *
Pytonius, 180 *

Quail, 299
Quezal, 307 *, 308

Raccoons, 368, 367 *
Racial senescence, structural criteria of, 20, 21, 163
Rag-fish, 111
Raia batis, 124 *; eggs of, 164
Rana pipiens, 200 *, *R. catesbiana*, 200 *, 203; *R. esculenta*, .200 *; *R. temporaria*, 203
Ranidæ, 202, 203
Raninæ, 203
Ranzania makua, 153 *
Ratitæ (see Neornithes Ratitæ)
Ratite birds (see Neornithes Ratitæ)
Recapitulation theory, 23, 24
Recurvirostra, foot of, 280 *; beak of, 282 *
Remora brachyptera, 148 *, 149
Reptilia, 13, 14, 69, 210–259; characters of, 225, 226
Respiratory system, of Amphioxus, 38; of *Balanoglossus*, 64; of lamprey, 92; of Pisces, 106 *, 107 *, 108, 109; of *Squalus*, 116; of turtle, 231; of bird, 270
Rhabdopleura, 31, 32, 65, 66, 67 *
Rhacophorus pardalis, 200 *, 201
Rhamphastus ariel, 307 *
Rhea, 284, 285 *; *R. americana*, 284, 285 *
Rheiformes, 284
Rhina squatina, 122 *, 123
Rhinoceros bicornis, 393 *
Rhinocerotidæ, 393, 394
Rhinodontidæ, 121, 122, 123
Rhodeus amarus, embryos of, 135 *, 137
Rhynchocephalia, 216, 232, 233
Rhynchops, beak of, 282 *
Rhynchotus rufescens, 303 *
Rhytidoceros undulatus, 303 *
Ribbon-fish, 150
Road runner, 305
Rodentia, 346, 370–373

Roller, 306; common, 307 *
Rorquals, 404
Ruminantia, 389–392
Running birds (see Neornithes Ratitæ)

Salamandra maculosa, 189, 190; *S. atra*, 191
Salamandridæ, 188–192
Salmo fario, 139 *
Salmon, 111, 138, 139 *
Salmonidæ, 138, 139
Salpa, 58
Salpians (see Thaliacea)
Sand-eel, 111
Sapsucker, 309
Sarcophilus ursinus, 355 *, 356
Sargassum fish, 151
Sauria, 246–259
Sauripterus taylori, pectoral fin of, 178 *
Saurischia, 219
Sauropoda, 220
Saw-fish, 111, 123, 125
Scales, types of in fishes, 109, 110
Sceloporus spinosus, 248 *, 249
Sciuromorpha, 371, 372
Sciuropterus volucella, 372 *
Schizocardium, 62 *, 63
Schuchert, G., 335
Scolopax rustica, 303 *
Scombridæ, 147
Scorpœnidæ, 149
Screamer, 296; horned, 296
Scyllium canescens, 122; eggs of, 164
Scymnognathus, 215 *, 216
Scyphophori, 137
Sea-squirts (see Ascidiacea)
Sea-horses, 111, 143, 144 *, 145 *
Seal, 368, 369 *, 370
Sea-lion, 368, 369 *, 370
Sea-moth, 144, 145
Secretary bird, 298
Selachii, 111, 121–123
Semon, 167, 168
Senescence, racial, 20; internal causes of, 22
Senescent types of vertebrates, 12, 14, 15
Serranus, eggs of, 164
Seymouria, 215 *, 216
Shark, 13, 17 *, 115 *
Shark-sucker, 111, 148 *, 149
Sheep, 392
Shrews, 360, 361 *, 362
Siluridæ, 140
Siluroid fish, 140 *
Simia satyrus, 382, 383 *
Simiidæ, 378, 381–384
Simplicidentata, 371
Siredon axolotl, 189, 190 *
Siren lacertina, 193 *
Sirenia, 346, 396, 397 *
Sirenidæ, 194
Skate, 123, 124 *
Skeletal rods of branchial basket of *Amphioxus*, 38
Skeletal system, of *Squalus*, 114; of Amphibia, 179; of turtle, 227–229; of bird, 267, 268 *
Slow loris, 380
Smilodon, 370 *
Sminthopsis, 356
Snakes (see Ophidia); rattle, 256 *, 258; sea, 256 *, 259; venom of, 257; coral, 257
Solenocytes, 41 *
Sparrow, English, 312, 313
Spawning, of *Amphioxus*, 42; of lamprey, 94 *; of *Amia*, 134; of Stickleback, 144; of sea-horse, 144
Specialized types of verbetrates, 13–15
Specializations and adaptations, 15
Spelerpes fuscus, 188 *; *S. bilineatus*, 189
Spermaceti, 402
Sphenisciformes, 289, 291, 292
Sphenodon, 232, 233 *
Sphyrna zygæna, 122 *
Sphyrnidæ, 121, 122
Spinescence, a criterion of racial senescence, 21
Spiracles, of *Squalus*, 114; of *Polypterus*, 131; of frog larva, 207
Spiral valve, of lampreys, 92, of fishes, 101; of *Squalus*, 115; of *Polypterus*, 131
Spoon-bill, 132 *; 295, 296
Squalus acanthias, 113 *–119
Squamata, 216, 246–259
Stapes, 325
Stegocephali, 180–183
Stegocephali Leptospondyli, 181
Stegocephali Temnospondyli, 181
Stegocephali Stereospondyli, 181–182

Stegodon, 396, 397 *
Stegosauria, 221, 222
Stegosaurus, 222 *
Stickle-back, 111, 143, 144 *
Sting rays, 123, 124 *
Stoasodon narinari, 124 *
Stork, white, 293 *, 295
Striges, 306
Struthio, foot of, 280 *; *C. camelus*, 283, 285 *; *S. molybdophanes*, 283; *S. australis*, 283
Struthiomimus, 218 *, 221
Struthioniformes, 283, 284
Sturgeons, 111, 132 *, 133
Sudoriparous glands, 327
Suina, 389
Sun-bitterus, 301
Sun-fishes, 112, 152, 153; fresh water, 146
Susceptibility method of demonstrating axial gradient, 10
Swallows, 311 *
Swans, 296
Swifts, 308
Swim-bladder, in fishes, 102
Symbranchii, 111, 140
Synapsida, 214
Synotus barbastellus, 365 *
Tæniodonta, 340
Tamandua, 374
Tapiridæ, 392
Tapirus, 393 *
Tarentola mauritanica, 248 *
Tarpon, 111, 138 *
Tarsipes, 357
Tarsius spectrum, 380
Tashiro biometer, 10
Taxidea taxus, 367 *
Teeth, of *Squalus*, 114; of dog, 329 *; various types of, 329 *; bunodont, 329; lophodont, 329; diphyodont, 329; monophyodont, 329; structure of typical, 329, 330; development of, 330, 331; of elephants, 396
Teleostei, 106, 111, 135–153
Teleostomi, 106, 107, 111, 127–153; characters of, 127, 128
Tench, 111
Terrapene (see *Cistudo*), 238, 239
Testudinidæ, 238–239
Testudo græca, 239; *T. elephantica*, 240 *
Tetrapteryx, 274 *
Thaliacea, 33, 56, 57 *, 58
Thecophora, 234–242
Thelodus, 169 *
Thinopus antiquus, 174, 175 *, 179
Thread-eel, 141
Thresher shark, 122 *
Thylacinidæ, 356
Thylacinus cynocephalus, 355 *, 356
Tinamiformes, 289
Tinamou, 287, 288, 303 *
Toad, Surinam, 196, 197 *, 198; fire-bellied, 197 *, 198; midwife, 197 *, 198, 199; spade-foot, 197, 199; common, 197 *, 199; tree, 199, 200 *; narrow-mouthed, 202; horned, 250 *, 251
Toad-fish, 150
Tolypeutes, 376
Toothed Diving Birds (see Neornithes Odontolcæ)
Tornaria larva of *Balanoglossus*, 64, 84, 85 *
Torpedinidæ, 123
Torpedo ocellata, 124 *
Tortoises (see turtles)
Tosa fowl, Japanese, 300
Toucan, 307, 308, 309
Trachodon, 221
Träger, of mammal, 408
Tragulina, 389
Trematosaurus, 182 *
Triænops persicus, 365 *
Triceratops, 220, 223 *
Trichechidæ, 368
Trichechus latirostris, 398 *
Trichosurus, 357
Triconodonta, 337, 338 *
Triconodon ferox, jaw of, 338 *
Trigger-fish, 151 *
Trilophodon, 396, 397 *
Trinacromerion, 218 *, 219
Trionychidæ, 241–242
Tritemnodon, 341 *
Triton tæniatus, foot development of, 178 *; *T. cristatus*, 190 *, 191, 192; *T. torosus*, 192; *T. virescens*, 192
Trituberculata, 338
Trogon, 307 *, 308
Tropic birds, 295
Trunk-fish, 112, 151, 152 *
Tuatara (see *Sphenodon*)
Tubulidentata, 346, 377, 378

Tunicates (see Ascidiacea)
Tupaia, 336, 360
Turdus, foot of, 280 *; beak of, 282 *
Turkeys, 299
Turnices, 299
Turtle, anatomy of, 226–231; external characters of, 226–227, 234–242; soft-shelled, 240 *, 241; snapping, 235 *, 236; pond, 238; sea, 239; land, 239; snake-necked, 240 *, 241; musk, 237
Tylopoda, 389, 390, 391 *
Tympanic membrane, of Amphibia, 179; of turtle, 226; of alligator, 243
Types of body form in fishes, 97, 98, 99 *
Typhleps, 258
Typhlomolge rathbuni, 193 *, 194
Tyrannosaurus, 218 *, 219, 220

Unguiculata, 346, 360–378
Ungulata, 389–400
Urochordata, 33, 48, 49 *, 50 *, 51 *, 52, 53, 54 *, 55 *, 56 *, 57 *, 58, 59, * 60; classification of, 60; characterization of, 48, 49
Urodela, 186–194
Urogenital system, of fishes, 102; of Amphibia, 33; of turtle, 231; of mammals, 331
Ursidæ, 368
Ursus, 368, *U. gyas*, 367 *

Vampires, 364, 365 *
Varanops, 215 *, 216
Varanus albigulares, 250 *, 252, 253
Vertebræ, of Cyclostomes, 88; of myxinoids, 90; of lampreys, 92; of *Squalus*, 114; of mammals, 326
Vertebrate phylogeny, 22–30; embryological aspects of, 23–25; geological aspects of, 26–30
Vertebrates, definition of, 1; chronological succession of, 27 *; classification of, 33
Vertebral theory of the head, 7
Viverra civetta, 367 *
Viverridæ, 366
Vultures, American, 297, 298; Old World, 298

Wallace, A. R., 203
Walrus, 369 *
Wapiti, 391 *
Water moccasin, 257
Whalebone, 403 *, 404
Whales, 400–404; toothed, 400–403; beaked, 402; sperm, 400, 401 *, 402; killer, 401 *; southern right, 401 *, 402; balleen, 403; whalebone, 403, 404; Greenland right, 404
Whale-sharks, 121, 122, 123
Wheeler, W. M., 19
Whip-poor-will, 308
Whip-tailed rays, 123
Wiedersheim, R., viii, 103
Wilder, H. H., viii, 77, 78, 79, 84, 85, 189
Willey, A., 35, 42
Williston, S. W., 182, 214, 215, 216
Wolf, timber, 367 *
Wombat, 359 *, 360
Woodcock, American, 302, 303 *
Woodpeckers, 309

Xantharpyia, 363, 365

Yolk-sac, in reptiles, 313 *; in birds, 320 *, 321

Zanclus, 98, 104, 146, 151, 159; *X. canescens*, 147 *
Zebra, 392; Burchell's, 393 *

www.ingramcontent.com/pod-product-compliance
Ingram Content Group UK Ltd.
Pitfield, Milton Keynes, MK11 3LW, UK
UKHW021430280726
14060UKWH00001BA/1